Gustav Kortüm

Reflectance Spectroscopy
Principles, Methods, Applications

Translated from the German by
James E. Lohr, Philadelphia

With 160 Figures

Springer-Verlag New York Inc. 1969

Professor Dr. *Gustav Kortüm*
Institut für Physikalische Chemie der Universität
74 Tübingen, Wilhelmstraße 56

All rights reserved. No part of this book may be translated or reproduced in any form without written permission from Springer-Verlag. © by Springer-Verlag Berlin-Heidelberg 1969. Printed in Germany. Library of Congress Catalog Card Number 79-86181

The use of general descriptive names, trade marks, etc. in this publication, even if the former are not especially identified, is not be taken as a sign that such names, as understood by the Trade Marks and Merchandise Act, may accordingly be used freely by anyone

Title-No. 1605

Preface

Reflectance spectroscopy is the investigation of the spectral composition of surface-reflected radiation with respect to its angularly dependent intensity and the composition of the incident primary radiation. Two limiting cases are important: The first concerns *regular* (specular) reflection from a smooth surface, and the second *diffuse* reflection from an ideal matte surface. All possible variations are found in practice between these two extremes. For the two extreme cases, two fundamentally different methods of reflectance spectroscopy are employed:

The first of these consists in evaluating the optical constants n (refractive index) and $\varkappa$ (absorption index) from the measured regular reflection by means of the *Fresnel* equations as a function of the wavelength λ. This rather old and very troublesome procedure, which is incapable of very accurate results, has recently been modified by *Fahrenfort* by replacing the air-sample phase boundary by the phase boundary between a dielectric of higher refractive index (n_1) and the sample (n_2). If the sample absorbs no radiation and the angle of incidence exceeds a certain definite value, total reflection occurs. On close optical contact between the two phases, a small amount of energy is transferred into the less dense phase because of diffraction phenomena at the edges of the incident beam. The energy flux in the two directions through the phase boundary caused by this is equal, however, so that total reflection takes place. On the other hand, if the sample absorbs light, a fraction of the radiation energy transferred is lost, and the total reflection is attenuated. This phenomenon has proved extremely valuable in the evaluation of the absorption spectra of liquid and solid materials, especially in the infrared region ("internal" reflectance spectroscopy). It is described in Chapter VIII.

The second method, diffuse reflectance from matte surfaces, assumes that the angular distribution of the reflected radiation is *isotropic*; that is, that the density of the reflected radiation (surface brightness) is directionally independent. This involves the validity of the Lambert Cosine Law. Two conceptions are possible for the occurence of this isotropic angular distribution: For particle diameters much greater than the wavelength, the radiation is partly reflected by means of regular reflection at elementary mirrors (crystal surfaces) inclined statistically at all possible angles to the macroscopic surface. It also partly penetrates into the inside of the sample where it then undergoes numerous reflections, refractions, and diffractions at the irregularly located particles, finally emerging diffusely from the surface. For particle diameters of the order of the wavelength, scattering occurs. This scattered radiation is angularly distributed according to *Mie's* theory of single scattering at gases or colloids and is by no means isotropic. For multiple scattering, there is no theory at present. It can be shown qualitatively, however, that for a sufficiently large number and a sufficiently thick layer of closely packed particles, an isotropic scattering distribution can still be anticipated. This too can be confirmed by measurements made under suitable conditions.

As no rigorous theory of multiple scattering exists, many authors have attempted to develop a phenomenological theory of absorption and scattering of such tightly packed particle layers. To describe radiation transfer in a simultaneously scattering and absorbing medium, one divides the radiation field into two or more opposing radiation paths and describes the change in radiation intensity in an increment ds by means of two constants, the absorption coefficient and the scattering coefficient. These constants are regarded as characteristic properties of the layer per unit thickness and may be determined from measurements of reflectance and transmittance on such layers. These two-constant theories lead in general to related formulas, the most general and comprehensive of which being that of *Kubelka* and *Munk*, which will be thoroughly treated in the course of this book. Under suitable conditions, this theory is particularly useful for representing experimental results.

The importance of reflectance spectroscopy for quantitative and qualitative analysis of crystalline powders, pigments, and solid dyes, as well as for numerous problems concerning molecular structure, adsorption, catalysis at surfaces, Ligand Field Theory, kinetics of reactions at surfaces or of solid materials, photochemistry, the determination of optical constants (n and $\varkappa$), etc., has increased enormously in recent years. Actually, measurements of diffuse reflectance have been carried out over several decades, but the method was only rarely fully exploited, and in many cases was employed rather uncritically.

The purpose of the present book is to investigate theoretically and demonstrate by means of experimental results the capabilities of diffuse and internal reflectance spectroscopic techniques. The effects of external parameters such as particle size, regular parts of reflection, concentration of absorbing components, moisture content, effect of additional phase boundaries such as covering glasses, etc, on the results obtained from these techniques will also be demonstrated. The conclusion must be that reflectance spectroscopy deserves its proper place among other spectroscopic methods and is capable of solving numerous problems only difficultly or not at all soluble by other available methods.

Tübingen, August 1968   G. Kortüm

Contents

Chapter I. Introduction 1

Chapter II. Regular and Diffuse Reflection 5
   a) Regular Reflection at Non-Absorbing Media 5
   b) Total Reflection 13
   c) Regular Reflectance at Strongly Absorbing Media 21
   d) Definition and Laws of Diffuse Reflection 25
   e) Experimental Investigation of Diffuse Reflection at Non-Absorbing Materials 33
   f) Diffuse Reflectance at Absorbing Materials 55
   g) Dependence of Remission Curves on Particle Size 58

Chapter III. Single and Multiple Scattering 72
   a) Rayleigh Scattering 73
   b) Theory of Scattering at Large Isotropic Spherical Particles 81
   c) Multiple Scattering 94
   d) The Radiation-Transfer Equation 100

Chapter IV. Phenomenological Theories of Absorption and Scattering of Tightly Packed Particles 104
   a) The Schuster Equation for Isotropic Scattering 104
   b) The Kubelka-Munk Exponential Solution 106
   c) The Hyperbolic Solution Obtained by Kubelka and Munk . 116
   d) Use of Directed Instead of Diffuse Irradiation 127
   e) Consideration of Regular Reflection at Phase Boundaries . . 130
   f) Absolute and Relative Measurements 137
   g) Consideration of Self-Emission or Luminescence 150
   h) Attempts at a Rigorous Solution of the Radiation-Transfer Equation 156
   i) Discontinuum Theories 163

Chapter V. Experimental Testing of the "Kubelka-Munk" Theory . 170
   a) Optical Geometry of the Measuring Arrangement 170
   b) The Dilution Method 175
   c) Concentration Dependence of the "Kubelka-Munk" Function $F(R_\infty)$ 178
   d) The Typical Color Curve 186
   e) Influence of Cover Glasses 189
   f) Scattering Coefficients and Absorption Coefficients 191

g) Influence of Scattering Coefficients on the "Typical Color" Curve 210
h) Particle-Size Dependence of the Kubelka-Munk Function. . 213

Chapter VI. Experimental Techniques 217
  a) Test of the Lambert's Cosine Law 217
  b) The Integrating Sphere 219
  c) Measuring Apparatus 221
  d) Measurements with Linearly Polarized Radiation 231
  e) The Measurement of Fluorescent Samples 232
  f) Influence of Moisture on Reflectance Spectra 234
  g) Preparation of Samples for Measurement 237
  h) Adsorption from the Gas Phase and from Solution . . . 242
  i) Measurements in the Infrared 245
  k) Discussion of Errors 250

Chapter VII. Applications 253
  a) The Spectra of Slightly Soluble Substances, or Substances that are Altered by Dissolution 253
  b) Spectra of Adsorbed Substances 256
    Acid-Base Reactions between Adsorbed Substance and the Adsorbent 257
    Charge Transfer Complexes 262
    Redox Reactions 265
    Reversible Cleavage Reactions 266
    Surface Area Determination of Powders 270
    Establishment of Equilibria and Orientation at Surfaces . . 273
    Photochemical Reactions 278
  c) Kinetic Measurements 281
  d) Spectra of Crystalline Powders 285
  e) Dynamic Reflectance Spectroscopy 288
  f) Analytical Photometric Measurements 290
  g) Color Measurement and Color Matching 301

Chapter VIII. Reflectance Spectra Obtained by Attenuated Total Reflection 309
  a) Determination of the Optical Constants n and $\varkappa$ 309
  b) Internal Reflection Spectroscopy 313
  c) Methods 319
  d) Applications 329

Appendix: Tables of the Kubelka-Munk-Function 337
        Tables of $\sinh^{-1} x$; $\cosh^{-1} x$; $\coth^{-1} x$ 350

Subject Index 357

Chapter I. Introduction

The term absorption spectroscopy is used to denote the qualitative or quantitative measurement of the absorbance of a material as a function of the wavelenght or wavenumber. With quantitative measurements using a parallel beam of light the so-called "transmittance" of a plane-parallel layer,

$$T(\lambda) \equiv \frac{I}{I_0} \qquad (1)$$

is measured, where I and I_0 denote the radiation flux after and before the transmission of the radiation through the absorbing layer[1]. Even a measurement of this order of simplicity, however, involves certain complications. If a continuous beam of light enters into a homogeneous medium bounded by plane-parallel windows, it is partially *reflected* at each phase boundary, while within the medium it is partially *absorbed* and partially *scattered*.

If we are dealing with gases or dilute solutions, the energy loss due to *reflection* at the phase boundaries can be eliminated to a large extent by suitable experimental measures: the radiation is allowed to pass successively through two identical cells, one of which contains the absorbing solution or gas, and the other the non-absorbing solvent, or air at the same pressure. The reflection losses are then identical to within a very small difference due to the different refractive indices of the dilute solutions and the solvent, or of the two gases. This small difference usually lies well within the limits of errors of the measuring methods. For concentrated solutions, pure liquids or transparent solids (crystals, glasses, foils, etc.) for which the foregoing is not true, the reflection losses at the phase boundaries can be eliminated by irradiating the absorbing material in various layer thicknesses[2]. The intensity ratio $\Delta I/I_0$ then, corresponding to the difference in the layer thickness, gives the true transmittance, since the reflection losses for both measurements are identical. In this case, when the reflection losses have been eliminated, the value of T defined by means of Eq. (1) is called the "true" or "internal" transmittance.

It is not possible, on the other hand, to eliminate the radiation loss caused by scattering by the dissolved molecules or by scattering differ-

---

[1] For methods of spectroscopic measurement see *Kortüm, G.*: Kolorimetrie, Photometrie und Spektrometrie (Colorimetry, photometry and spectrometry), 4th edition. Berlin-Göttingen-Heidelberg: Springer 1962.

[2] *Schachtschabel, K.*: Ann. Physik **81** (4), 929 (1926).

ences by different gases, but these errors also lie within the limits of error of the measuring methods, provided that we are concerned with genuine molecularly dispersed solutions, and provided that the transmittance does not drop below ca. 0.03%. Very large errors arise, however, through scattering in colloidal solutions, where radiation losses due to absorption and scattering cannot be separated because the scattering depends on the shape, and particularly on the size and concentration of the collodial particles as well as on the wavelength of the radiation, and can only be evaluated theoretically in simple cases (see p. 81 ff.).

The difficulties become still greater when attempts are made to obtain from scattered transmission the absorption spectrum of solid powdered materials such as pigments, suspensions, substances adsorbed at solid surfaces, etc. Recently, infrared spectra of adsorbed molecules have often been obtained directly in this way. Because of the indefinite layer thickness, however, only qualitative results can be expected from such experiments. That satisfactory results can be obtained in this way depends upon the scattering coefficients being relatively low and the particles being small compared with the wavelength of the radiation used. This is particularly true in the range of medium and long-wave infrared radiation[3]. By immersing the powder to be investigated in a suitable non-absorbing liquid of similar refractive index, it is possible to largely remove the radiation loss due to scattering. This procedure has often been employed in the infrared using such immersion liquids as paraffin, perfluorokerosine, Nujol, and others. Apart from the frequently encountered difficulty in finding suitable liquids, the indefinite thickness of the layer traversed remains a problem, so that this method also only gives qualitative results. The so-called KBr-method[4] brought some progress. This involves grinding the finely divided powder under investigation with an excess of solid potassium bromide (or silver chloride) and pressing the mixture under high pressure into transparent plates whose absorbance is measured relative to a plate of the same thickness made from the pure dilution material. The limitation of this procedure for quantitative evaluation arises from radiation losses which are not completely removable, partly due to the difference in refractive index in the sample and dilution medium, partly to the inadequate distribution of the sample, and partly also to traces of moisture which are very difficult to remove[5].

---

[3] This is the reason for the well-known ability of infrared radiation to penetrate through mist and haze.

[4] *Stimson, M. M.*, and *M. J. O'Donnell:* J. Am. Chem. Soc. **74**, 1805 (1952). — *Schiedt, U.*, and *H. Reinwein:* Z. Naturforsch. **7b**, 270 (1952); **8b**, 66 (1953).

[5] See, for example, *Lejeune, R.*, and *G. Duyckaerts:* Spectrochim. Acta **6**, 194 (1954).

I. Introduction

The condition that the particle dimensions must be small compared with the wavelength is no longer true in the short-wave infrared and especially not in the visible and ultraviolet region. Scattering then exceeds absorption so greatly that it is no longer possible to obtain a useful transmission spectrum without taking account of scattering. The question therefore arises whether it is possible from the measured transmittance or diffuse reflectance of a simultaneously scattering and absorbing layer, or from one of these measurements alone, to obtain the absorption spectrum of the material under investigation.

This question is not merely of theoretical interest, but of the greatest practical importance for the characterizing and standardizing of industrial products of all kinds (pigments, synthetic plastics, textiles, papers, paints, etc.); that is, for the quantitative physical analysis of "color". In addition, the solution of this problem could facilitate the investigation of numerous other problems for which hitherto suitable methods have been lacking. Examples which may be mentioned are: the kinetic investigation of reactions between solid materials, color-changes of solid materials under the influence of temperature and pressure (thermochromism and piezochromism), the influence of the adsorption of molecules on solid phases on their spectrum, the catalytic influence of solid surfaces on the reactivity of adsorbed molecules with gases or with the surface itself, the photochemistry of adsorbed materials, the quantitative analysis of mixed solid materials, the quantitative analysis of paper and thin-layer chromatograms, etc.

To separate the effect of scattering from absorption on the spectroscopic composition of the radiation flux reflected or transmitted by a scattering and absorbing layer, a number of theories have been developed, all of which commence with the transmittance and reflectance of monochromatic radiation in an infinitesimal layer. The differential equations used are then integrated over the total thickness of the layer. This leads to suitable formulas which may be tested experimentally, provided that the initial equations are not too complicated, that is, do not contain so many constants which must later be evaluated from the measurements as to be impracticable. This means that simplifying assumptions which can never be exactly fulfilled in practice must be made. This necessarily limits the applicability of the formulas developed. In the majority of the theories, only two constants are introduced, the *absorption coefficient* and the *scattering coefficient*, by means of which transmittance and reflectance of a layer of finite thickness may be expressed. These so-called *two-constant* theories lead in general to similar formulas which may frequently be converted into each other by interrelating the parameters used. They are intended to provide a solution for the problem mentioned above: that is, to determine from the reflectance and scattered transmittance of a simultaneously absorbing and scattering layer the absorption spectrum of the substance in question. They prove in practice to be useable in this way when suitable measuring conditions are maintained. In particular success has been achieved in obtaining the so-called *"typical color curve"* of the material in question from the diffuse reflectance of an "infinitely thick" layer (semi-infinite medium) which is no longer transparent. The curve obtained is frequently in agreement with the true spectrum after a parallel-displacement in the ordinate. By this means, it is possible not only to obtain the quantitative physical analysis of pigments, but also to approach the solution of numerous problems of the kind already referred to above.

Chapter II. Regular and Diffuse Reflection

If a parallel beam of light is allowed to fall on the smooth surface of a solid material, two limiting cases arise for the reflected portion of the radiation: it is either reflected "specularly" (that is, as from a mirror), or it is reflected in all directions of the hemisphere uniformly. In the first case, the surface is an ideal reflecting (polished) surface, while in the second it is an ideal matte (scattering) surface. These two ideal limiting cases for a surface are never attained in practice. Even on the best mirrors, such as a mercury surface the place where a beam of light impinges (even outside the incident plane) can be seen. It is desirable, however, to consider the behaviour of such ideal surfaces in a theoretical manner, and then to proceed to the behaviour of real surfaces.

a) Regular Reflection at Non-Absorbing Media

In Fig. 1 we see a parallel monochromatic beam of light fall at an angle α on the interface of two media with refractive indices n_0 and n_1, where $n_1 > n_0$. We assume further that neither medium has any appreciable absorbance. According to the law of Conservation of energy the incident radiation flux per unit area of the beam cross section I_e (measured in watts/cm^2) must be equal to the sum of the flux of the reflected light,

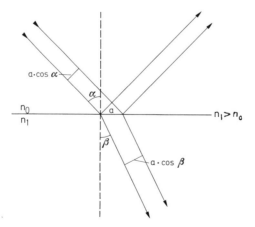

Fig. 1. Mirror reflection of a parallel beam of rays on a plane phase boundary

II. Regular and Diffuse Reflection

I_r, and that of the refracted light, I_g. That is:

$$I_e = I_r + I_g \cdot \frac{\cos\beta}{\cos\alpha}. \tag{1}$$

The factor $\cos\beta/\cos\alpha$ takes into account the increase in the cross-section of the beam due to refraction, which can be seen from Fig. 1. Since the radiation flux is proportional to the square of the electric vector $\vec{E}$ and to the velocity $v = c/n$ of the radiation in the appropriate medium,

$$I = \frac{\varepsilon}{4\pi} E^2 v \quad \text{or} \quad I = \frac{n^2}{4\pi} E^2 v, \tag{2}$$

Eq. (1) can be written in the form:

$$n_0 E_e^2 = n_0 E_r^2 + n_1 E_g^2 \cdot \frac{\cos\beta}{\cos\alpha}. \tag{3}$$

If the electric vector is now resolved into the two components $\vec{E}_\perp$ and $\vec{E}_{\|}$, respectively perpendicular and parallel to the incident plane, then for the perpendicular oscillating component, because of the constancy of the tangential components at the phase boundary,

$$E_{e\perp} + E_{r\perp} = E_{g\perp}. \tag{4}$$

Since Eq. (3), as follows from the law of Conservation of energy, is naturally valid for both components we must have for the vertical component:

$$n_0(E_{e\perp}^2 - E_{r\perp}^2) = n_0(E_{e\perp} + E_{r\perp})(E_{e\perp} - E_{r\perp}) = n_1 E_{g\perp}^2 \cdot \frac{\cos\beta}{\cos\alpha},$$

or, taking into account Eq. (4)

$$n_0(E_{e\perp} - E_{r\perp}) = n_1 E_{g\perp} \cdot \frac{\cos\beta}{\cos\alpha}. \tag{5}$$

From (4) and (5), by elimination of $E_{g\perp}$ we obtain

$$E_{r\perp} = -E_{e\perp} \cdot \frac{n_1 \cos\beta - n_0 \cos\alpha}{n_1 \cos\beta + n_0 \cos\alpha}. \tag{6}$$

Then, taking into account the Snell Law of Refraction, $\frac{\sin\beta}{\sin\alpha} = \frac{n_0}{n_1}$. we have for the vertical component of the reflected radiation

$$E_{r\perp} = -E_{e\perp} \cdot \frac{\sin(\alpha - \beta)}{\sin(\alpha + \beta)}. \tag{7}$$

If we, however, eliminate from (4) and (5) the term $E_{r\perp}$, we obtain similarly for the refracted component

$$E_{g\perp} = E_{e\perp} \frac{2n_0 \cos\alpha}{n_1 \cos\beta + n_0 \cos\alpha} = E_{e\perp} \frac{2\cos\alpha \sin\beta}{\sin(\alpha+\beta)}. \tag{8}$$

For the oscillating component parallel to the incident plane, we have the equation analogous to (4)

$$E_{e\|} - E_{r\|} = E_{g\|} \cdot \frac{\cos\beta}{\cos\alpha}, \tag{9}$$

as we can see directly from Fig. 2.

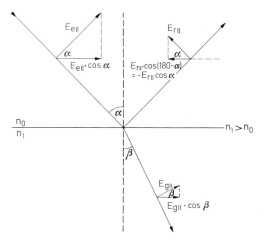

Fig. 2. Parallel components of the incident, the reflected, and the refracted beam in specular reflection

From (3) and (9) we obtain, in a similar way to (5)

$$n_0(E_{e\|} + E_{r\|}) = n_1 E_{g\|} \tag{10}$$

and by elimination first of $E_{g\|}$ and then of $E_{r\|}$ we obtain from Eq. (9) and (10) the two relationships

$$E_{r\|} = E_{e\|} \cdot \frac{n_1 \cos\alpha - n_0 \cos\beta}{n_1 \cos\alpha + n_0 \cos\beta} = E_{e\|} \frac{\tan(\alpha-\beta)}{\tan(\alpha+\beta)}, \tag{11}$$

and

$$E_{g\|} = E_{e\|} \cdot \frac{2n_0 \cos\alpha}{n_1 \cos\alpha + n_0 \cos\beta} = E_{e\|} \frac{2\cos\alpha \sin\beta}{\sin(\alpha+\beta)\cos(\alpha-\beta)}. \tag{12}$$

Eqs. (7), (8), (11) and (12) are the well-known *Fresnel equations*, which describe the reflection, refraction, and polarization of non-absorbing media. The ratio of the radiation fluxes $I_{r\perp}/I_{e\perp}$ and $I_{r\parallel}/I_{e\parallel}$ are measured as a function of the angle of incidence α. The roots of this give the "reflection coefficients", $E_{r\perp}/E_{e\perp}$, and $E_{r\parallel}/E_{e\parallel}$. If these are plotted against α, we obtain, according to the Fresnel equations (7) and (11), the two curves of Fig. 3, in which $n_0 = 1$ and $n_1 = 1.5$ have been employed, corresponding approximately to the phase boundary between air and

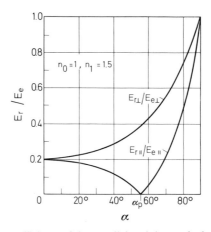

Fig. 3. Reflection coefficients of the parallel and the vertical component of the beam, dependent on the angle of incidence α in specular reflection for $n_0 = 1$, $n_1 = 1.5$

crown glass. It is seen, in accordance with the formulas, that with perpendicular and grazing incidence the values of $E_{r\perp}/E_{e\perp}$, and $E_{r\parallel}/E_{e\parallel}$ are equal, and that $E_{r\parallel}/E_{e\parallel}$ for a given angle of incidence α_P is equal to zero, when according to equation (11), $\alpha_P + \beta = \pi/2$ (that is, the reflected and refracted beams are perpendicular to each other), and so $\sin\beta = \cos\alpha_P$. Since in addition, according to the Law of Refraction $\dfrac{\sin\alpha_P}{\sin\beta} = n_1$, it follows that

$$\tan\alpha_P = n_1 . \tag{13}$$

The quantity α_P is called the "*angle of polarization*", and Eq. (13) represents the well-known Brewster Law used, for example, for the production of polarized infrared radiation by reflection at selenium mirrors. In a similar way, if the "transmission coefficients" $E_{g\perp}/E_{e\perp}$, and $E_{g\parallel}/E_{e\parallel}$ are plotted against the angle of incidence α, taking into account (as in

Eq. (3)) the altered cross-section of the beam of light and its refractive index n_1 [6], we then obtain the curves of Fig. 4.

The *regular reflectance* of a non-absorbing medium is given according to Eq. (7) for the component polarized perpendicular to the incident plane, by

$$R_{\text{reg}\perp} = \frac{\sin^2(\alpha - \beta)}{\sin^2(\alpha + \beta)} \qquad (14)$$

and for the component polarized parallel to the incident plane, according to Eq. (11) by

$$R_{\text{reg}\|} = \frac{\tan^2(\alpha - \beta)}{\tan^2(\alpha + \beta)}. \qquad (15)$$

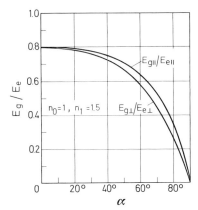

Fig. 4. Transmission coefficients of the parallel and the vertical component of the refracted beam, dependent on the angle of incidence α for $n_0 = 1$, $n_1 = 1.5$

The reflectance of natural radiation is, therefore, given by

$$R_{\text{reg}} = \frac{1}{2}\left[\frac{\sin^2(\alpha - \beta)}{\sin^2(\alpha + \beta)} + \frac{\tan^2(\alpha - \beta)}{\tan^2(\alpha + \beta)}\right]. \qquad (16)$$

For perpendicular incidence ($\alpha = \beta = 0$) the difference between the perpendicular and parallel polarized components disappears, since the incident plane is no longer defined. We then obtain from Eq. (6)

$$R_{\text{reg}} = \left(\frac{n_1 - n_0}{n_1 + n_0}\right)^2. \qquad (16a)$$

---

[6] The radiation flux of the refracted beam is, according to Eq. (3), given by $E_g^2 \dfrac{n_1 \cos\beta}{n_0 \cos\alpha}$.

The regular reflectance of non-absorbing materials is small, since the majority of these have refractive indices of less than 2. Thus, for crown glass ($n_1 = 1.5$) and air ($n_0 = 1$), $R_{reg} \cong 0.04$, that is, only 4% of the radiation is reflected in the case of perpendicular incidence. With increasing angle of incidence, R_{reg} increases, as can be seen from Fig. 3. For the transmitted fraction of the radiation we obtain from Eq. (8) and (12) for perpendicular incidence, taking into account the correction factor $n_1 \cos\beta / n_0 \cos\alpha$,

$$T = n_1 \frac{4 n_0^2}{(n_0 + n_1)^2}. \tag{17}$$

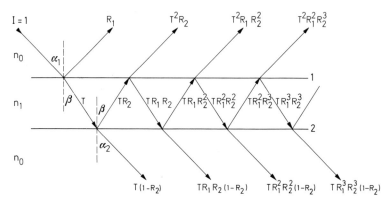

Fig. 5. Multiple reflection between the phase boundaries of a plate bounded by parallel planes. R = reflection; T = transmittance

For the example just given, using the phase boundary between air and crown glass ($n_0 = 1$; $n_1 = 1.5$) we obtain $T = 0.96$, so that $R_{reg} + T = 1$, as is necessarily the case.

Finally we will calculate the regular reflectance of a plane parallel plate of refractive index n_1 which is surrounded by a medium of refractive index n_0. In the simplest case this will consist of air ($n_0 = 1$). In this case, the multiple reflection between the two phase boundaries must be taken into account as is indicated in Fig. 5. According to the Snell Law, when $n_0 = 1$, $\sin\alpha = n_1 \sin\beta$. Also, from the Helmholtz Reciprocity Law, $\alpha_1 = \alpha_2$. That is, a parallel beam of light is displaced parallel to itself on passing through a plane parallel plate. In addition, at the internal surface 2 exactly the same amount of light will be reflected at the angle β as is reflected at the outer surface 1 at the angle α. This follows from the Fresnel Equations (14) and (15). Since $\sin^2(\alpha - \beta) = \sin^2(\beta - \alpha)$, and $\tan^2(\alpha - \beta) = \tan^2(\beta - \alpha)$. With a prearranged angle of incidence, the angles α and β are maintained during multiple reflection. The sum of all the fractions reflected upwards, representing the transmission by T, gives, therefore:

$$\begin{aligned}R_{12} &= R_1 + T^2 R_2 + T^2 R_1 R_2^2 + T^2 R_1^2 R_2^3 + \cdots \\ &= R_1 + T^2 R_2 (1 + R_1 R_2 + R_1^2 R_2^2 + \cdots) \\ &= R_1 + \frac{T^2 R_2}{1 - R_1 R_2},\end{aligned} \qquad (18)$$

if the summation formula for the geometrical series is employed. Then, using the relation already demonstrated that $R_1 = R_2 \equiv R(\alpha)$, we obtain,

$$R_{12}(\alpha) = R(\alpha) + \frac{T^2 R(\alpha)}{1 - R^2(\alpha)}. \qquad (19)$$

Writing $T = 1 - R$, we have finally

$$R_{12}(\alpha) = R(\alpha) + \frac{R(\alpha)(1 - R(\alpha))}{1 + R(\alpha)} = \frac{2R(\alpha)}{1 + R(\alpha)}. \qquad (20)$$

From (16) we have

$$R(\alpha) = \frac{1}{2}\left[\frac{\sin^2(\alpha - \beta)}{\sin^2(\alpha + \beta)} + \frac{\tan^2(\alpha - \beta)}{\tan^2(\alpha + \beta)}\right].$$

We see from (20) that the two boundary surfaces together do not reflect twice as much as the upper surface alone. For the total transmission of the plate by similar summation we obtain the transmitted fraction:

$$T_{12} = \frac{T(1 - R_2)}{1 - R_1 R_2}, \qquad (21)$$

which, when we write $R_1 = R_2 \equiv R(\alpha)$ is converted into:

$$T_{12} = \frac{T(1 - R(\alpha))}{1 - R^2(\alpha)} = \frac{T}{1 + R(\alpha)}. \qquad (22)$$

R_{12} plus T_{12} are equal to 1, as should be expected.

Of particular interest for problems which will arise later is the regular reflectance for *diffuse incident radiation*, which can be obtained by means of an integrating sphere (see p. 219). The radiation then impinges on the phase boundary surface at all angles α between 0 and $\pi/2$, so that integration must be carried out over all these angles and the average obtained. Since the radiation flux impinging on a unit surface is proportional to the cosine of the angle of incidence (see p. 28), for the regular reflectance with diffuse incident radiation we have from the mean value theorem:

$$R_{\text{reg(diff. inc.)}} = \frac{2\pi \int_0^{\pi/2} \sin\alpha \cos\alpha f(\alpha, n)\, d\alpha}{2\pi \int_0^{\pi/2} \sin\alpha \cos\alpha\, d\alpha}$$

$$= 2 \int_0^{\pi/2} \sin\alpha \cos\alpha f(\alpha, n)\, d\alpha, \tag{23}$$

where $f(\alpha, n)$ is given by Eq. (16). The integration has been carried out by *Walsh*[7], and gave the relation:

$$R_{\text{reg(diff. inc.)}} = \frac{(n-1)(3n+1)}{6(n+1)^2} + \left[\frac{n^2(n^2-1)^2}{(n^2+1)^3}\right]\log\frac{n-1}{n+1}$$
$$- \frac{2n^3(n^2+2n-1)}{(n^2+1)(n^4-1)} + \left[\frac{8n^4(n^4+1)}{(n^2+1)(n^4-1)^2}\right]\log n. \tag{24}$$

Table 1 gives R_{reg} according to Eqs. (16) and (24) for different values of n[8]. For these tabulations, n_1 and n_0 were set equal to n and 1, respectively.

Table 1. R_{reg} and $R_{\text{reg(diff. inc.)}}$ *for various values of n*

n	$R_{\text{reg(perp. inc.)}}$	$R_{\text{reg(diff. inc.)}}$	n	$R_{\text{reg(perp. inc.)}}$	$R_{\text{reg(diff. inc.)}}$
1.00	0.0000	0.000	1.45	0.033	0.085
1.10	0.0023	0.026	1.50	0.040	0.092
1.15	0.0049	0.035	1.55	0.047	0.100
1.20	0.0083	0.045	1.60	0.053	0.107
1.25	0.012	0.053	1.65	0.060	0.114
1.30	0.017	0.061	1.70	0.067	0.121
1.35	0.022	0.069	1.80	0.082	0.134
1.40	0.028	0.077	1.90	0.096	0.146

It is seen that the regular reflectance for diffuse incidence is always larger than that for perpendicular incidence, in the case of crown glass with $n = 1.5$ amounting to more than double this. The regular reflection losses therefore are also appreciably greater on passage through phase boundary surfaces.

---

[7] *Walsh, J. W. T.*: Dept. Sci. Ind. Res. Illum. Res. Techn. Pap. **2**, 10 (1926); compare *Judd, D. B.*: J. Res. Natl. Bur. Std. **29**, 329 (1942).

[8] See *Ryde, J. W.*, and *B. S. Cooper*: Proc. Roy. Soc. London A **131**, 464 (1931).

If we insert in Eq. (23) for $f(\alpha, n)$ expression (20) for the regular reflectance $R_{12}(\alpha)$ of a plane parallel plate, we obtain the reflectance of such a plate for diffuse incidence as

$$R_{12\,\text{reg.(diff.inc.)}} = 2 \int_0^{\pi/2} \frac{2R(\alpha)}{1+R(\alpha)} \sin\alpha \cos\alpha \, d\alpha. \qquad (25)$$

Here also $R(\alpha)$ is obtained from Eq. (16). A graphical integration, taking $n=1.5$, gave the value of $R_{12\,\text{reg(diff.inc.)}}$ as 0.155, which is also smaller than twice the reflection at the upper surface which according to Table 1 amounts to 0.092 for diffuse incidence.

b) Total Reflection

We now consider the reverse procedure when a beam of light passes from the optically more dense medium (n_1) into an optically less dense medium (n_0), such as from glass to air (Fig. 6). In this case, the beam is deflected away from the perpendicular, and from the Law of Refraction we have

$$\frac{\sin\beta}{\sin\alpha} = \frac{n_1}{n_0}. \qquad (26)$$

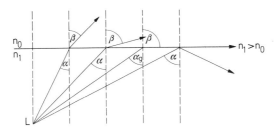

Fig. 6. Critical angle α_g of total reflection when a beam of rays passes from a denser to a rarer medium (emerges from an optically more dense medium and enters an optically less dense medium). $n_1 > n_0$

For a given angle of incidence α_g, called the limiting or critical angle, the beam emerges tangential to the boundary surface, so that $\beta = 90°$, and $\sin\beta = 1$. For this limiting angle we have, therefore,

$$\sin\alpha_g = \frac{n_0}{n_1}. \qquad (27)$$

For an even greater angle of incidence α, the radiation is totally reflected, so that no radiation energy passes into the thinner medium.

For $\alpha \leq \alpha_g$, the Fresnel formulas remain valid. If we plot $E_{r\perp}/E_{e\perp}$, or $E_{r\|}/E_{e\|}$, respectively against α, we obtain curves similar to those in Fig. 3 (see Fig. 7). Here the parallel component of the reflected beam also vanishes at the polarization angle α_p, which is always smaller than α_g; and when $\alpha = 0$ and $\alpha = \alpha_g$, $E_{r\perp}/E_{e\perp}$ and $E_{r\|}/E_{e\|}$ are equal, as the Fresnel equations require. Eq. (16) is also valid for the regular reflectance.

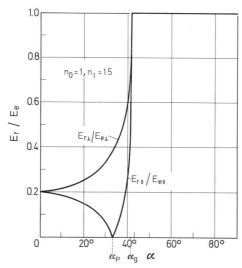

Fig. 7. Reflection coefficients of the parallel and the vertical component of the beam, dependent on the angle of incidence α in the case of reflection at the rarer medium ($n_1 = 1.5$; $n_0 = 1$)

Similarly, for the interval $0 \leq \alpha \leq \alpha_g$ we can determine the transmission coefficients, which here, unlike Fig. 4, increase with α, and at α_g acquire the maximum values $E_{g\perp}/E_{e\perp} = 2$, and $E_{g\|}/E_{e\|} = 3$ (cf. Fig. 8, left-hand side). In this case also the modified propagation velocity and the changed cross-section of the beam must be taken into account by using the correction factor $\dfrac{n_0 \cos\beta}{n_1 \cos\alpha}$. Thus, for example, if $\alpha = \beta = 0$, the transmitted fraction of the radiation is given, according to Fig. 8, by $T = (1.2)^2 \cdot \dfrac{1}{1.5} \cdot \dfrac{\cos\beta}{\cos\alpha} = 0.96$, or, as predicted by the Reciprocity Law, exactly the same magnitude as when the incidence direction is reversed.

The transmittance and reflectance of *diffuse radiation* passing from more to less dense media are of interest for the problem of so called "internal reflection", to be treated later. Because the spatial energy density of an electromagnetic wave is, according to Eq. (2), given by $n^2 E^2/4\pi$, the radiation intensities in the two media are related as n_0^2/n_1^2. On the other hand, the radiation flux must be equal in both directions through the phase boundary. Therefore, if we set $n_0 = 1$, the transmitted fraction T for the diffuse radiation from more into less dense media will differ from

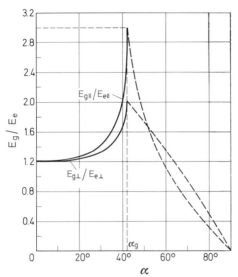

Fig. 8. Transmission coefficients of the parallel and the vertical component of the refracted beam, dependent on the angle of incidence α in case of transition from an optically denser medium to an optically rarer medium ($n_1 = 1.5$; $n_0 = 1$)

that in the opposite direction by the factor $1/n_1^2$. The reflectance [9] is thus given by

$$R_{n_0 \to n_1} = 1 - T_{n_0 \to n_1}; \quad R_{n_1 \to n_0} = 1 - \frac{T_{n_0 \to n_1}}{n_1^2}. \qquad (28)$$

The "internal reflectance" at the phase boundary with the less dense medium can be obtained from the values in Table 1. Thus, for example, if $n_1 = 1.5$ and $n_0 = 1$,

$$R_{1.5 \to 1} = 1 - \frac{0.908}{2.25} = 0.596.$$

This treatment is valid for planar phase boundaries, although internal reflectance is largely independent of surface character [10].

---
[9] See *Judd, D. B.*: J. Res. Natl. Bur. Std. **29**, 329 (1942).
[10] *Giovanelli, R. G.*: Opt. Acta **3**, 127 (1956).

In the region $\alpha_g \leq \alpha \leq \pi/2$, the total reflection is always $E_{r\perp} = E_{e\perp}$ and $E_{r\|} = E_{e\|}$, so that the energy of the reflected radiation is equal to that of the incident radiation[11]. Further it is observed that the totally reflected light is no longer linearly polarized but elliptically polarized if the incident light itself was linearly polarized. That is, a phase difference Δ arises between $E_\|$ and $E_\perp$ which is itself dependent on α. This can also be obtained from the Fresnel equations:

Since from Eq. (27), when $\alpha = \alpha_g$, $\sin\beta = 1$, then when $\alpha > \alpha_g$, it follows from the Law of Refraction (Eq. (26)) that $\sin\beta = \sin\alpha \, \dfrac{n_1}{n_0} > 1$, so that there no longer exists a real refractive index. It follows that

$$\cos\beta = \pm\sqrt{1 - \sin^2\beta} = \pm\sqrt{1 - \sin^2\alpha \cdot \left(\dfrac{n_1}{n_0}\right)^2}$$

is complex, so that it may be written in the form

$$\cos\beta = \pm i \dfrac{n_1}{n_0} \sqrt{\sin^2\alpha - \left(\dfrac{n_0}{n_1}\right)^2}. \tag{29}$$

Only the negative sign in front of the root leads to physically signficant consequences. If this is inserted in the Fresnel formula (6), which for the reverse path of the radiation from the more dense into the less dense medium must be written in the form

$$\dfrac{E_{r\perp}}{E_{e\perp}} = -\dfrac{n_0 \cos\beta - n_1 \cos\alpha}{n_0 \cos\beta + n_1 \cos\alpha}, \tag{6a}$$

we obtain after a simple rearrangement the relation

$$\dfrac{E_{r\perp}}{E_{e\perp}} = \dfrac{\left[\cos\alpha + i\sqrt{\sin^2\alpha - \left(\dfrac{n_0}{n_1}\right)^2}\right]^2}{1 - \left(\dfrac{n_0}{n_1}\right)^2}, \tag{30}$$

Similarly, we obtain from (11)

$$\dfrac{E_{r\|}}{E_{e\|}} = \dfrac{\left[\dfrac{n_0}{n_1}\cos\alpha + i\dfrac{n_1}{n_0}\sqrt{\sin^2\alpha - \left(\dfrac{n_0}{n_1}\right)^2}\right]^2}{\left(\dfrac{n_0}{n_1}\right)^2 \cos^2\alpha + \left(\dfrac{n_1}{n_0}\right)^2 \left[\sin^2\alpha - \left(\dfrac{n_0}{n_1}\right)^2\right]}. \tag{31}$$

---

[11] A totally reflecting phase boundary represents an ideal mirror. This may be compared with a polished metal mirror which perhaps reflects 98 % of the incident radiation. With such a system, the radiation intensity would be reduced to $(0.98)^{10} = 0.82$ its original intensity after only 10 reflections, while in so called optical fibers, thousands of reflections are practically possible without loss of energy.

The reflection coefficients themselves are therefore complex quantities and we can in the usual way write them as complex numbers of the following form:

$$E_{r\perp} = |A| \cdot e^{i\delta_\perp} \quad \text{and} \quad E_{r\|} = |B| \cdot e^{i\delta_\|}.$$

In the brackets, $|A|$ and $|B|$ represent the absolute contributions of the reflected amplitudes, and δ is a quantity which represents the phase shift with respect to the incident wave. In our case, $|E_{r\perp}| = E_{e\perp}$, and $|E_{r\|}| = E_{e\|}$, so that

$$\left. \begin{array}{l} E_{r\perp} = E_{e\perp} \cdot e^{i\delta_\perp} = E_{e\perp}(\cos\delta_\perp + i\sin\delta_\perp) \\ E_{r\|} = E_{e\|} \cdot e^{i\delta_\|} = E_{e\|}(\cos\delta_\| + i\sin\delta_\|) \end{array} \right\}. \quad (32)$$

The phase angle δ is obtained in the following way: from Eqs. (30) and (32) we have for the perpendicular component

$$\cos\delta_\perp + i\sin\delta_\perp = \frac{\cos^2\alpha - \sin^2\alpha + \left(\dfrac{n_0}{n_1}\right)^2}{1 - \left(\dfrac{n_0}{n_1}\right)^2}$$

$$+ i\frac{2\cos\alpha\sqrt{\sin^2\alpha - \left(\dfrac{n_0}{n_1}\right)^2}}{1 - \left(\dfrac{n_0}{n_1}\right)^2},$$

from which,

$$\tan\frac{\delta_\perp}{2} \equiv \frac{\sin\delta_\perp}{1 + \cos\delta_\perp} = \frac{\sqrt{\sin^2\alpha - \left(\dfrac{n_0}{n_1}\right)^2}}{\cos\alpha}. \quad (33)$$

Correspondingly, from Eqs. (31) and (32) we obtain

$$\tan\frac{\delta_\|}{2} = \frac{\sqrt{\sin^2\alpha - \left(\dfrac{n_0}{n_1}\right)^2}}{\left(\dfrac{n_0}{n_1}\right)^2 \cos\alpha}. \quad (34)$$

Eqs. (32) indicate that because $|E_{r\perp}| = E_{e\perp}$, and $|E_{r\|}| = E_{e\|}$, the absolute contributions of the reflection coefficients are equal to unity, the reflection is *total*; they also show that between the reflected and incident waves there will always arise a phase difference $\delta_\perp$ or $\delta_\parallel$. The values of $\delta_\perp$ and $\delta_\parallel$ are different, so that between the reflected waves themselves a phase difference, $\delta_\parallel - \delta_\perp \equiv \Delta$, arises, which indicates that the totally reflected light will be elliptically polarized, which is confirmed by experiment[12]. From Eqs. (33) and (34) we obtain

$$\tan\frac{\Delta}{2} = \tan\left(\frac{\delta_\parallel}{2} - \frac{\delta_\perp}{2}\right) \equiv \frac{\tan\frac{\delta_\parallel}{2} - \tan\frac{\delta_\perp}{2}}{1 + \tan\frac{\delta_\parallel}{2} \cdot \tan\frac{\delta_\perp}{2}}$$

$$= \frac{\cos\alpha\sqrt{\sin^2\alpha - \left(\frac{n_0}{n_1}\right)^2}}{\sin^2\alpha} = \frac{\sqrt{(1-\sin^2\alpha)\left(\sin^2\alpha - \left(\frac{n_0}{n_1}\right)^2\right)}}{\sin^2\alpha}. \quad (35)$$

For $\alpha = \pi/2$, and also for $\alpha_g = n_0/n_1$, the phase difference according to Eq. (27) is zero, and the totally reflected light will remain linearly polarized. For all angles of incidence lying between it is elliptically polarized and for a given value of n_0/n_1, Δ must pass through a maximum, for which $\frac{d}{d\sin^2\alpha}\left(\tan\frac{\Delta}{2}\right) = 0$. We obtain then

$$\sin^2\alpha_{max} = \frac{2\left(\frac{n_0}{n_1}\right)^2}{1 + \left(\frac{n_0}{n_1}\right)^2}.$$

If this is inserted in Eq. (35) we obtain for the maximum phase difference

$$\tan\frac{\Delta_{max}}{2} = \frac{1 - \left(\frac{n_0}{n_1}\right)^2}{2\frac{n_0}{n_1}}. \quad (36)$$

Thus, the smaller n_0/n_1, the greater will be the maximum phase-difference. This has also been confirmed experimentally, and shows that the Fresnel equations are also valid for total reflection.

In the region of total reflection, $\alpha_g \leq \alpha \leq \pi/2$, if the complex expression given by Eq. (29) for $\cos\beta$ is inserted into the transmission coefficients

---

[12] The value of Δ can be brought to zero by means of the "Babinet compensator", that is, elliptically polarized light can be reconverted into linearly polarized light, so that Δ is made accessible to measurement.

defined by (8) and (12), we find the following expressions

$$\frac{E_{g\perp}}{E_{e\perp}} = \frac{2\cos\alpha\left(\cos\alpha + i\sqrt{\sin^2\alpha - \left(\frac{n_0}{n_1}\right)^2}\right)}{1 - \left(\frac{n_0}{n_1}\right)^2} \qquad (37)$$

and

$$\frac{E_{g\|}}{E_{e\|}} = \frac{2\cos\alpha\left(\frac{n_0}{n_1}\cos\alpha + i\frac{n_1}{n_0}\sqrt{\sin^2\alpha - \left(\frac{n_0}{n_1}\right)^2}\right)}{\left(\frac{n_0}{n_1}\right)^2\cos^2\alpha + \left(\frac{n_1}{n_0}\right)^2\left[\sin^2\alpha - \left(\frac{n_0}{n_1}\right)^2\right]} . \qquad (38)$$

If we then write the complex amplitudes in the normal form

$$E_{g\perp} = E_{e\perp}\,\frac{2\cos\alpha}{\sqrt{1 - \left(\frac{n_0}{n_1}\right)^2}} \cdot \frac{\cos\alpha + i\sqrt{\sin^2\alpha - \left(\frac{n_0}{n_1}\right)^2}}{\sqrt{1 - \left(\frac{n_0}{n_1}\right)^2}}$$

$$\equiv |A|e^{i\delta_\perp} = |A|(\cos\delta_\perp + i\sin\delta_\perp)$$

and

$$E_{g\|} = E_{e\|}\,\frac{2\cos\alpha}{\sqrt{\left(\frac{n_0}{n_1}\right)^2\cos^2\alpha + \left(\frac{n_1}{n_0}\right)^2\left[\sin^2\alpha - \left(\frac{n_0}{n_1}\right)^2\right]}}$$

$$\cdot \frac{\frac{n_0}{n_1}\cos\alpha + i\frac{n_1}{n_0}\sqrt{\sin^2\alpha - \left(\frac{n_0}{n_1}\right)^2}}{\sqrt{\left(\frac{n_0}{n_1}\right)^2\cos^2\alpha + \left(\frac{n_1}{n_0}\right)^2\left[\sin^2\alpha - \left(\frac{n_0}{n_1}\right)^2\right]}}$$

$$\equiv |B|e^{i\delta_\|} = |B|(\cos\delta_\| + i\sin\delta_\|),$$

we see directly that

$$|E_{g\perp}| = E_{e\perp} \cdot \frac{2\cos\alpha}{\sqrt{1 - \left(\frac{n_0}{n_1}\right)^2}}, \qquad (39)$$

$$|E_{g\|}| = E_{e\|} \cdot \frac{2\cos\alpha}{\sqrt{\left(\frac{n_0}{n_1}\right)^2\cos^2\alpha + \left(\frac{n_1}{n_0}\right)^2\left[\sin^2\alpha - \left(\frac{n_0}{n_1}\right)^2\right]}}, \qquad (40)$$

whereas for $\tan\delta = \dfrac{\sin\delta}{\cos\delta}$

we obtain

$$\tan\delta_\perp = \frac{\sqrt{\sin^2\alpha - \left(\dfrac{n_0}{n_1}\right)^2}}{\cos\alpha} \tag{41}$$

and

$$\tan\delta_\parallel = \frac{\sqrt{\sin^2\alpha - \left(\dfrac{n_0}{n_1}\right)^2}}{\left(\dfrac{n_0}{n_1}\right)^2 \cos\alpha}. \tag{42}$$

We therefore obtain the result, which at first appears surprising: In spite of the total reflection, radiation is present even in the optically less dense medium and a phase difference arises between the components polarized parallel or perpendicular to the plane of incidence, so that the radiation shows an elliptical polarization dependent on the value of α.

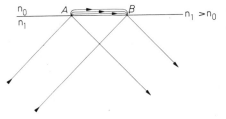

Fig. 9. Influence of diffraction at the edges of the incident beam on the total reflection

If the absolute values (39) and (40) of the complex amplitudes are plotted as a function of α, we obtain the broken curve of the right-hand side of Fig. 8.

This radiation in the less dense medium arises from diffraction phenomena at the edges of the incident beam (see Fig. 9) so that in the region of A, energy penetrates into the less dense medium, and at B returns to the more dense medium. Between A and B there proceeds, therefore, a transversely damped surface wave whose amplitude is already completely vanished within a few wavelengths. Although, between A and B, no energy penetrates into the less dense medium (total reflection), at the edges of the beam, because of the diffraction, energy is in fact transported through the phase boundary, and can be detected directly by the use of suitable means. This, of course, disturbs the total reflection phenomenon. That this interpretation is correct follows from the fact that the incident beam is advanced laterally about a fraction of a wavelength[13].

We shall return to this phenomenon as a basis of absorption measurements at a later stage (see page 309 ff.).

c) Regular Reflectance at Strongly Absorbing Media

The refractive index of absorbing materials can be formally expressed in the same way as for nonabsorbing materials, by replacing the real n by the complex quantity

$$n' = n(1 - i\varkappa). \tag{43}$$

In this, $\varkappa$ is the "absorption index" which arises from dispersion theory, and is defined by the Lambert Absorption Law:

$$I = I_0 \cdot \exp\left[-\frac{4\pi n \varkappa s}{\lambda_0}\right]. \tag{44}$$

This equation indicates that the radiation flux in passing through a medium with the absorption coefficient $n\varkappa$ diminishes along the layer thickness $s = \lambda_0$ (vacuum wavelength) by the fraction[14] $e^{-4\pi n \varkappa}$.

The complex expression in Eq. (43) combines the two quantities which are connected in the dispersion theory, that is, the refractive index n and the absorption coefficient $n\varkappa$. The Fresnel equations (6) and (11) for the reflection of non-absorbing materials also become complex by the introduction of the complex refractive index of (43), as in the case of total reflection, and describe the reflection behavior of absorbing materials.

Let us consider now the most important case, in which linearly polarized radiation falls from a transparent medium with the real refractive index n_0 and the azimuth $\varphi = 45°$ ($E_{r\perp} = E_{r\|}$) on the boundary surface of a strongly absorbing medium with the complex refractive index $n'_1 = n_1(1 - i\varkappa_1)$. The reflected components must then be expressed

---

[13] *Goos, F.,* and *H. Hähnchen:* Ann. Physik **1**, 333 (1947); **5**, 251 (1949). See also *Fragstein, C. v.:* Ann. Physik **4**, 271 (1949); — *Renard, R. H.:* J. Opt. Soc. Am. **54**, 1190 (1964).

[14] If, for example, for $\lambda_0 = 500$ mµ $= 5 \cdot 10^{-5}$ cm, the absorption coefficient $n\varkappa = 0.08$, then I decreases within the range $s = \lambda_0$ by a factor e. This corresponds to a natural extinction modulus m_n, defined by $I = I_0 e^{-m_n s}$, of $2 \cdot 10^4$; since $m_n = \dfrac{4\pi n \varkappa}{\lambda_0}$ has the dimension cm^{-1}, $\varkappa$ itself is dimensionless.

in a manner similar to (32) in the complex form

$$E_{r\perp} = |E_{r\perp}| e^{i\delta_\perp} \quad \text{and} \quad E_{r\|} = |E_{r\|}| e^{i\delta_\|}. \tag{45}$$

A phase difference Δ between $E_{r\perp}$ and $E_{r\|}$ must also arise in this case, which indicates that the reflected radiation is elliptically polarized except when $\alpha = 0$ and $\alpha = \pi/2$. There is in this case, however, no polarization angle for which $E_{r\|}$ is equal to zero, but the reflection simply passes through a minimum with increasing angle of incidence α. The corresponding angle is usually designated as the "major angle of incidence" α_H. For this, $\Delta = 90°$. Finally, the refractive index n_1 is no longer a constant, but is dependent on the angle of incidence, which is also the case in the region of total reflection at non-absorbing media. For measurement of n_1 and $\varkappa_1$ we therefore restrict ourselves in general to the determination of the major angle of incidence α_H at which $\Delta = \pi/2$. The phase difference, Δ, is compensated by means of a compensator, so that the elliptically polarized light is again converted into linearly polarized light, and its azimuth, which after reflection is no longer 45°, is measured. From the so-called "major azimuth" φ_H and the "major angle of incidence" α_H, it is possible to evaluate n_1 and $\varkappa_1$, but it is necessary to introduce a number of approximations so that the equations obtained are only approximately valid [15].

It is simpler, for the determination of n_1 and $\varkappa_1$, to consider the case of perpendicular incidence ($\alpha = 0$). It is of course no longer possible in such a case to distinguish between perpendicular and parallel polarized components in relation to the incident plane, since there is no longer any defined incident plane. If we therefore permit natural radiation with amplitude E_e to impinge perpendicularly, the reflected complex amplitude may be written in a manner similar to (45)

$$E_r = |E_{r0}| e^{i\delta}, \tag{46}$$

where $|E_{r0}|$ is the absolute value of the amplitude and δ the phase difference arising on reflection. Using (43) we then obtain from the Fresnel equations (6) or (11) with $\alpha = \beta = 0$,

$$\frac{|E_{r0}|}{E_e} e^{i\delta} = \frac{n_1' - n_0}{n_1' + n_0} = \frac{n_1 - n_0 - in_1\varkappa_1}{n_1 + n_0 - in_1\varkappa_1}. \tag{47}$$

---

[15] See Textbooks of Optics. In general the optical constants n_1 and $\varkappa_1$ may be determined by two measurements, such as those of R_{reg} at two different angles of incidence, or of $R_\perp$ and $R_\|$ at a particular angle of incidence. A review of the methods is given, for example, by *Wendtlandt, M. M.*, and *H. G. Hecht*: In: Reflectance Spectroscopy. Chap. II; New York: John Wiley&Sons 1966. — *Nassenstein, H.*: Ber. Bunsenges. **71**, 303 (1967).

If both sides of the equation are multiplied by the conjugate complex quantities, we obtain for the *regular reflectance* of the absorbing material 1:

$$R_{reg} \equiv \frac{|E_{ro}|^2}{E_e^2} = \frac{(n_1 - n_0 - in_1\varkappa_1)(n_1 - n_0 + in_1\varkappa_1)}{(n_1 + n_0 - in_1\varkappa_1)(n_1 + n_0 + in_1\varkappa_1)}$$
$$= \frac{(n_1 - n_0)^2 + (n_1\varkappa_1)^2}{(n_1 + n_0)^2 + (n_1\varkappa_1)^2}, \quad (48)$$

which corresponds to Eq. (16a) for the regular reflectance of non-absorbing materials. In spectral regions for which the absorbance of the material 1 is great ($n_1\varkappa_1 \gg n_1$), the reflectance must also present high values, a fact which is well-known from the reflectance of metals, or from the so-called "residual radiation method" in the infrared.

Regular reflectance R_{reg} may be easily measured photometrically, though values of α somewhat different from zero must here be chosen. Eq. (48) remains valid in this case to a good approximation. Since E_e^2 is accessible in the same way, we then obtain the absolute value $|E_{ro}|$ as an experimental quantity for the determination of n_1 and $\varkappa_1$. As a second measured quantity, the phase displacement δ must be determined. It follows from (47) that

$$\frac{|E_{ro}|}{E_e}(\cos\delta + i\sin\delta) = \frac{(n_1 - n_0 - in_1\varkappa_1)(n_1 + n_0 + in_1\varkappa_1)}{(n_1 + n_0)^2 + n_1^2\varkappa_1^2}$$
$$= \frac{n_1^2 - n_0^2 + n_1^2\varkappa_1^2 - i2n_0n_1\varkappa_1}{(n_1 + n_0)^2 + n_1^2\varkappa_1^2}$$

and hence

$$\left.\begin{array}{l}\dfrac{|E_{ro}|}{E_e}\cos\delta = \dfrac{n_1^2 - n_0^2 + n_1^2\varkappa_1^2}{(n_1 + n_0)^2 + n_1^2\varkappa_1^2}\\[2mm] \dfrac{|E_{ro}|}{E_e}\sin\delta = \dfrac{-2n_0n_1\varkappa_1}{(n_1 + n_0)^2 + n_1^2\varkappa_1^2}\end{array}\right\} \quad (49)$$

or
$$\tan\delta = \frac{2n_0n_1\varkappa_1}{n_0^2 - n_1^2 - n_1^2\varkappa_1^2}. \quad (50)$$

From (48) and (49) the desired quantities n_1 and $\varkappa_1$ may be calculated:

$$n_1 = n_0 \cdot \frac{1 - \dfrac{|E_{ro}|^2}{E_e^2}}{1 + 2\dfrac{|E_{ro}|}{E_e}\cos\delta + \dfrac{|E_{ro}|^2}{E_e^2}}, \quad (51)$$

$$\varkappa_1 = n_0 \cdot \frac{2\dfrac{|E_{ro}|}{E_e}\sin\delta}{1 + 2|E_{ro}|\cos\delta + |E_{ro}|^2}. \quad (52)$$

For the measurement of the phase difference δ the interference method of *Th. Young* is used, as shown schematically in Fig. 10. Two coherent rays of light, 1 and 2, are allowed to fall vertically on a glass wedge, at the back of which they are reflected at the small angle of incidence ε. As long as they are both reflected at the glass-air phase boundary, no phase shift occurs, and an interference pattern caused by the interference of two rays 1' and 2' may be recorded on a photographic plate. If the wedge is now displaced parallel to itself to such an extent that the ray 1

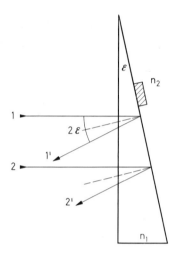

Fig. 10. Interference method for measuring the phase shift in case of reflection on metals

is reflected at the more dense medium of the evaporated metal layer, so that at reflection a phase shift δ arises, then the interference pattern of the two reflected rays 1' and 2' is displaced by a quantity from which δ may be determined, since a displacement by the width of a whole band would correspond to a phase shift of the ray 1 of π, or $\lambda/2$.

In Table 2 a number of measurements at metals are summarized. It is seen from the Table that the refractive index, because of the anomalous dispersion in the absorption region can fall below the value of unity, and that for $n\varkappa \geq n$ the reflectance may rise to the neighborhood of 1. It is further seen that the refractive indices are only given to the second decimal place, instead of the five places usually seen for non-absorbing materials. This is partly due to the observation being carried out in reflection, and partly to the fact that pure metal surfaces are almost impossible to prepare, being always covered with an external oxide layer. Because the reflection process takes place within thin surface

Table 2. *Regular reflectance of metals*

Metal	$\lambda[\mu]$	n	$n\varkappa$	R_{reg} %
Silver	4.04	2.98	28.8	99.5
	2.10	1.00	14.3	98.0
	1.00	0.24	6.96	98.1
	0.578	0.106	3.59	97.0
	0.546	0.108	3.25	96.3
	0.4916	0.123	2.72	94.3
	0.4358	0.149	2.16	90.0
	0.302	1.2	0.7	12.0
	0.2653	1.1	1.3	20.4
Copper	4.20	1.92	22.8	98.7
	1.03	0.43	5.6	94.5
	0.660	0.996	3.70	77.7
	0.5461	0.74	3.19	65.0
	0.520	1.434	2.40	51.2
	0.3656	1.58	1.17	37.0
	0.3129	1.71	0.91	30.0
	0.2536	1.96	0.80	25.0
Gold	4.13	1.60	28.8	99.2
	1.07	0.25	7.1	98.0
	0.870	0.21	5.4	97.0
	0.680	0.617	3.859	85.3
	0.5893	0.469	2.826	81.5
	0.520	1.104	2.817	53.0
	0.3611	1.300	1.750	37.7
	0.2573	0.918	1.142	27.6

layers, the results obtained by different observers differ considerably. In general the absorption coefficients observed in reflection are also very inaccurate, which indicates that weak absorptions are practically incapable of being determined by this method (see page 309).

d) Definition and Laws of Diffuse Reflection

While the reflection and refraction of a light beam at macroscopically flat phase boundaries may be described completely by means of geometrical optics, a new phenomenon arises if the wave front is not subject to the relevant optical procedure over its entire extent. In a crystal, for example, whose dimensions are large compared with the wavelength, but small compared with the cross-section of the beam of radiation, it is possible to distinguish between rays striking the crystal and those passing by it. The former are partly reflected and partly refracted, in many cases repeatedly, so as to give a definite angular distribution of the corresponding radiation flux. In the latter case an incomplete wave front is formed, and because of the Huygens principle, interference of the elementary waves leads to diffraction phenomena, which give a quite different angular distribution of the corresponding radiation intensity depending on the form and dimensions of the crystal, though not on the nature of the crystal and its surface.

If the wavelength of the radiation is slowly diminished at constant crystal size, the intensity distribution of the diffracted portion contracts continuously into a narrow angular range in the forward direction of the beam of radiation, while the intensity distribution of the reflected and refracted portion gradually approaches that determined by geometrical optics. If $\lambda \to 0$ there would be no diffraction at all, so that for vanishingly small wavelength the geometrical optics would be strictly valid.

On the other hand, if λ is gradually increased so that it becomes comparable to or greater than the dimensions of the crystal, the intensity and angular distribution of the diffracted radiation on the one hand, and of the reflected or refracted portion on the other, would become comparable to each other, so that it would no longer be possible to separate the two portions. Thus, the phenomena of reflection, refraction and diffraction are no longer separately definable under these conditions. In this case, we speak of "scattering" of the wave at the particle (see page 72).

We will first consider the case where the dimensions of the particles upon which the beam of radiation impinges (such as a crystal powder) are large compared with the wavelength, so that the phenomena of reflection, refraction and diffraction for the individual particles are still well-defined. Because the crystal surfaces can be oriented in all possible directions, the incident radiation is now reflected at all angles into the hemisphere from which the incident radiation originates. In such cases, we speak of "*diffuse reflection*" in contrast to regular (directional) reflection from a plane phase boundary. Ideal diffuse reflection is thereby defined by the condition that the angular distribution of the reflected radiation is independent of the angle of incidence.

The first law of diffuse reflection was proposed by *Lambert*[16] on the basis of observations that a white wall illuminated by sunlight appears equally bright at all observation angles. He assumed, therefore, that such a surface behaves as though itself were radiating independently of the angle of incidence of the radiation. Consider the "matte" macroscopically plane surface of a solid body (such as a disc of compressed

---

[16] *Lambert, J. H.:* Photometria Augsburg 1760; German edition by Anding Leipzig 1892.

Definition and Laws of Diffuse Reflection

magnesium oxide) uniformly irradiated by a parallel beam of light at angle of incidence α. If S_0 is the *irradiation intensity* in watt/cm² for normal incidence, the radiation flux per unit surface area is then given by

$$\frac{dI_e}{df} = S_0 \cos\alpha. \tag{53}$$

The remitted radiation flux per unit surface area of the irradiated disc depends on the cosine of the angle ϑ at which the radiating surface is observed (Fig. 11) and is proportional to the solid angle, that is:

$$\frac{dI_r}{df} = B \cos\vartheta\, d\omega. \tag{54}$$

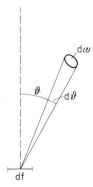

Fig. 11. Concerning the definition of the radiation density of an area

The proportionality factor is given by

$$B = \frac{dI_r/df}{\cos\vartheta\, d\omega} \tag{55}$$

and will be called the *radiation density* (surface brightness), having the dimension [watt/ω · cm²], if the radiation flux is again measured in watts. If the solid angle $d\omega$ is replaced by the surface element of a unit sphere,

$$d\omega = \sin\vartheta\, d\vartheta\, d\varphi \tag{56}$$

and integrated over the angular range of ϑ from 0 to $\pi/2$, and the azimuth φ from 0 to 2π (that is, over the total hemisphere of the remitted radiation), we obtain for the total (integral) remitted radiation strength from (54)

$$\int_0^{\pi/2} \int_0^{2\pi} B \cos\vartheta \sin\vartheta\, d\vartheta\, d\varphi = \pi B. \tag{57}$$

Since the incoming and outgoing radiation intensities must be proportional to each other, it follows from both (53) and (57) that

$$\pi B = \text{Const} S_0 \cos\alpha \tag{58}$$

or, introducing (55)

$$\frac{dI_r/df}{d\omega} = \frac{\text{Const} S_0}{\pi} \cos\alpha \cos\vartheta = B \cos\vartheta. \tag{59}$$

The remitted radiation flux per square centimetre and unit solid angle is proportional to the cosine of the incident angle α and to the cosine of the angle of observation ϑ (Lambert Cosine Law). The constant gives the fraction of the incident radiation flux which is remitted, and is always less than 1, since some portion of the radiation is always absorbed. This constant is often called "Albedo". The Lambert Cosine Law is rigorously valid for the emission of a black body radiator. It is, however, no mere empirical law for reflection but can be derived from the Second Law of Thermodynamics for an "ideal diffuse reflector" as defined above [17]. More or less large deviations from the Lambert Law are always found in practice, and actually there is no such thing as an "ideal diffuse reflector".

v. Seeliger [18] derives a law for diffuse reflection on the assumption that the radiation penetrates into the inside of the powder where it is partly absorbed and partly returned through the surface after numerous reflections, refractions and diffractions. This emitted radiation is the sum of the contributions of all the elements dV of the penetrated volume. If we sum over all these elements of volume, we obtain for the remitted radiation flux, per square centimeter and unit solid angle, the expression

$$\frac{dI_r/df}{d\omega} = \text{Const} \frac{\cos\alpha \cos\vartheta}{\cos\alpha + \cos\vartheta}, \tag{60}$$

where the constant includes both S_0 and the absorption constant k. Comparing with (59) we see that the radiation density defined by (55) for the remitted radiation according to *Lambert* is independent of the observation angle ϑ, while according to *v. Seeliger* this is not so. If, for example, we put $\alpha = 0$ (for perpendicular incidence) and $\vartheta = 0$, or $\vartheta = 90°$, the radiation density according to *v. Seeliger* is in the first case, Const/2, and in the second, Const. Thus, according to *v. Seeliger*, the radiation density B increases with increasing ϑ. The Seeliger formula is, however, only a first approximation, since the irradiation of the volume element dV through the neighboring elements is neglected. Actually, each element

---

[17] *Witte, W.*: in press.
[18] *v. Seeliger, R.*: Münch. Akad. II. Kl. Sitzungsber. **18**, 201 (1888).

of volume will not only be reached by the radiation flux due to direct irradiation, but also by the diffusely reflected radiation from all other volume elements[19]. If account is taken of this, however, the expression obtained is so complicated that it is of little practical value.

Eqs. (59) and (60) may be experimentally tested by two different methods of measurement:

1. Holding a uniformly illuminated surface constant and measuring the reflected radiation flux (constant $d\omega$) at different angles ϑ to the normal, the measured beam narrows proportionally to $\cos\vartheta$. This also means that the measured radiation intensity must decrease proportionally to $\cos\vartheta$ with increasing ϑ.

2. Using for measurement a variable surface $df = df_0/\cos\vartheta$ and constant cross section of the measured beam, the measured quantity is proportional to the radiation density B according to Eq. (59) and should be independent of ϑ, while, according to Eq. (60), it should increase with increasing ϑ.

For graphical presentation of the observed radiation intensity as a function of ϑ, we can plot the radiation density B as a function of ϑ and, according to *Lambert*, obtain a line parallel to the abscissa. Likewise, we can employ a suggestion of *Bouguer* by plotting the measured radiation flux under otherwise constant conditions as radius vector in the relevant direction ϑ to obtain a polar curve designated as the *indicatrix*. For a Lambert-type radiator, this indicatrix according to (59) is a sphere in contact with the radiating surface (Fig. 12a). The diameter of the indicatrix is proportional to the irradiation strength. The radiation reflected from the surface in the unit solid angle is thus angularly isotropic only in relation to the azimuth φ and not in relation to ϑ. We therefore sometimes speak also of "circularly diffuse" reflection. In that case, the surface brightness B is naturally independent of both ϑ and φ. For a Seeliger-type radiator, the indicatrix is an ellipsoid (Fig. 12b), which becomes flatter with larger incidence angles.

The first investigation to theoretically explain the diffuse reflection of macroscopic surfaces was made by *Bouguer*[20]. He assumed that the diffuse reflection occurred by regular reflection from the elementary mirrors of the crystal faces whose surface planes are statistically distributed at all angles. Because of refraction, however, light will also penetrate inside the crystals and will, to the extent it is not absorbed, eventually exit through the surface after numerous reflections and diffractions. *Bouguer* did not take this into account at all while *v. Seeliger* assumed it was solely responsible for diffuse reflection (see p. 28). The same re-

---

[19] *Lommel, E.:* Münch. Akad. II. Kl. Sitzungsber. **17**, 95 (1887).
[20] *Bouguer, P.:* Traité d'optique sur la gradation de la lumière. Paris 1760.

presentation had earlier been developed by Zöllner [21] who assumed, in contrast to v. Seeliger, that an isotropic angular distribution of the remitted radiation density corresponding to Lambert's Law would result.

In measurements on paper and MgO according to method 1, Pokrowski [22] found deviations from the Lambert Cosine ϑ Law. In order

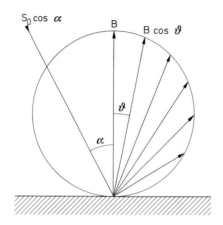

Fig. 12a

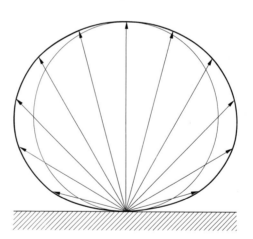

Fig. 12b

Fig. 12. a Angular distribution of a Lambert radiator (Indicatrix); b Angular distribution of a Seeliger radiator in case of vertical incidence ($\alpha = 0$)

---

[21] *Zöllner, J. C. F.*: Photometrische Untersuchungen (Photometric investigations). Leipzig 1865. His assumption went back to a consideration originating with *Fourier, J. B.* [Ann. Chim. Phys. **4**, 128 (1817)].

[22] *Pokrowski, G. I.*: Z. Physik **30**, 66 (1924); **35**, 35 (1926); **36**, 472 (1926).

to explain these deviations, he divided the reflected radiation into two parts

$$I(\vartheta) = I_{\text{reg}} + I_{\text{diff}} \qquad (61)$$

and tried to calculate the two parts separately. For I_{diff} he used the Lambert expression

$$I_{\text{diff}} = \frac{I(\vartheta)}{I(0)} \cos\vartheta, \qquad (62)$$

and for I_{reg}, the Fresnel formula (16)

$$I_{\text{reg}} = \frac{a}{2}\left[\frac{\sin^2(i-\beta)}{\sin^2(i+\beta)} + \frac{\tan^2(i-\beta)}{\tan^2(i+\beta)}\right] \equiv \frac{a}{2} f(i,\beta). \qquad (63)$$

In this expression, i is the angle of incidence of the incident beam on an individual crystal surface (elementary mirror) and, according to Snell's Law of Refraction, $\sin\beta = \dfrac{\sin i}{n}$. a should be a constant depending only upon the number of elementary mirrors and the intensity of irradiation. This means that the statistical distribution of the mirror-like crystal surface facets is not taken into account and thus, a will have to be taken from measurement. It is not clear whether *Pokrowski* attributes the diffuse part I_{diff} as originating in the interior or at the surface of the sample. We almost always observe deviations from the Lambert Law with increasing angle of incidence. Because the function $f(i,\beta)$ increases with increasing i, we can at least qualitatively understand these deviations.

As *Schulz*[23] has shown with the aid of his own measurements, the correlation with experimental results cannot be sufficient because a cannot be constant. On the contrary, it must be supposed that there exists a distribution function for the inclination angles of the elementary mirrors such that, with reference to the macroscopic surface, flat angles of inclination are encountered more frequently than steep angles. Thus, according to the distribution function chosen, various values for the regular part of the reflection are obtained and it is difficult to derive a generally valid law.

The *Pokrowski* conception according which the reflected radiation must consist of regular and diffuse parts was later again taken up by *Barkas*[24]. Here again, in contrast to the representation of *v. Seeliger*, the diffuse part is assumed to originate, not from the interior of the powder but, just as the regular part, from the surface of the

---

[23] Schulz, H.: Z. Physik **31**, 496 (1925); — Z. Techn. Physik **5**, 135 (1924); see also *Middleton, W.E.*, and *A. G. Mungall*: J. Opt. Soc. Am. **42**, 572 (1952).

[24] Barkas, W. W.: Proc. Phys. Soc. London **51**, 274 (1939).

sample. This means that any given real surface can be described by a model consisting partly of diffuse reflecting and partly of regular reflecting elementary surfaces which, in their turn, can be inclined against the macroscopic surface statistically. Thus this somewhat arbitrary supposition ignores the energy reflected from the interior of a powder.

In spite of not accounting for the reflected light originating in the interior of a powder, the *Bouguer* elementary mirror hypothesis has, until recently, been employed to explain "diffuse" reflection from surfaces. This has been the case even though *Grabowski*[25] a half century ago had proved through theoretical considerations that a circularly diffuse reflection which obeys the Lambert Law cannot in principle be explained by the Bouguer hypothesis, independently of the distribution function used for the orientation of the elementary mirrors. If we assume that this orientation is statistically symmetrical and independent of the azimuth, the Bouguer hypothesis for the reflection coefficients leads to an operating equation whose general solution is given by

$$R(i) = \frac{A}{\cos^{2n} i}, \qquad (64)$$

where i is again the angle of incidence at an elementary mirror and A and n stand for suitably chosen constants. This reflection coefficient is different from that in the Fresnel equations. Since $R(i)$ should increase with i, n must be positive. Then, however, Eq. (64) contradicts the requirement that the reflection coefficient reach the limiting value of 1 when $i = 90°$. Instead, the reflection coefficient becomes greater than 1 and this is a physical impossibility. *Grabowski* concludes that isotropic diffuse reflection can in no case be explained by the *Bouguer* elementary mirror hypothesis.

Berry[26] attempted another interpretation by calculating the reflectance of matte surfaces assuming different distribution functions for the Bouguer elementary mirrors and comparing the results with the Lambert Law. If we designate by dF the surface of the mirrors whose normals lie in the solid angle $d\omega$ such that their inclinations lie between p and $p + dp$, the Gaussian distribution of these inclinations is given by

$$dF = k \cdot e^{-a^2 p^2} dp. \qquad (65)$$

In place of (59) we thus obtain for the reflected radiation flux per cm² and unit solid angle

$$\frac{dI_r/df}{d\omega} = \frac{\text{Const } e^{-a^2 \tan^2 \frac{1}{2}(\alpha - \vartheta)}}{\cos^2 \frac{1}{2}(\alpha - \vartheta)}. \qquad (66)$$

---

[25] *Grabowski, L.*: Astrophys. J. **39**, 299 (1914).
[26] *Berry, E. M.*: J. Opt. Soc. Am **7**, 627 (1923).

This law agrees very well with the Lambert Law if $\alpha = 0$; i.e., at perpendicular incidence. On the contrary, at other incidence angles (e.g., when $\alpha = -\vartheta$, or if the incidence and observation directions are coincident), large differences between the two laws occur. If we use in place of the Gaussian probability distribution the similar distribution function

$$dF = \frac{k}{a^2 + p^2} dp, \qquad (67)$$

we obtain in place of (66) the equation

$$\frac{dI_r/df}{d\omega} = \frac{\text{Const}\, a^2 [1/\cos^2 \frac{1}{2}(\alpha - \vartheta)]}{a^2 + \tan^2 \frac{1}{2}(\alpha - \vartheta)}. \qquad (68)$$

By judicious choice of a^2 we can obtain curves that agree with those of the Lambert Law at least at small observation angles if $\alpha = 0$. At higher ϑ-values and greater angles of incidence, however, considerable differences between the two laws arise. Nevertheless, *Berry* concludes from his calculations that *Grabowski's* view of *Bouguer's* elementary mirror hypothesis as untenable is not entirely justified.

It is nevertheless not to be perceived why the Fresnel formulas should no longer be valid for the elementary mirror so long as the condition that the crystal dimensions greatly exceed the wavelength remains fulfilled.

Rense[27] made measurements of the intensity relationships of beams reflected from matted or etched glass by comparing the parts polarized parallel and perpendicular to the plane of incidence. The results showed that for observation and incidence angles less than 45°, the diffuse reflection could be nearly completely described by the regular reflection of elementary mirrors whose orientations obey a Gaussian distribution function. To be sure, this represents a confirmation of *Berry's* views, but again is valid only in a limited angular region of α and ϑ. Thus, a solution of the problem of "diffuse reflection" cannot be asserted.

e) Experimental Investigation of Diffuse Reflection at Non-Absorbing Materials

We shall not dwell on the earlier information [28]. From the measurements of *Bouguer*[29] and *Ångström*[30], however, it has been already seen

---

[27] *Rense, W. A.:* J. Opt. Soc. Am. **40**, 55 (1950).

[28] Compare the comprehensive summary of *Schoenberg, E.:* Handb. der Astrophysik, Vol. II, T. 1. Berlin: Springer-Verlag 1929; — *Falta, W.:* Photometrische Untersuchungen an photographischen Papieren verschiedener Oberfläche. Jenaer Jahrbuch **1954**, 91.

[29] *Bouguer, P.:* Traité d'optique sur la gradation de la lumière. Paris 1760.

[30] *Ångström, K.:* Wied. Ann. **26**, 253 (1885).

that at least for small angles of incidence, α, the Lambert Law (59) is superior to that of *Seeliger* (60). A survey of the experimental material as a whole shows that the Lambert formula fits the observations better than the Seeliger relation. The latter will therefore be omitted from now on. Special attention should be given to the careful measurements by *Wright*[31] on plates made from powders (such as magnesium carbonate, calcium sulphate, and others) of particle size around 2μ, com-

Table 3. *Verification of the Lambert cosine law by Wright. Reflecting medium: calcium sulphate;* $\lambda = 589$ mμ; *azimuth* $= 180°$

ϑ \ α	0°	20°	40°	60°	80°
0°	—	—	-1.3	$+4.9$	-6.3
20°	—	-2.2	$+3.5$	$+5.5$	-4.0
40°	-1.9	$+0.3$	$+1.4$	$+8.0$	-1.9
60°	0	-1.6	$+1.1$	$+16.0$	$+39.0$
80°	0	$+0.3$	$+0.6$	$+47.0$	$+212.0$

pressed under high pressure. The plates were prepared between fine drawing boards at pressures of 4 to $20 \cdot 10^3$ atm. and were largely matte (gloss-free). Measurements of relative radiation density were carried out at different values of the angles α and ϑ, and at the azimuth value $\varphi = 180°$. Table 3 presents one of *Wright's* series of measurements; it shows the percentage deviations of the measured values from those evaluated from equation (59) for different values of α and ϑ. The reference point adopted was the measured value for $\alpha = 0°$ and $\vartheta = 30°$. The mean error amounted to $\pm 2\%$.

It is seen from the table that, for angles of incidence between 0° and 40°, and angles of observation between 0° and 80°, the Lambert Law is completely confirmed within the accuracy of the measurement, whereas for larger angles of incidence, the errors lie far outside the accuracy of measurement. This is also true for the remaining examples investigated by *Wright*. The good proportionality of the measured radiation flux to cosϑ likewise represents a complete contradiction with the *Seeliger* formula. Later authors[32] have also reported similar observations, though the samples and their surfaces were often not so well-defined as in the case of *Wright*.

[31] *Wright, H.*: Ann. Physik **1**, 17 (1900).

[32] *Woronkoff, G. P.*, and *G. J. Pokrowski*: Z. Physik **20**, 358 (1924); compare also *Henning, F.*, and *W. Heuse*: Z. Physik **10**, 111 (1922); — *Schulz, H.*: Z. Physik **31**, 496 (1925).

As a result of *Wright's* work, the following conclusions can be drawn:

a) The reflected radiation flux is not symmetrical in relation to the angle of incidence α and observation angle ϑ. This is in contrast to the *Lambert* Law and also the *Helmholtz* reciprocity law (see p. 172).

b) The irradiation intensity does not vary proportionally to cos α as *Lambert* assumed.

c) On the other hand, the Lambert Law is quite valid for matte surfaces with regard to the remitted energy flux if we exclude the cases of large angles of incidence α and simultaneously large angles of observation ϑ.

d) At large incidence angles and equally large observation angles we observe even from apparently completely matte surfaces noticeable glossy peaks, indicating large deviations from the Lambert Law.

The influence of the azimuth between the incident and observed radiation has been investigated more closely by *Thaler*[33]. In general for an azimuth of 180°, when α = ϑ, that is, in the direction of the macroscopic reflection angle, a reflection which is appreciably too high is observed, which indicates that a contribution of regular reflection has become considerable. This is particularly true for large angles of incidence.

Finally, the results of this kind of reflection measurements on powders are also affected by the size of the particles and their packing density, as well as by the condition of the surface. Examples of this kind of measurement on barium sulphate carried out by *Budde*[34] show that the so-called goniophotometric curve ($B = f(\vartheta)$ for a given value of α) is not parallel to the abscissa, as the Lambert Law requires, but depends upon both particle size and the condition of the surface. Rough surfaces show falling and difficultly reproducible curves in constrast to polished ones. From the more recent measurements by *Höfert* and *Loof*[35], it is likewise evident that the deviations from the goniophotometric curves increase with increasing ϑ. Besides the shape of the curves depends upon the nature of the material (being characteristically different for magnesium oxide and barium sulphate).

A characteristic example[36] of the effect of packing density is shown in Fig. 13, in which the indicatrix for the reflection of infrared radiation at powdered germanium (particle size <0.15 mm) is given. The upper portion is concerned with densely packed powder with a smoothly pressed surface, and the lower with loosely scattered powder. The angle of incidence was 45° in both cases. The outer curves deal with the total radiation of a "Globar" at a temperature of 1330 °K, while the inner

---

[33] *Thaler, F.*: Ann. Physik **11** (4), 996 (1903).
[34] *Budde, W.*: J. Opt. Soc. Am. **50**, 217 (1960).
[35] *Höfert, H. J.*, and *H. Loof*: Die Farbe **13**, 53 (1964).
[36] *Agnew, J. T.*, and *R. B. McQuistan*: J. Opt. Soc. Am. **43**, 999 (1953).

curves refer to the same radiation from which the waves of $\lambda > 4\,\mu$ have been filtered out by means of an inserted quartz filter. It is seen that in the case of the densely packed sample a very substantial fraction of regularly reflected radiation is found when $\alpha = \vartheta$ and $\varphi = 180°$. The same is found in the case of silica, graphite, and glass powder, the effect increasing with increase in particle size. Individual samples (e.g. metal powders

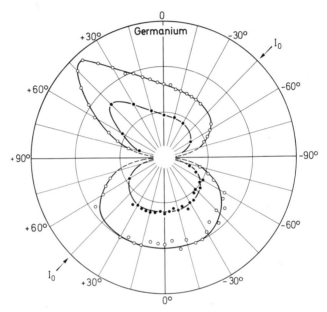

Fig. 13. Indicatrix for the reflection of infrared radiation on germanium powder, densely packed with level pressed surface (top) and loose (bottom); grain size < 0.15 mm

of Te, Al and brass) also show "reverse gloss" indicating additional regular reflection at azimuth $\varphi = 0°$. Apart from this effect, the indicatrix when $\alpha = 45°$ is in all cases a flattened ellipsoid instead of a sphere, as already indicated qualitatively by the observations of earlier workers. Thus, in the IR also it is difficult to speak of the validity of the Lambert Law.

The deviations from the spherical indicatrix are naturally even more pronounced, if a regular fraction is reflected directly in the form of "gloss" from the surface when observed under larger values of ϑ. This is the case, for example, with papers. Fig. 14 shows reflection measurements of a white drawing paper[37] for various angles of incidence and observation (indicated by arrows). For $\alpha = \vartheta$ and $\varphi = 180°$, that is, for the macroscopic

---

[37] Barkas, W. W.: Proc. Phys. Soc. **51**, 274 (1939).

angle of reflection, the radiation density is higher with increasing angles of incidence. It is also noteworthy that, not only when $\alpha = \vartheta$, but also for a larger range of angles on both sides of specular reflection, the radiation density found is too high. The fibre structure in textiles [38], or the internal surface of cell walls [39] revealed by the cutting of wood, show an appreciable amount of specular reflection which reveals itself in deviation from the spherical indicatrix.

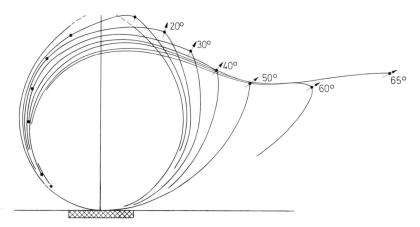

Fig. 14. Indicatrix for the reflection of white light on drawing paper for various incidence angles (marked by arrows)

Measurements of the state of polarization of "diffusely" reflected radiation lead to the same results. For ideal diffuse reflection it would be expected that incident natural light would not be partly polarized, while incident linear polarized light would be fully depolarized. Measurements [40] of polarization of natural light on diffuse reflection at magnesium oxide for various angles of incidence and reflection have, according to the observations of *Wright*, shown that only for very large values of α and ϑ is there a slight linear polarization of the reflected light.

A few measurements have been reported of the depolarization of linear polarized light reflected at the diffusely reflecting surface of magnesium oxide [41]. It has been found that light polarized either perpendicular

---

[38] *S: son Stenius, Å.:* Svens. Papperstidning **54**, 663 (1951).

[39] *Barkas, W. W.:* Proc. Phys. Soc **51**, 274 (1939).

[40] *Woronkoff, G. P.,* and *G. I. Pokrowski:* Z. Physik **30**, 139 (1924).

[41] *Umov, N.:* Physik Z. **6**, 674 (1905); **13**, 962 (1912); — *Návrat, V.:* Wiener Ber. II a, **120**, 1229 (1911); — *Pokrowski, G. I.:* Z. Physik **32**, 563 (1925); — *Woronkoff, G. P.,* and *G. I. Pokrowski:* Z. Physik **30**, 139 (1924); **33**, 860 (1925); — *Gorodinskii, G. M.:* Opt. Spectr. **16**, 59 (1964).

or parallel to the plane of incidence is never completely depolarized on reflection, and that the fraction of polarized light after reflection increases with increasing angles of incidence and reflection. This further supports the view that a certain fraction of regularly reflecting surface elements exist for which the Fresnel equations (14) and (15) should be used, while the radiation returning from the inside of the medium to the surface will always be depolarized. The minimum expected at the polarization angle for light polarized parallel to the plane of incidence (see Fig. 3) was not observed, which could be attributed to inadequate measurement accuracy.

With the help of a modern photoelectric method, systematic goniophotometric measurements testing the Lambert Cosine Law on powders of magnesium oxide, barium sulfate, rutile and Aerosil have been recently made. The object was to test the influence of such parameters as particle size, packing density, method of sample preparation, wavelength, etc., in such a way that not only could more precise measurements be made, but also to provide a comprehensive explanation of these factors on experimental results [42]. The measured quantity was the relative radiation density $(B(\alpha, \vartheta)/B(0°, 45°))$ normalized to $\alpha = 0°$ and $\vartheta = 45°$. If the measured quantity is divided by $\cos\alpha$ (i.e., related to constant illumination intensity), $B_{rel}/\cos\alpha$ should, according to the Lambert Cosine Law, be a constant equal to 100, or lie on the abscissa as a function of ϑ. Deviations from the constant represent deviations from the $\cos\vartheta$-Law, while deviations from 100 represent the total deviation from Lambert behavior.

It appears that the method of preparation of the sample has a decisive influence on the validity or non-validity of the Lambert Law. Also, specific material properties such as refractive index and crystal form have an influence.

We can survey the influence of various parameters on the validity of the Lambert Law with the aid of Fig. 15a to 15g. The figures all represent measurements on the same purified sample of $BaSO_4$ dried at 500° C, and $B_{rel}/\cos\alpha$ is plotted as a function of ϑ. The azimuth is always 180°.

Fig. 15a shows that the reduced radiation density of sample 1 falls for all chosen incidence angles (with the exception of $\alpha = 0$) with increasing observation angle. At the macroscopic reflection angle $\vartheta = \alpha$, we observe weak glossy peaks that become more intense with increasing α. This must be explained as a Fresnel effect: The reflection coefficient increases with increasing α. If we do not consider these regions of specular reflection, the average deviation from the $\cos\vartheta$-Law on this

---

[42] *Kortüm, G.*, and *R. Hamm*: Ber. Bunsenges. **72**, 1182 (1968); see further: *Torrance, K. E., E. M. Sparrow*, and *R. C. Birkebak*: J. Opt. Soc. Am. **56**, 916 (1966); – *Torrance, K. E.*, and *E. M. Sparrow*: J. Opt. Soc. Am. **57**, 1105 (1967).

sample, related to the average, amounts to about 5.8%. On the other hand, the average deviation from the cosα-Law of 11.2% is almost twice as large as that predicted by the Helmholtz Reciprocity Law. All curves lie below the Lambert Constant 100 and the total deviations from Lambert behavior likewise increase with α.

The sample 2 of Fig. 15b was prepared no differently from that of Fig. 15a except merely that particles from a sieve fraction with diameter less than 50 microns were used. The reduced radiation density falls, however, only slightly at the beginning and then increases again with increasing observation angle ϑ. At $\alpha = 60°$, it even climbs over the

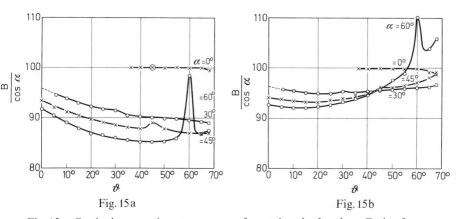

Fig. 15a.

Fig. 15b.

Fig. 15a. Goniophotometric measurements for testing the Lambert Cosine Law: reduced radiation density $B_{rel}/\cos\alpha$ as a function of the observation angle ϑ for BaSO$_4$ powder
Sample 1. Particle diameter d 74 to 90 μ. Surface leveled with a glass plate under low pressure; $\lambda = 450$ mμ

Fig. 15b. Sample 2. Particle diameter $d < 50$ μ, otherwise the same as Fig. 15a; $\lambda = 450$ μ

reference value 100; i.e., the curves cross one another at intermediate angles ϑ. The curves do not lie so far under the reference value 100 as those of Fig. 15a and thus the deviations from complete Lambert behavior are considerably smaller. Because the deviations from the cosϑ-Law are of the same magnitude as those of Fig. 15a, this behavior must be due to smaller deviations from the cosα-Law. In any case, the glossy peaks are somewhat smaller than with a coarser powder.

Sample 3 (Fig. 15c) was prepared from still finer powder and under high pressure. It exhibited extraordinarily high and wide glossy peaks that largely determined the shape of the curves and increased with increasing α. The Richter "Gloss Number", defined as

$$\eta \equiv \frac{B(\alpha = 22.5°;\ \vartheta = 22.5°)}{\cos 22.5°} \bigg/ \frac{B(\alpha = 45°;\ \vartheta = 0°)}{\cos 45°}$$

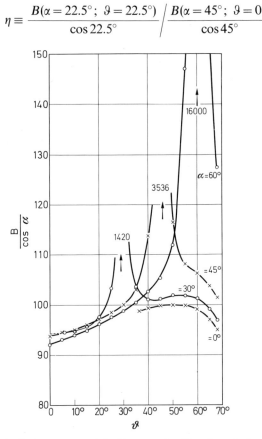

Fig. 15c. *Sample 3.* Average particle diameter $d \ll 42\ \mu$. Sample hard pressed with a polished die stamp at about 1000 atm. Surface smooth and strongly specular; $\lambda = 450\ m\mu$

(i.e., the quotient of the reduced radiation density of a specularly reflecting spot to the reduced radiation density of a non-specularly reflecting spot with a value of 1 for a completely matte surface) has a value here of 10.6 at 450 mµ. In addition, except at small and large angles ϑ the curves lie above the reference value 100, and the total deviation from Lambert behavior is positive here instead of negative when we exclude the specularly reflecting areas. At large observation angles ϑ the curves again fall rather sharply in contrast to those of samples 1 and 2.

If we sieve a thin layer of the same powder on the glossy surface (sample 4, Fig. 15d), the glossy peaks become smaller and narrower and the ascending tendency of the curves of Fig. 15c at small angles ϑ

again reverts to a falling tendency. The curves run almost exclusively below the reference value 100 and the total deviations from Lambert behavior are again negative.

In sample 5, the sieved layer of $BaSO_4$ with $d \ll 42\,\mu$ was increased until the glossy peaks disappeared, η fell from 10.6 to 1.005 at 450 mµ. With the exception of the $\alpha = 60°$ curve, the curves have a descending tendency throughout the entire ϑ-range investigated and are similar to those of Fig. 15a if we ignore the glossy peaks of the latter. Only

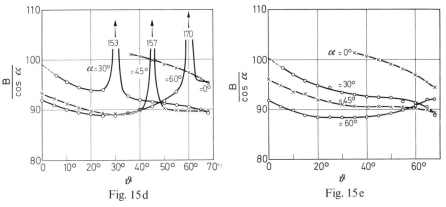

Fig. 15d. *Sample 4.* A thin layer of $BaSO_4$ was sieved through a 42 µ sieve on the specular layer of sample 3; $\lambda = 450$ mµ

Fig. 15e. *Sample 5.* Increasing of the sieved layer until the gloss disappears; $\lambda = 450$ mµ

the curve for $\alpha = 0$ is not constant here as in sample 1, but likewise descends strongly with ϑ. All curves lie below the reference value 100 and the total deviations from Lambert behavior are negative. The deviations from the $\cos\vartheta$-Law come to between 4 and 10% of the reference value, and are, to be sure, smaller at large α-values than at small α-values, analogous to Fig. 15a.

If the glossy surface of the strongly pressed sample 3 is roughened with blotting paper (sample 6, Fig. 15f), we obtain curves similar to those in Fig. 15c but with much smaller glossy peaks. η falls from 10.6 to 1.08 at 450 mµ. What is more important, for $\alpha > 0$ the curves exceed the reference value 100 at relatively small observation angles ϑ indicating a much steeper ascent than in Fig. 15c if we continue to ignore the glossy peaks present. The deviation from the $\cos\vartheta$-Law at $\alpha = 60°$ amounts to about 44% of the relative value without considering the glossy peaks. At small α- and large ϑ-angles, a falling tendency again occurs in the curves. The total deviation from Lambert behavior is nevertheless positive.

Sample 7 was pressed under about 1000 atm pressure as sample 3, but between smooth sheets of paper that could be easily torn away later. The sample was not specular in contrast to sample 3 and the gloss number $\eta = 1.04$. As Fig. 15g shows, there were almost no glossy peaks, all curves, including that for $\alpha = 0$, climb with increasing ϑ, and all curves lie such that the total deviation from Lambert behavior approaches zero if we average over all points[43]. Nevertheless, consider-

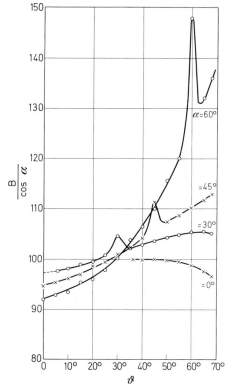

Fig. 15f. *Sample 6.* The hard pressed surface of sample 3 was carefully rubbed and thereby roughened with blotting paper; $\lambda = 450$ mµ

---

[43] By such an averaging over all measured incidence angles α and observation angles ϑ, we can, omitting the specularly reflecting spots, give a single number as the value for the total deviation from Lambert behavior, as *Wright* has done. The example of this sample shows that such a number says nothing about the true course of the goniophotometric curves because the observed deviations mutually can almost completely compensate themselves. Such an average value would be characteristic at best for a specially prepared sample surface and would be strongly dependent on the preparative method of the sample.

able deviations from the $\cos\vartheta$-Law occur and become greater with increasing α. On the average, these deviations amount to 16.7%. The deviations from the $\cos\alpha$-Law are small at small ϑ-angles but increase more and more above $\vartheta = 20°$ so that their average value is 8.6%. Thus the view of *Wright* that the $\cos\vartheta$-Law is always fulfilled better than the $\cos\alpha$-Law does not always hold true.

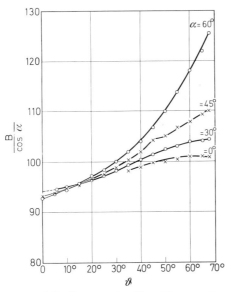

Fig. 15g. *Sample 7.* Particle diameter $d \ll 42\,\mu$. The sample was hard pressed as sample 3 at about 1000 atm. except that a piece of smooth paper was placed between the die and the powder. After removal of the paper, the surface appeared smooth but not specular

In a similar manner the goniophotometric curves of the following white standards were investigated in dependence on the parameters named above:

Rutile (TiO_2): most pure from Farbenfabriken Bayer, Uerdingen, was dried at 500 to 600° C and fractionated according to powder size with sieves. Measurements at $\lambda = 450$, 550, and 649 mμ.

Magnesium oxide (MgO): p.a. from Merck was dried at 600° C. The material was such a fine powder that no fractionation was possible with sieves ($\bar{d} \cong 2\,\mu$). Measurements at $\lambda = 450$ and 550 mμ.

Aerosil (SiO_2): from Degussa was dried at 600° C. Samples were prepared from material of 106 and 38 m^2/g specific surface area, respectively. The powder particles were considerably smaller ($\bar{d} \leq 20$ mμ) than the wavelength of the light ($\lambda = 401$ mμ) used in the measurements.

Excluding the influence of specific material characteristics (crystal form, compressibility, etc.) that cause certain small differences in the behavior of the various white standards, we see in the goniophotometric curves of all four materials, in dependence on the named parameters, an analogous very characteristic pattern which can be read from the Fig. 15a to 15g [44].

With the *application of pressure* with a flat surface (a glass plate or polished metal die), we always obtain glossy peaks at $\alpha = \vartheta$ that become higher with increasing pressure and increasing incidence angle α. At very high pressures, they increase considerably in breadth; i.e., they extend themselves over a much larger range of ϑ (Fig. 15c). We can trace these glossy peaks back to the elementary mirrors of *Bouguer*. If we assume regular reflection from the crystal faces, the elementary mirrors must be largely oriented in the macroscopic surface because of the applied pressure. Only in this way is the angle of incidence at the elementary mirror equal to the macroscopic incidence angle α and with it $\alpha = \vartheta$. Furthermore, because the frequency distribution of the oriented elementary mirrors is equal for all α, the same number of elementary mirrors always contributes to the specular reflection. The increase in intensity of the glossy peaks with α may thus be informally explained as the increase of the reflection coefficient with increasing angle of incidence according to the Fresnel formula (16). With increasing pressure, the number of elementary mirrors oriented in the surface will increase and with it, the intensity of the glossy peaks [45].

It is noteworthy that we observe an analogous behavior even with aerosil where the particle dimensions are somewhat less than one tenth the wavelength used for the measurements. We should thus expect that in this case perhaps no fraction of regular reflection from elementary mirrors in the surface would be possible. Nevertheless, we see here also discernable glossy peaks at high pressures. Their intensity increases as α increases while they can scarcely be recognized at low pressures. We must evidently assume that the very small and closely packed aerosil particles can partly constitute closed surface regions

---

[44] In this connection, compare also the p. 38 cited work of *Kortüm* and *Hamm* and the dissertation of *R. Hamm*, Tübingen 1968.

[45] The widening of the glossy peaks with increasing incidence angle α, which is chiefly observed with high glossy peaks, cannot be traced back to the distribution function of the elementary mirrors, because this is fixed and independent of α. The widening must arise from elementary mirrors that are only slightly inclined against the macroscopic surface and still are relatively frequent. But because their number is also independent of α, the widening of the glossy peaks with increasing α must depend on how many elementary mirrors send light into the detector of given $d\omega$. As Eq. (71) shows, this depends on the angle i or $\frac{\alpha + \vartheta}{2}$ (see p. 46).

where the single particles are held together by hydrogen bonds in such a way that real specular reflection can be observed[46]. The widely held view that remission becomes more diffuse as particle size decreases is thus not valid in this case.

If we press the samples at high pressure between smooth papers that may later be removed, the glossy peaks almost completely disappear. In contrast, the goniophotometric curves still climb with increasing ϑ and the rate of change becomes greater with increasing incidence angle α (Fig. 15g). The same thing happens with magnesium oxide and rutile but the rate of change in this series decreases in spite of the identical method of preparation of the samples. On the whole, rutile obeys the Lambert Law the best. At small incidence angles α, and especially at $\alpha = 0$, we almost always see a small decrease in the goniophotometric curves with increasing ϑ. This is easily recognized in Fig. 15f. We can extract several clues from these observations:

a) The piece of paper between die and sample hinders the parallel orientation of the elementary mirrors to the surface largely through its characteristic structure to which the surface adjusts itself. In this way, the glossy peaks disappear, i.e., the distribution function of the elementary mirrors relative to their position in the macroscopic surface is broadened and steeper angles ε also occur. The same effect can be obtained by roughening a hard pressed glossy surface instead of pressing the sample between sheets of paper. Comparison of Figs. 15c and 15f shows this.

b) The ascent of the goniophotometric curves with increasing ϑ, and especially at large incidence angles α, increases in the order rutile, magnesium oxide, barium sulfate to aerosil under otherwise equal conditions. This is, however, the same series progression as the increase of the reflectance of these materials according to the Fresnel formula (16). These relationships are again shown in Fig. 16 where, according to *Snell*'s law of refraction, $\sin\beta = \sin\alpha/n$ [47]. The reflectance thus becomes dependent on the refractive index of the reflecting medium. We see that R_{reg} for rutile has the most constant behavior. The steep ascent in the series MgO, BaSO$_4$ to SiO$_2$ begins at smaller and smaller incidence angles and is undoubtedly a result of the refraction indices which decrease in the same order.

---

[46] Compare discussion meeting: Physik. Grundlagen der anwendungstechn. Eigenschaften von Pigmenten. Ber. Bunsenges. **71**, 239 (1967).

[47] The refractive indices for $\lambda = 450$ mµ were taken from the literature and, because the crystals lie distributed in all directions, averaged when several indices of refraction were existing. Since it is a question of a half-quantitative consideration, the *n*-values can be excepted from great exactitude.

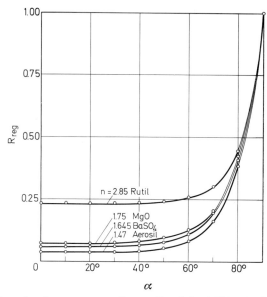

Fig. 16. Fresnel reflectance according to Eq. (16) of TiO_2 (rutile), MgO, $BaSO_4$, and SiO_2 (Aerosil) as a function of the angle of incidence α of the radiation

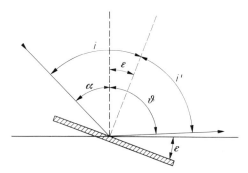

Fig. 17. Orientation of elementary mirrors to the macroscopic surface

The angle of incidence i on an elementary mirror with the orientation angle ε (compare Fig. 17) is determined through the macroscopic angle of incidence α and the observation angle ϑ. The elementary mirrors whose surface normals halve the angle between the incidence and observation direction reflect regularly in the direction ϑ. This means that $i + i' = 2i = \alpha + \vartheta$ or

$$i = i' = \frac{\alpha + \vartheta}{2}. \tag{69}$$

The expression for the orientation angle ε is

$$\varepsilon = \vartheta - i = \vartheta - \frac{\alpha + \vartheta}{2} = \frac{\vartheta - \alpha}{2}. \tag{70}$$

At a given incidence direction α, all elementary mirrors whose surface normals lie in a solid angular element $d\omega_\varepsilon$ around the orientation angle ε contribute to the regular reflection in the solid angle $d\omega$ around the observation direction ϑ. Between both solid angular elements we find the relationship [48]

$$d\omega_\varepsilon = \frac{d\omega}{4\cos i}. \tag{71}$$

The number of reflecting elementary mirrors in the surface element dF under measurement is given by

$$Z(i) = f(\varepsilon)\, d\omega_\varepsilon\, dF, \tag{72}$$

where $f(\varepsilon)$ is the distribution function of the elementary mirrors over the orientation angles. As we can see from (71) and (72), in the case of an equal distribution, the light arrives in the detector from more elementary mirrors the greater the angle i; i.e., the greater the angle between the incident and observed beam.

If $\bar{q}$ is the average area of a mirror,

$$dQ(i) \equiv \bar{q} Z(i) = \bar{q} f(\varepsilon) \frac{d\omega}{4\cos i} dF \tag{73}$$

is the resultant reflecting surface in the solid angular element $d\omega$.

Now, if we assume an equal distribution of the orientation angles according to *Bouguer*, the distribution function is given by

$$f(\varepsilon) = f\left(\frac{\vartheta - \alpha}{2}\right) = \text{const}.$$

In this case, the projection of the surface normals of dQ in the direction of the incident beam becomes independent of all angles; i.e.,

$$dQ_n = dQ(i)\cos i = \text{const}\, \bar{q}\, \frac{d\omega}{4}\, dF = \text{Const}. \tag{74}$$

This projected plane dQ_n lies parallel to the cross section of the incident light beam. Because the intensity of the light source is constant and because the beam cross section f is held constant, the radiation flux falling on the plane dQ_n is constant. It is given by

$$dI_e = dQ_n \cdot \frac{dI_e}{df} = \text{Const}\, \frac{dI_e}{df}. \tag{75}$$

---

[48] *Rense, W.:* J. Opt. Soc. Am. **40**, 55 (1950).

The reflected radiation flux is, according to Fresnel,

$$dI_r = R_{reg}(i)\, dI_e \tag{76}$$

where $R_{reg}(i) = R_{reg}\left(\dfrac{\alpha + \vartheta}{2}\right)$ is the regular reflectance of an elementary mirror.

From (76) and (75), the reflected radiation flux per unit surface area is

$$\frac{dI_r}{df}(\vartheta) = \text{Const}\, \frac{dI_e}{df}\, R_{reg}\left(\frac{\alpha + \vartheta}{2}\right). \tag{77}$$

If we relate the reflected radiation flux at the angle ϑ from any incidence angle α to the reflected radiation flux at the angle $\vartheta = 45°$ from perpendicular incidence, we obtain

$$\frac{dI_r(\vartheta)}{dI_r(\alpha = 0°;\, \vartheta = 45°)} = \frac{R_{reg}\left(\dfrac{(\alpha + \vartheta)}{2}\right)}{R_{reg}(22.5°)} \equiv F_{n,\alpha}(\vartheta). \tag{78}$$

The angle of incidence α plays the roll of a parameter here. The function $F_{n,\alpha}(\vartheta)$ is dependent on the refractive index and is shown in Fig. 18 for rutile ($n \cong 2.85$), MgO ($n \cong 1.75$), BaSO$_4$ ($n \cong 1.645$), and SiO$_2$ ($n \cong 1.47$) at α-value parameters of $0°$, $30°$, and $60°$. We see that the curves ascend more steeply with the observation angle ϑ the greater the incidence angle α and the smaller the refractive index n. This is unquestionably a result of the Fresnel relationships.

According to the Lambert Law, the radiation flux remitted from a unit surface area is proportional to the cosine of the observation angle ϑ and to the solid angle $d\omega$ (Eq. (54)), whereas the radiation density B should be independent of ϑ. If we leave open for the moment whether or not this is compatible with the Bouguer elementary mirror hypothesis and set B as a function of ϑ, according to (54)

$$\frac{dI_r}{df} = B(\vartheta)\cos\vartheta\, d\omega$$

should be valid. If we now relate again to the value for $\alpha = 0°$ and $\vartheta = 45°$, the result is

$$\frac{dI_r(\vartheta)}{dI_r(\alpha = 0°;\, \vartheta = 45)} = \frac{B(\vartheta)\cos\vartheta}{B(\alpha = 0;\, \vartheta = 45)\cos 45°}. \tag{79}$$

If we now equate the left sides of (78) and (79), i.e., assume that the Lambert Law is compatible with the Bouguer hypothesis and the Fresnel formulas, we obtain

$$F_{n,\alpha}(\vartheta) = \frac{B(\vartheta)}{B(\alpha=0;\vartheta=45)} \cdot \frac{\cos\vartheta}{\cos 45°}. \quad (80)$$

The normalized surface brightness was measured directly and from (80) is given by

$$\frac{B(\vartheta)}{B(\alpha=0;\vartheta=45)} = \frac{\cos 45°}{\cos\vartheta} \cdot F_{n,\alpha}(\vartheta) \equiv G_{n,\alpha}(\vartheta). \quad (81)$$

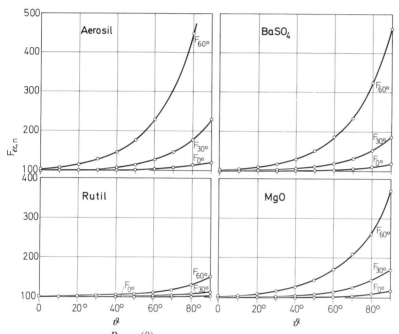

Fig. 18. $F_{n,\alpha}(\vartheta) \equiv \dfrac{R_{regn,\alpha}(\vartheta)}{R_{regn,0}(45°)} \cdot 100$ as a function of ϑ for rutile, MgO, BaSO$_4$, and SiO$_2$ at $\alpha = 0°$, $\alpha = 30°$, and $\alpha = 60°$

The functions $G_{n,\alpha}(\vartheta)$ resulting from the Bouguer hypothesis with equally distributed elementary mirrors and the Fresnel Laws upon which the functions $F_{n,\alpha}(\vartheta)$ depend are the normalized radiation densities. The geometric factor $\cos 45°/\cos\vartheta$ amplifies still the steepness of the $F_{n,\alpha}(\vartheta)$-curves of Fig. 18 [49]. The $G_{n,\alpha}(\vartheta)$ functions are given in Fig. 19 for the white standards under investigation. They increase more steeply with ϑ the greater the angle of incidence α and the smaller the refractive index n.

[49] Actually $\lim\limits_{\vartheta \to 90} G_{n,\alpha}(\vartheta) = \infty$ is valid; i.e., we come to the same conclusion as *Grabowski* (compare p. 32).

This shows that the radiation density $G_{n,\alpha}(\vartheta)$, calculated according to the Bouguer elementary mirror hypothesis using the Fresnel formulas, can never be constant in the case of equal distribution of the elementary mirrors. For this case, the Lambert Law and the Bouguer representation are not compatible. It remains an open question if there exists a distribution function $f(\varepsilon) = f\left(\dfrac{\vartheta - \alpha}{2}\right)$ for which the expression

$$G_{n,\alpha}(\vartheta) \cdot f\left(\frac{\vartheta - \alpha}{2}\right) = \text{const} \tag{82}$$

at all angles of incidence α, so that the Lambert Law remains valid. *Grabowski* answers this question in the negative (compare p. 32).

Comparison of the calculated $G_{n,\alpha}(\vartheta)$-curves with the measured normalized radiation densities is only permissable for those samples with constant elementary mirror distribution functions. This requirement is most nearly fulfilled for samples prepared under high pressure between paper sheets (e.g., Fig. 15g). Samples with specular glossy peaks as well as loose samples whose radiation densities fall with increasing ϑ are naturally not suitable for this comparison.

If the comparison is restricted to samples of the sort illustrated in Fig. 15g, we find that the experimentally observed radiation density in the series rutile, MgO, BaSO$_4$, and SiO$_2$ ascend more and more steeply in dependence on ϑ. This behavior is expected from the Bouguer hypothesis. In the region of observation angles ϑ, where measurements were made, the inconstancy of $B(\vartheta)/\cos\alpha$ with rutile is least, with MgO and BaSO$_4$ almost equal, and with aerosil considerable. On the basis of the differences in the refractive indices of these materials, this result corresponds completely to the expected results according to the Fresnel equations. These samples do not fulfill the requirement of a constant distribution function for the elementary mirrors. Thus, a *quantitative* correspondence between the measured and calculated normalized radiation densities cannot be expected. To be shure, these samples were prepared under high pressure, but between paper sheets, so that very large angles of inclination ε will occur much more seldom than smaller angles. Such an assuredly wide, although not constant, distribution function will give a flat maximum at $\varepsilon = 0$. This could perhaps explain the experimental radiation density curves for $\alpha = 0°$ (Fig. 15g) that fall somewhat at high ϑ-angles instead of increasing steadily. When large inclination angles ε are relatively rare, the radiation density reflected at large angles ϑ must be too small compared to the case where there is an equal distribution of the elementary mirrors over all possible angles ε. Furthermore, the calculated radiation density curves take a steeper course also at large incidence angles α than the measured curves. This

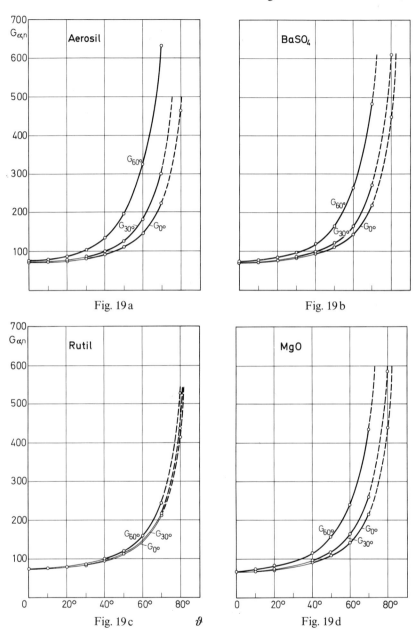

Fig. 19. Normalized radiation densities, calculated on the basis of the Bouguer elementary mirror hypothesis, as a function of ϑ for rutile, MgO, BaSO$_4$, and SiO$_2$ at $\alpha = 0°$, $\alpha = 30°$, and $\alpha = 60°$. A statistically homogeneous distribution of the elementary mirrors is assumed

points out that the Bouguer elementary mirror hypothesis does not alone suffice to explain the observations. For example, the law of multiple scattering should be applied to the radiation that penetrates into the powder and reemerges from the surface after numerous reflections (compare p. 94). However, samples that exhibit a broad distribution function for the orientation angles of the elementary mirrors show qualitatively similar deviations between the measured and calculated radiation densities. This would indicate that the Bouguer conception of elementary mirrors essentially correctly describes the processes of reflection of radiation at powder surfaces. We must, in addition, assume an ideal diffuse scattered fraction from the interior of the sample.

Finally, it is noteworthy that with the finely powdered aerosil, the goniophotometric curves likewise ascend steeply with increasing observation angle ϑ and incidence angle α even when the samples have not been pressed. Here it is sufficient to sinter the samples by firmly pounding the bottom of the sample dish against something solid. Presumably regularly reflecting surface areas with a definite distribution function that obey the Fresnel Laws can form themselves by sintering. Besides, the radiation density ascends more steeply for a fine powder than for a coarse powder. In this case the reflectance by no means becomes "more diffuse" with decreasing grain size.

Although it was to be expected that the *wavelength* λ would have no great influence on the radiation density of the white standards as a function of the observation angle, the λ-dependence with $BaSO_4$, MgO, and rutile was nevertheless investigated in the region from 450 to 650 mμ. The only wavelength influence observed concerned the constancy of the radiation density for rutile which was noticeably better at 450 mμ than at 649 mμ under otherwise identical conditions. The greater constancy of the radiation density of rutile compared to the two other standards can be traced back to the larger refractive index of rutile, so that the dispersion of n will influence this constancy too. The refractive index of rutile, in the given wavelength region, decreases by about 0.2 with increasing λ while the indices of the other two standards decrease by only about 0.02. Therefore, this observation also can be traced back to the Fresnel equations.

A completely different behavior is shown by *loose* or sieved samples of the same materials. The radiation density curves $B/\cos\alpha$ fall with ϑ at small observation angles ϑ and also frequently at larger angles (Figs. 15d and 15e). The curves furthermore always lie under the reference value 100. Such samples appear evidently darker to the eye from an oblique view than by perpendicular observation. The reason lies in the rough condition of the surface which leads to a shadowing of the deeper lying zones by the higher lying zones. Thus by oblique

Diffuse Reflection at Non-Absorbing Materials 53

observation we see preponderantly the dark zones. This effect also commonly increases with increasing angle of incidence α.

A comparison of the experimental results with the law (60) according to v. Seeliger-Lommel gives the following: According to Eq. (60), the radiation density of the remission is given by

$$B = \text{Const} \frac{\cos\alpha}{\cos\alpha + \cos\vartheta}. \tag{83}$$

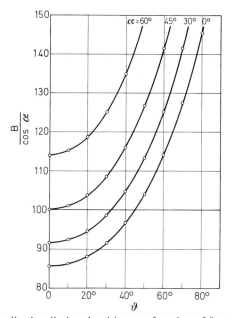

Fig. 20. Normalized radiation densities as a function of ϑ according to v. Seeliger

By normalization to $B(\alpha = 0°; \vartheta = 45°) = 100$, we obtain

$$\text{Const} = 100(1 + \tfrac{1}{2}\sqrt{2}) = 170.7.$$

Thereby we obtain for the reduced radiation density

$$\frac{B}{\cos\alpha} = \frac{170.7}{\cos\alpha + \cos\vartheta}.$$

This function is shown for $\alpha = 0°$, $30°$, $45°$, and $60°$ in Fig. 20 as a function of the observation angle ϑ, as compared with the α- and ϑ-independent Lambert value 100. The comparison shows that the v. Seeliger Law represents the results worse than the Lambert Law. The radiation density in particular ascends much too steeply with the observation angle ϑ at all incidence angles α. The v. Seeliger curve for $\alpha = 60°$ lies far above that for $\alpha = 0°$, while the experimental curves for large α instead begin under the curve for $\alpha = 0°$. The deviations from the Lambert Law are minimized for media with large refractive indices as long as the observation and incidence angles are relatively small. This is not the case at all in the v. Seeliger Law case.

We can summarize the results of this investigation as follows:

1. The Bouguer elementary mirror hypothesis is not compatible with the Lambert Law if an equal distribution of the orientation angles is assumed.

2. None of the samples investigated exactly fulfilled the requirements of the Lambert Law. With increasing incidence and observation angles systematically increasing deviations appear. Earlier contrary findings are traced back to inadmissable averageing.

3. The Bouguer hypothesis cannot explain the ideally diffuse reflectance, but instead makes it possible to understand the observed deviations of the radiation densities from the Lambert law and the influence of the refractive index of the material.

4. Hard pressed samples without glossy peaks show the smallest deviations from constant radiation density. Those with the largest refractive index n also come closest to a Lambert radiator. This can be understood on the basis of the Fresnel formulas and the Bouguer hypothesis.

5. Even very fine powders such as aerosil cannot show constant radiation density if n is small. The widely held view that small particle size favors the validity of the Lambert Law does not, in general, hold true.

6. Loose powder surfaces show shadowing effects. The radiation density decreases more with increasing observation angle the larger the particles.

7. Mutual shadowing and Fresnel effects work contrary to one another and can so interact that, under suitable conditions of particle size and refractive index, the radiation density is to some extent constant in large regions of α and ϑ. In such cases the Lambert Law is apparently valid.

8. The degree of the shadowing effect depends on the surface roughness. Because the goniometric course of the Fresnel curves depends on the refractive index, the regularly reflected parts are material specific. The Bouguer hypothesis therefore can explain this material influences.

9. With nonabsorbing materials, the radiation density is dependent on the wavelength only if the material in question possesses a significant dispersion of n.

10. The v. Seeliger-Lommel Law is inferior to the Lambert Law and can not be theoretically derived even for an ideal diffuse reflector.

f) Diffuse Reflectance at Absorbing Materials

The experimental results of the last section concern themselves with nonabsorbing materials. Providing we disregard low levels of unavoidable contaminants, this means that for all practical purposes the incident radient energy is completely reflected, in part diffusely and in part regularly. However, for the case where the reflecting medium selectively absorbs, the spectral composition of the reflected radiation must deviate from that of the incident beam when it is a continuum. This is based on two processes: Radiation in a certain wavelength region is selectively absorbed in the interior of the sample. However, on the basis of the Fresnel Laws, this same radiation will be preferentially reflected from the sample surface at the Bouguer elementary mirrors. The two processes work against one another and determine in common the spectral composition of the remitted radiation.

Let us assume that the radiation density coming from the interior of the sample is "diffuse," that is, has an isotropic angular distribution, and obeys the Lambert Law. (We will show later that this is quite probable.) In this case, the observed angular dependence of the remitted radiation density will be again practically exclusively determined by the Bouguer elementary mirrors and the distribution function of their inclination angles ε to the macroscopic surface. We should thus expect radiation density curves similar to those of nonabsorbing media in their dependence on the angle of incidence α and the angle of observation ϑ. There will be differences, however, because the deviations from the Lambert Law in the region of selective absorption are essentially greater and because they begin at smaller ϑ-values than in the spectral regions where the illuminated material does not absorb.

This expectation has already been confirmed by older measurements. *Woronkoff* and *Pokrowski*[50] made measurements on paper treated with rhodamine B. The measurements were made at wavelengths of 650 mµ, where rhodamine absorbs only slightly, and 550 mµ, where rhodamine absorbs very strongly. The deviations from the Lambert Law increase with increasing observation angle ϑ and incidence angle α as a parameter, as has been already observed with nonabsorbing materials. Under otherwise equivalent conditions, however, the deviations at $\lambda = 550$ mµ are much greater than those at $\lambda = 650$ mμ and also begin at smaller ϑ-angles. Recent measurements [51] with a photoelectric method on powders of K_2CrO_4, $CuSO_4 \cdot 5H_2O$, chrome alum and naphthol yellow adsorbed on aerosil have yielded analogous results. An example of this behavior appears in Fig. 21 which shows the normalized goniophoto-

---

[50] *Woronkoff*, G. P., and G. I. *Pokrowski:* Z. Physik **20**, 358 (1924).
[51] *Kortüm*, G., and R. *Hamm:* Ber. Bunsenges. **72**, 1182 (1968).

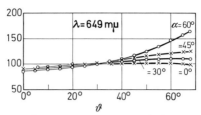

Fig. 21a. Goniophotometric measurements on K_2CrO_4 powder at 649 mμ beyond selective absorption

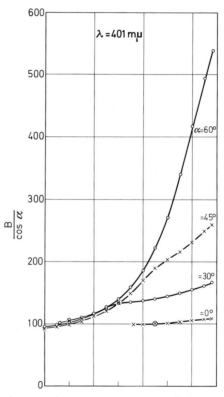

Fig. 21b. Goniophotometric measurements on K_2CrO_4 powder at 401 mμ in the region of selective absorption

metric curves of K_2CrO_4 as a function of observation angle ϑ at different incidence angles α. The measurements at $\lambda = 649$ mμ where K_2CrO_4 practically does not absorb at all (reflectance at 649 mμ is (97.2%) shows Fig. 21a. The sample was prepared under light pressure and the results correspond somewhat with those given in Fig. 15g for $BaSO_4$. Fig. 21b shows curves for the same sample investigated at $\lambda = 401$ mμ where

K_2CrO_4 has a reflectance of only 5.2%. We immediately recognize the following characteristic differences: a) The curve for $\alpha = 0$ increases slightly with increasing ϑ instead of falling; b) The deviations from Lambert behavior are 4–5 times greater and indicate weak glossy areas; c) The deviations are already beginning at very small ϑ-angles and climb

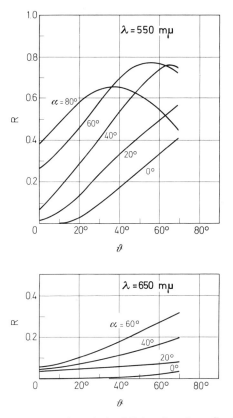

Fig. 22. Fraction R of linearly polarized light after the reflection on paper dyed with rhodamin, for various angles of incidence α as parameter and dependent on the observation angle ϑ. Lower part: $\lambda = 650$ mµ; upper part: $\lambda = 550$ mµ (in the absorption range)

much more steeply than outside the region of absorption. This all indicates that these experimental results can be at least qualitatively explained by a combination of the Bouguer elementary mirror hypothesis and the Fresnel equations. According to Eq. (48) in the simplest case of perpendicular incidence, we can write for the refractive index $n_1 = n$

and (using air as the surrounding medium) $n_0 = 1$, so that

$$R_{reg} = \frac{(n-1)^2 + n^2\varkappa^2}{(n+1)^2 + n^2\varkappa^2}. \tag{84}$$

The larger the absorption index, $\varkappa$, the larger becomes R_{reg}. The regular part of the reflection at the Bouguer elementary mirrors must already for perpendicular incidence on absorbing materials be more noticeable than on nonabsorbing materials. At large angles of incidence the influence of $\varkappa$ on R_{reg} must greatly increase so that R_{reg} eventually approaches the limiting value of 1.

Finally, similar results were shown by measurements of the polarization of natural radiation on diffuse reflection at materials with different absorbance under otherwise equivalent conditions [52]. Fig. 22 shows the fraction R of linearly polarized radiation after reflection from paper dyed with rhodamine B for different angles of incidence, as a function of the angle of observation ϑ, using wavelengths of 650 mμ (where there was practically no absorption) and 550 mμ (where absorption was strong). In the first case the polarization is almost the same as for white paper, while in the second it is very great, and may even pass through a maximum, corresponding to the Brewster Law (see p. 8), at $\alpha_P = \dfrac{\alpha + \vartheta}{2}$.

In the case of the depolarization of linearly polarized radiation by diffuse reflection, it is also found that this is smaller the more strongly the reflecting material absorbs [53].

g) Dependence of Remission Curves on Particle Size

From what has been said in the previous paragraph and confirmed by experiment we must bear in mind that in reflection of radiation at solid matte surfaces, diffuse and regular fractions will be superimposed, the intensity ratio depending not only on the angles of incidence and observation, but also on packing density, crystal form, refractive index, particle size, and the absorbance of the material used. It is clear that an ideal diffusely reflecting surface can only be approximately achieved in practice, even with the finest possible grinding of the material. The measurements of the goniophotometric curves of aerosil provide an example of this. The basis for this behavior lies in the fact that there are always coherently reflecting surface regions (elementary mirrors) whose reflection obeys the Fresnel formulas. Their radiation density, increasing with ϑ, may be partly compensated by shadowing effects.

---

[52] *Woronkoff*, G. P., and G. I. *Pokrowski*: Z. Physik **30**, 139 (1925).

[53] *Woronkoff*, G. P., and G. I. *Pokrowski*: Z. Physik **33**, 860 (1925); — *Umow*, N. A.: Physik Z. **6**, 674 (1905); **13**, 962 (1912).

Added to this is the radiation coming back through the surface of the sample from its interior. The density distribution of this radiation can be assumed as largely isotropic and should accordingly fulfill the requirements of the Lambert Law (compare p. 99).

The validity of these views is impressively demonstrated by measurements of the dependence of remission curves on particle size. In what follows, we shall consider the diffuse-regular reflectance $J/I_0 = R_\infty$ as determined at layer thicknesses sufficiently great to prevent transparency [54]. This quantity will be determined as a function of the wavelength λ or the wave number $\overset{*}{v}$. I_0 is the diffuse (see p. 219), or, with given angle of incidence α, directed incident radiation flux, and J the corresponding remitted radiation flux. In transmission measurements, the corresponding term is the transmittance, $I/I_0 = f(\lambda)$ of $f(\overset{*}{v})$. Instead of the so-called absolute remittance, R_∞, in practice the quantity most often measured is the relative remittance, R'_∞, referred to a non-absorbing standard, such as magnesium oxide (see p. 137). Instead of R'_∞ it is also possible to plot $\log(1/R'_\infty)$ against λ or $\overset{*}{v}$, which corresponds to absorbance curves in transmission measurements. For this reason, the quantity $\log(1/R'_\infty)$ is often called the "apparent absorbance".

In general, for not too strongly absorbing materials, it is found that R'_∞ (for example for a crystalline powder) increases with reduction in particle size, so that $\log(1/R'_\infty)$ diminishes. This means that grinding absorbing materials in general causes them to become paler. A well-known example is $CuSO_4 \cdot 5H_2O$ which appears practically white in a finely triturated form. This depends upon the fact, which we shall examine later (see p. 63), that the scattering coefficient increases with diminishing particle size, while the *depth of penetration* of the radiation diminishes. The mean layer thickness penetrated within the sample therefore becomes smaller, so that the absorbed fraction of the radiation is reduced.

The absorption coefficient itself is a function of the particle size in heterogeneous systems. This can be seen also in transmission if scattering is largely eliminated by making the refractive indices of the particles and of the surrounding medium as nearly alike as possible, which has been the general practice in infrared spectroscopy of heterogeneous systems for a long time. Assuming negligible scattering, the dependence of the absorption coefficient, K_T, on particle size in heterogeneous systems has been investigated by several authors [55–58]. The most general statistical de-

---

[54] This is usually the case when the layer thickness is a few millimeters or more.
[55] *Duyckaerts, G.:* Spectrochim. Acta **7**, 25 (1955).
[56] *Otvos, J. W., H. Stone,* and *W. R. Harp:* Spectrochim. Acta **9**, 148 (1957).
[57] *Gledhill, R. J.,* and *D. B. Julian:* J. Opt Soc. Am. **53**, 239 (1963).
[58] *Felder, B.:* Helv. Chim. Acta **47**, 488 (1964).

velopment by *Felder*[58] gives for a system of spherical monodisperse particles the absorption coefficient in transmission

$$K_T = -\frac{3}{2d}\varphi(P_m)\ln\left[1 - \frac{P}{\varphi(P_m)}(1 - T_d)\right] \quad (85)$$

in which d = particle diameter; P = packing density of particles, defined by $P = \frac{N}{V} \cdot \frac{\pi d^3}{6}$; $\varphi(P_m)$ is a function of a maximum possible packing density P_m, defined by the limit to which the particles can approach one another $[\varphi(P_m) \sim P_m < 1]$; T_d is the transmittance of a single particle, given according to *Duyckaerts*[55] by

$$T_d = \frac{2}{(kc_0 d)^2}[1 - (1 + kc_0 d) \cdot e^{-kc_0 d}], \quad (86)$$

in which $k = 2.303 \cdot \varepsilon$ is the natural extinction coefficient of the molecules, while c_0 is their concentration in moles per litre in the particle.

Eq. (85) indicates that the absorbance for constant concentration c_0 increases with diminishing particle size, and approaches a limiting value which corresponds to the absorbance of the molecularly dispersed dissolved material. It is, of course, important to remember that for $d \to 0$ the formula is no longer valid, since the extinction coefficient k may change in the transition from the crystal state to the solvated molecule. Ignoring this point, however, we obtain for the molecularly dispersed solution, according to the Lambert-Beer Law:

$$K_{(d \to 0)} = kc_0 P. \quad (87)$$

The ratio $K_T/K_{T(d \to 0)}$ is shown in Fig. 23 as a function of $\log kc_0 d$ for various packing densities. For $kc_0 d > 0.1$, a remarkable and rapidly increasing dependence of the absorption coefficient on the particle size is to be expected[59]. If $kc_0 d > 10$, the K_T is practically independent of the extinction coefficient k of the molecules.

If we assume that this result obtained for spherical monodispersed particles with negligible scattering can be approximately applied also to polydispersed scattering particles, we find the following: it may be expected, and is experimentally confirmed (see p. 195), that the absorption coefficient in reflection K_R is proportional to the absorption coefficient in transmission K_T. If the latter is dependent on the particle size, then K_R must behave in the same way, so that it should diminish with increasing values of d in accordance with Fig. 23. This change

---

[59] The values of the abscissae chosen in Fig. 23 correspond approximately to cases occurring in practice. In such a case c_0 is of the order of magnitude 10, d of 10^{-4} cm and k of 10^2 to 10^5.

should be more strongly observed at greater extinction modulus kc_0 of the molecules, so that the whole spectrum should become flattened. In practice, the opposite is found for weakly absorbing materials: with increasing particle size the absorbance increases (the powder becomes darker). This is due to the fact that, in addition to the absorption coefficient, the scattering coefficient S is a function of the particle size, increasing with particle sizes $d \geq 1\mu$ in proportion to d^{-1}, which com-

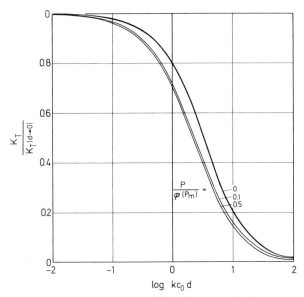

Fig. 23. Particle-size dependence of light absorption in heterogeneous systems

pensates the particle size dependence of K_R (see p. 213). The scattering therefore diminishes with increasing d, the radiation can penetrate deeper into the powder, the average path traversed $\bar{x}$ becomes greater, and the absorbance also increases. As we shall show later (see p. 214), for very large values of d the quotient K_R/S will be independent of d.

This is always found to be the case for not too strongly absorbing materials. Fig. 24 gives as an example the remission curves of glass powder of various particle sizes, measured against barium sulphate as standard, as well as the absorption coefficients k of the same glass obtained from transmission measurements[60], defined by $\log(I_0/I) = kx$.

[60] *Kortüm, G.*, and *P. Haug*: Z. Naturforsch. **8a**, 372 (1953). The regular reflection loss at the boundary surface in the transmission measurement was eliminated by obtaining results with two different layer thicknesses [see *Schachtschabel, K.*: Ann. Physik **81**, 929 (1926)].

It is seen that with diminishing particle size, the remittance increases, so that $\log(1/R'_\infty)$ diminishes, but that with increasing $\overset{*}{\nu}$ the remitted radiation flux does not diminish to the same extent as would have been expected from the steadily increasing intrinsic absorbance of the glass. This is to be attributed, at least partially, to the fact that at constant d

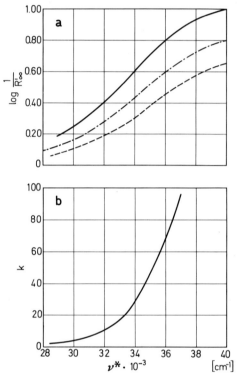

Fig. 24. a) Remission curves of glass powder of different grain sizes, measured against BaSO$_4$ as standard: $\sqrt{\overline{d^2}} \cong 100\,\mu$ ———; $\cong 35\,\mu$ —·—; $\cong 10\,\mu$ ---.
b) Absorption coefficients of glass in transmission

and increasing kc_0, the absorption coefficient according to Fig. 23 diminishes, so that the whole spectrum is flattened.

This connection is seen more clearly when the particle size dependence of the remission for selectively absorbing materials is investigated. Fig. 25a shows the remission curves of solid solutions of potassium permanganate in potassium perchlorate with a potassium permanganate content of 0.17 mole-% for three different mean particle sizes, measured against the purest barium sulphate as standard [61]. Here also

---

[61] Kortüm, G., and H. Schöttler: Z. Elektrochem. **57**, 353 (1953).

it is seen that with diminishing particle size, the remittance increases, so that the apparent absorbance $\log(1/R'_\infty)$ diminishes, a consequence of the diminished penetration depth of the radiation with reduction in particle size. If the same measurements are made on pure potassium permanganate, however, two remarkable differences are found (Fig. 25 b). In the first place,

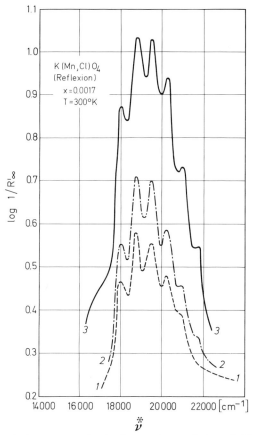

Fig. 25a. Remission curves of mixed crystals consisting of $KMnO_4$ and $KClO_4$ with a $KMnO_4$-content of 0.17 mol-%, measured against $BaSO_4$ as standard. Average grain size: 1) approx. 2μ; 2) approx. 20μ; 3) approx. 100 μ

the apparent absorbances are only about one unit larger than that of the solid solution in spite of the 500-fold increased concentration of the absorbing ions. For the same particle diameters, kc_0 is thus very much greater than in the solid solution, so that according to Fig. 23 the absorption coefficient K_R also diminishes, while S is not appreciably changed. The second difference is the increase in the apparent absorbance of the pure crystal with diminishing particle size, as compared with the decrease in the case of the solid solution. This can be explained by the fact that because of the strong absorbance the fraction of regular reflection becomes more evident, as the Fresnel equations would lead one to expect.

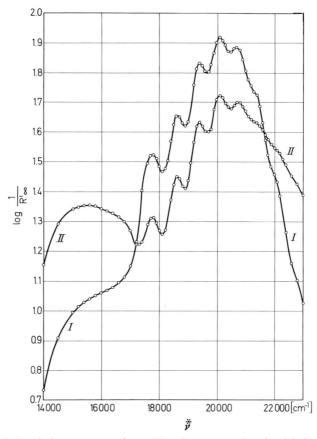

Fig. 25b. Remission spectrum of pure $KMnO_4$, measured against MgO as standard. I: $\bar{d} \cong 1$ to 2μ; II: $\bar{d} \cong 20\,\mu$

The fraction of regular reflection diminishes with decrease in particle size. Since the regular refraction diminishes the apparent absorbance, its decrease due to reduction in particle size must cause the apparent absorbance to appear greater. This inverse dependence on particle size of the apparent absorbance for large or small k-values, which has often been observed, shows just as the course of the goniophotometric curves (see p. 44) that even in the case of such fine powders the contribution of regular surface reflection cannot be neglected.

In addition to the particle size dependence of the absorption coefficient K_R, the fraction of regular surface reflection which exists causes the curves obtained from remission measurements at powders to be always greatly flattened compared with a spectrum obtained in transmission: that is, the apparent absorbance differences between maxima and minima are appreciably smaller than in the transmission spectrum, so that the remission spectrum appears less differentiated [62]. Fig. 26 gives $\log(1/R'_\infty)$ of powdered potassium ferricyanide for different values of the mean particle diameter $\bar{d}$ of particles homogenized by sieving, measured against magnesium oxide as standard [63].

Curves 1 to 4 show the reduction of the apparent absorbance with falling $\bar{d}$ due to diminishing penetration depth, and simultaneously the increasing differentiation of the remission spectrum. For large particles, (curves I and II) the fraction of surface reflection is relatively large, and can be recognized as "gloss" with the eyes. Corresponding to the strong damping of electronic vibrations assumed by the dispersion theory, the maxima are broad and indistinct. For very small particles (curve V) [64] the apparent absorbance again increases. This may be attributed in part to a more complete elimination of the fraction of regular reflection, since the particle size is now of the same order of magnitude as the wavelength. It is also possible, however, that in this particle size range the scattering coefficient, by analogy with simple scattering, is no longer inversely proportional to d, but increases with a higher power of d (see p. 201), so that again an increase in apparent absorbance with very fine particles is to be expected. The same effect has also been observed with potassium chromate [65].

---

[62] This is easily seen from the following estimate: for two neighboring bands measured in transmission, let the transmittance at the maximum $I_1/I_0 = 0.001$ and $I_2/I_0 = 0.1$, and the absorbances thus $E_1 = 3$, $E_2 = 1$ and $\Delta E = 2$. If now the surface reflection for corresponding remission measurements at the powder is regarded, for example, as being 4%, the apparent absorbances are given by

$$\log(1/R'_\infty)_1 = \log \frac{100}{4 + 0.096} = 1.388,$$

$$\log(1/R'_\infty)_2 = \log \frac{100}{4 + 9.6} = 1.867.$$

$\Delta \log(1/R'_\infty)$ is therefore 0.48 instead of 2 as in the case of measurements in transmission (see also p. 174).

[63] Kortüm, G., and J. Vogel: Z. physik. Chem. N. F. **18**, 110 (1958).
[64] A preparation with $\bar{d} \ll 1\,\mu$ was used [see Holzmann, G.: Science **11**, 550 (1950)].
[65] Kortüm, G., and P. Haug: Z. Naturforsch. **8a**, 372 (1953).

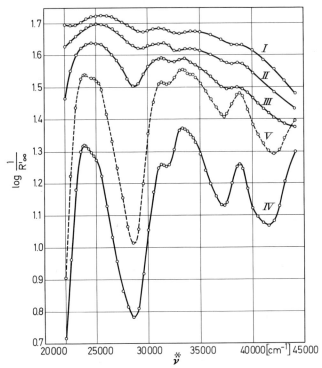

Fig. 26. Remission spectrum of pure $K_3[Fe(CN)_6]$, measured against MgO as standard. I: $\bar{d} \cong 200$ to $500\,\mu$; II: $\bar{d} \cong 60$ to $100\,\mu$; III: $\bar{d} \cong 20$ to $40\,\mu$; IV: $\bar{d} \cong 1$ to $2\,\mu$; V: $\bar{d} < 1\,\mu$

These observations on the particle size dependence of the remission spectrum taken together indicate strongly the existence of a superposition of regular and diffuse reflection components, but up to now there has been no direct evidence of this. This can, however, be obtained through measurements of the reflection of linearly-polarized radiation at powders of various particle sizes[66]. The diffusely reflected portion of the radiation is then completely depolarized and hence transmitted to the extent of 50% through a crossed analyzer attached to the polarizer. For the regularly reflected portion at crystal surfaces, two limiting cases must be distinguished:

With weak absorption, the radiation remains practically linearly polarized; depending on the size of the angle of incidence α and the azimuth ε (the angle between the incident plane and the oscillation plane of the radiation), the oscillation plane is rotated. This rotation has a

---

[66] *Kortüm, G.*, and *J. Vogel*: Z. physik. Chem. N. F. **18**, 230 (1958).

maximum when $\alpha = 90°$ and $\varepsilon = 45°$, but is zero for the limiting azimuths $\varepsilon = 0°$ and $\varepsilon = 90°$ for all angles of incidence. The same is true for all azimuths with perpendicular incidence ($\alpha = 0°$).

For a crystal powder with random positions of the reflecting surfaces, all possible azimuths and angles of incidence, and therefore all possible oscillation planes in the regularly reflected fraction are present. Under suitable geometrical conditions (see Fig. 27), it is possible, however, to

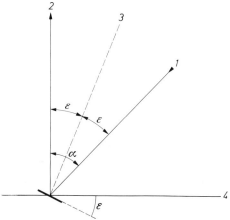

Fig. 27. Details of the method for measuring the regular reflection of linearly polarized radiation on crystal powders; *1* incident ray; *2* to receiver; *3* surface normal of a crystal face; *4* macroscopic surface of the sample

arrange that the radiation reaching the detector after regular reflection is practically completely polarized in the same plane as the incident primary radiation: the incident radiation impinges on the macroscopic surface of the sample at an angle $\alpha = 45°$, and the emergent radiation is measured in the direction of the normal to macroscopic surface (see also p. 232). The regular fraction is measured therefore only for those crystal surfaces on which the radiation impinges with an angle $\alpha_R = 45°/2$, whose surface normal is therefore inclined at $45°/2$ against the normal to the macroscopic surface. Because the incident radiation is not strictly parallel, the regular reflection also will be obtained from those surfaces which lie inclined to the surface normal within a certain small range of angles $\varepsilon \pm \Delta\varepsilon$. Therefore radiation will reach the detector from beneath an azimuth $\varepsilon \pm \Delta\varepsilon$ of the elementary mirrors. This $\Delta\varepsilon$ is small because of the optical arrangement used. If care is also taken to ensure that the "principal azimuth" ε_R is either $0°$ or $90°$, at which the radiation is reflected without rotation of the plane of polarization, then the plane of oscillation of the incident linearly polarized radiation, in so far as

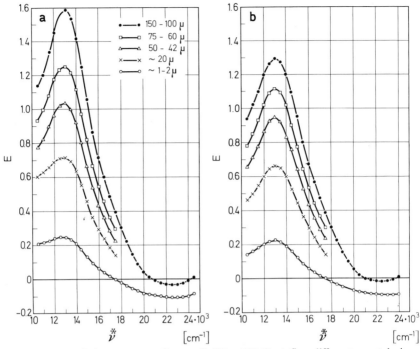

Fig. 28. Remission spectra of pure $CuSO_4 \cdot 5H_2O$ at five different crystal sizes, measured against the white Standard of Zeiss; a) taken with polarized radiation; b) taken with natural radiation

it reaches the detector by regular reflection, remains practically unchanged. If an analyzer is inserted in front of the detector, crossed with respect to the polarizer, then the regularly reflected fraction of the remitted radiation is eliminated to a large extent.

With stronger absorbance, the regularly reflected fraction of the linearly polarized incident radiation becomes elliptically polarized, depending on the angle of incidence and the azimuth of the primary radiation. With the limiting azimuths 0° and 90°, and all angles of incidence, as well as with perpendicular incidence and all azimuths, the radiation remains, however, linearly polarized in the original plane, so that in this case also the regular portion of the remitted radiation is practically completely eliminated by a crossed analyzer using the geometrical arrangement shown in Fig. 27.

If the remittance of powders of different particle sizes is measured, first with natural radiation, and then with linearly polarized radiation, using the same standard, then various remission spectra are obtained, as shown in Fig. 28 with respect to pure (weakly absorbing)

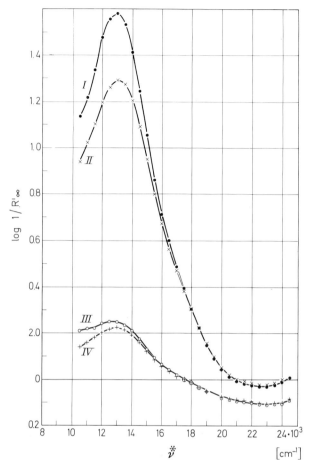

Fig. 29. Remission spectra of $CuSO_4 \cdot 5H_2O$ in cases of coarse and fine grains, taken with polarized and natural radiation. I: 150–100 μ, with polarized radiation; II: 150–100 μ, with natural radiation; III: 1–2 μ, with polarized radiation; IV: 1–2 μ, with natural radiation

$CuSO_4 \cdot 5H_2O$ [67]. For the polarized radiation measurements, the plane of the oscillations will lie either perpendicular or parallel to the plane of incidence, and an analyzer crossed with respect to the polarizer is inserted before the detector. Between the two measurements it is important that neither the surface of the sample nor the geometry of the measuring system, as in Fig. 27, should be altered. Spectrum (a) is obtained from linearly polarized radiation, and concerns the diffuse

---

[67] *Kortüm, G.*, and *J. Vogel:* Z. physik. Chem. N.F. **18**, 230 (1958).

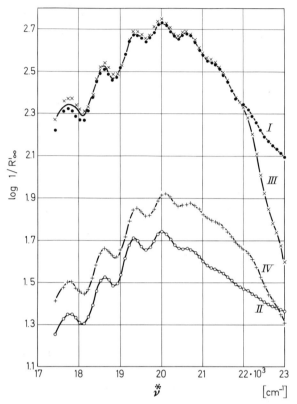

Fig. 30. Remission spectra of pure $KMnO_4$ in case of pure diffuse reflection (polarized radiation) and in case of superimposition of diffuse and regular reflection (natural radiation), for two very different grain sizes I: 150–100 µ, with polarized radiation; II: 150–100 µ, with natural radiation; III: 1–2 µ, with polarized radiation; IV: 1–2 µ, with natural radiation

fraction of the remission, while spectrum (b) was obtained under identical conditions with natural radiation, and concerns the total (regularly and diffusely reflected) radiation.

It is seen first that in both cases the apparent absorbance diminishes with reduction in particle size because the depth of penetration also diminishes, though with natural radiation it is always appreciably less than when the regularly reflected portion has been eliminated by crossed polarization plates. To show this clearly, the remission spectra in Fig. 29 are shown together for the largest and the finest particles with and without polarization plates. For the finest particles, the differences are very small, that is, with weak absorbance and the finest grinding, it is possible to get practically ideal diffuse reflection; with relatively large particles ($d \cong 150$–$100\,\mu$ diameter) the apparent absorbance difference at the maximum has already reached the value 20%.

Similar measurements of a strongly absorbing material (pure potassium permanganate) with the same large difference in particle size are shown in Fig. 30. Here, even with a particle size of 1–2 μ, the apparent absorbance depending on diffuse reflection is diminished by one-third through the regular reflection portion, when both portions are present together. The strong compression of the remission curve compared with the transmission spectrum of the same material with large absorption is in this case particularly well illustrated. It is also seen that even the finest grinding in this case by no means succeeds in eliminating the contribution of regular reflection, since the regular reflectance, even with particle sizes of around 2 μ, is still appreciable for high $\varkappa$-values, as the Fresnel equations would indicate.

It is further seen that the particle size dependence of pure diffuse reflection (using polarized radiation) is very small, unlike the case of the weakly absorbing copper sulphate, and indeed lies practically within the limits of error of the measurement. This indicates that diffuse reflection in the case of strong absorbance is largely independent of the particle size. We shall return to this at a later stage (see p. 213).

Chapter III. Single and Multiple Scattering

As we have seen, the mutual interaction of a plane electromagnetic wave with a particle of dimensions greater than the wavelength can be described, using the concepts of reflection, refraction, and diffraction, provided that absorption does not occur. The angular distribution of the density of radiation can in principle be calculated, and is by no means isotropic. Only with the simultaneous cooperation of numerous, densely packed particles (such as a crystal powder) is it possible in the most favorable cases to observe an approximately isotropic angular distribution (corresponding to the Lambert Cosine Law), which has been designated "diffuse reflection". If the dimensions of a particle are comparable with, or smaller than, the wavelength, it is no longer possible to separate the contributions to the intensity due to reflection, refraction, and diffraction, and the phenomenon is now described as "scattering". The angular distribution of scattering intensity at a single particle is also far from isotropic, depending on the size, state, and polarizability of the particle and on the direction of observation. This so-called "single scattering" corresponds therefore to reflection, refraction, and diffraction of a beam of rays at a single particle of dimensions very much greater than the wavelength. Since in fact it is impossible to investigate a single particle and a great number of particles always must be considered (as in gases, aerosols, or colloidal particles in suspension), it is necessary to assume that there is no fixed relation between the phases of the waves scattered from a single particle, so that interference cannot arise. This is the case when we are dealing with completely random and non-localized particles sufficiently separated from each other. Calculation indicates that a mean separation of twice the particle diameter is sufficient to provide independent scattering [68].

In this case we can assume that the scattering amplitudes of the waves scattered at single particles may be simply added to each other in each direction, without considering the phases. That is, the scattering intensity can be regarded as incoherent. A criterion as to whether single scattering is occurring is given by the apparent absorbance [69] of the sample for

---

[68] An dense mist of water drops of 1 mm diameter only contains one particle per cm^3; in general, the particles are thus separated by distances much greater than twice their diameters.

[69] We distinguish between apparent or "conservative", and true absorbance. The scattering of the particles simulates an absorption of the radiation in the direction of the emergent beam.

measurements in transmission. If the incident radiation flux is designated as I_0, and the transmitted, non-scattered, radiation flux by I^*, then in analogy to the Lambert-Beer Law, we have:

$$I^* = I_0 e^{-S'x} \quad \text{or} \quad \ln\frac{I_0}{I^*} = S'x, \tag{1}$$

in which S' is the scattering coefficient, also called the "turbidity" and x is the thickness of the illuminated layer. If $S' < 0.1$, the scattering is essentially single, while with $S' > 0.3$, multiple scattering predominates. For opaque samples, for which the radiation is practically completely remitted, multiple scattering is always predominant.

A rigorous theory exists only for single scattering at molecules which are small compared with the wavelength (*Rayleigh, Gans, Born*), and for isotropic spherical particles of any size (*Mie*). A detailed presentation of these theories is not possible within the scope of this book [70], and it will only be possible to indicate individual results of calculations important for our purposes.

a) Rayleigh Scattering

The theory of molecular light scattering by gases was first developed by *Rayleigh* [71]. It starts with the assumption that the electric vector of the incident monochromatic light wave excites the electrons of a molecule into forced oscillations of the same frequency, the frequency v of the radiation being supposed to be very much smaller than the natural frequency v_0 of the electrons (absence of absorption). If the dimensions of the molecule considered are small compared with the wavelength, it can be assumed that all the electrons oscillate in phase [72], so that the molecule forms an oscillating dipole, or a kind of antenna of molecular dimensions. This dipole itself radiates secondary electromagnetic waves of the same frequency in all directions, so that the primary wave is scattered. The amplitude of the scattered wave depends on the magnitude of the electrical moment, μ_i, induced in the molecule by the primary

---

[70] See the following comprehensive surveys: *Van de Hulst, H. C.*: Light scattering by small particles. New York: John Wiley & Sons 1957. — *Stuart, H. A.*: Molekülstruktur, 3. Aufl. Berlin-Heidelberg-New York: Springer 1967. — *Edsall J. T., and W. B. Dandliker*: Light scattering in solutions. Fortschr. chem. Forschg. **2**, 1 (1951). — *Oster, G.*: The scattering of light and its applications to chemistry. Chem. Rev. **43**, 319 (1947).

[71] *Rayleigh, J. W.*: Phil. Mag. **12**, 81 (1881); **47**, 375 (1899).

[72] We shall for simplicity subsequently disregard the number f of virtual oscillators.

wave, which itself depends upon the charge displacement, and is proportional to the instantaneous field strength $\vec{E}$:

$$\mu_i = \alpha \vec{E}. \tag{2}$$

The proportionality factor α is called the polarizability of the molecule. If the molecule is isotropic α is identical in all directions, and $\vec{E}$ and μ are parallel. If the molecule is anisotropic, α represents a mean value derived from the three principle polarizabilities in the three dimensions in space (or polarizability ellipsoid).

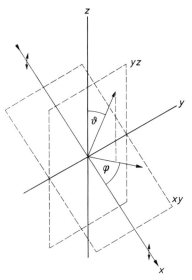

Fig. 31. Scattering of a light wave on a dipole in the origin; the light wave is linearly polarized in the z-direction, and the electric moment of the dipole oscillates in the z-direction

Consider a molecule, capable of scattering, to be in the origin of a coordinate system. A wave incident in the x-direction (see Fig. 31), whose electric vector $\vec{E}$ vibrates in the z-direction (corresponding to vertically polarized light) excites a dipole oscillating in the z-direction, and will be scattered uniformly in all directions of the xy plane, independently of the azimuth φ. On the other hand, the amplitude A of the scattered wave diminishes in the polar direction proportionally to sine ϑ, where ϑ represents the angle between the dipole-axis and the radius vector (the so-called zenith-distance). Thus, the dipole does not radiate in the direction of its axis, the z-direction. The scattered intensity, I_s, is proportional to the square of the amplitude (Eq. (II, 2)). A plot

Rayleigh Scattering

of I_s in a polar diagram from the middle of the dipole against ϑ is represented by a circular doughnut, shown in cross-section in Fig. 32a (representing the so-called radiation characteristic)[73]. The maximum of the scattered radiation lies in the equatorial plane.

The amplitude A of the scattered wave emitted by the oscillating dipole is given according to the electromagnetic theory of radiation[74] by

$$A = \frac{4\pi^2 v^2}{c^2 R} \cdot \mu \sin \vartheta, \tag{3}$$

where v is the frequency, R the distance from the dipole, and μ the electrical moment of the dipole defined by (2).

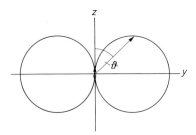

Fig. 32a. Radiation characteristic of a dipole when vertically polarized light is incident in the direction of the x-axis (vertical to the plane of the paper)

If we write the electric vector of the incident wave in the wellknown form:

$$\vec{E} = \vec{E}_0 \cdot \exp\left[i 2\pi v\left(t - \frac{R}{c}\right)\right], \tag{4}$$

where R/c represents the phase constant, then

$$\mu = \alpha \vec{E}_0 \cdot \exp\left[i 2\pi v\left(t - \frac{R}{c}\right)\right] \equiv \mu_0 \cdot \exp\left[i 2\pi v\left(t - \frac{R}{c}\right)\right]. \tag{5}$$

If this is inserted in (3),

$$A = \left(\frac{2\pi v}{c}\right)^2 \frac{\mu_0}{R} \sin \vartheta \cdot \exp\left[i 2\pi v\left(t - \frac{R}{c}\right)\right] \tag{6}$$

or, by converting into the real function:

$$A = \left(\frac{2\pi v}{c}\right)^2 \frac{\mu_0}{R} \sin \vartheta \cos 2\pi v\left(t - \frac{R}{c}\right). \tag{7}$$

---

[73] The figure should be rotated around the z-axis.

[74] See, for example, Joos, B.: Lehrbuch der theoretischen Physik. 11. Aufl. Leipzig: Akad. Verlagsges. 1959.

Since, According to Eq. (II, 2) when $\varepsilon = 1$ and $v = c$, the scattered intensity is

$$I = \frac{c}{4\pi} A^2, \tag{8}$$

we obtain for the scattered intensity in the direction ϑ

$$\begin{aligned} I_\vartheta &= \frac{c}{4\pi} \cdot \left(\frac{2\pi v}{c}\right)^4 \frac{\mu_0^2}{R^2} \sin^2\vartheta \cos^2 2\pi v\left(t - \frac{R}{c}\right) \\ &= \frac{c}{4\pi} \left(\frac{2\pi}{\lambda_0}\right)^4 \frac{\mu_0^2}{R^2} \sin^2\vartheta \cos^2 2\pi v\left(t - \frac{R}{c}\right), \end{aligned} \tag{9}$$

where λ_0 represents the wavelength in vacuum. I_ϑ is the intensity at a particular time t. Since I_ϑ varies with time, to find the mean value $\overline{I_\vartheta}$ we must integrate the intensity over the entire period T, and then divide by T, thus:

$$\overline{I_\vartheta} = \frac{1}{T} \int_0^T I_\vartheta \, dt.$$

$\overline{I_\vartheta}$ is thus the radiation flux per second. Since the mean value of $\cos^2 2\pi v(t - R/c)$ over a period is equal to $\frac{1}{2}$, we then obtain:

$$\overline{I_\vartheta} = \frac{c}{8\pi} \left(\frac{2\pi}{\lambda_0}\right)^4 \frac{\mu_0^2}{R^2} \sin^2\vartheta. \tag{10}$$

If we now replace μ_0, the maximum induced moment, using Eq. (5), by $\alpha \vec{E}_0$, then the radiation flux in the direction ϑ becomes

$$\overline{I_\vartheta} = \frac{c}{8\pi} \left(\frac{2\pi}{\lambda_0}\right)^4 \frac{\alpha^2 \vec{E}_0^2}{R^2} \sin^2\vartheta. \tag{11}$$

Since the incident radiation flux of the primary wave per second is

$$I_0 = \frac{c}{8\pi} \vec{E}_0^2, \tag{12}$$

we obtain finally for the intensity ratio of scattered to incident light:

$$\frac{\overline{I_\vartheta}}{I_0} = \frac{1}{R^2} \left(\frac{2\pi}{\lambda_0}\right)^4 \alpha^2 \sin^2\vartheta. \tag{13}$$

If, instead of vertically polarized light, we irradiate with horizontally polarized light, whose electrical vector oscillates in the y-direction, no alteration in principle occurs, and the induced dipole also oscillates in the y-direction. If we retain the coordinate system of Fig. 32a, the amplitude of the scattered radiation is now a maximum in the xz-plane, and is the same throughout independent of the value of ϑ, while it becomes a function of the angle φ. When $\varphi = 0$, and $\varphi = 180°$, reckoned from the emergent primary wave, it is obviously also a maximum, since these directions also lie in the xz-plane. For all values of φ lying between these values, the intensity depends on $\cos\varphi$, and for $\varphi = 90°$ and $\varphi = 270°$ is equal to zero (the direction of the dipole axis). The intensity distribution of the scattered radiation in a polar diagram is consequently proportional to $\cos^2\varphi$, and again is represented by a circular doughnut[75]. The corresponding radiation characteristic of Fig. 32b is rotated by 90° in the plane of the paper with respect to that of Fig. 32a.

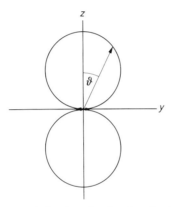

Fig. 32b. Radiation characteristic of a dipole when horizontally polarized light is incident in the direction of the x-axis (vertical to the plane of the paper)

If illumination is by means of natural light, which may be split up into independently vibrating vertical and horizontal components, then the intensity of the scattered radiation is clearly the mean of the intensities calculated above. In the xy-plane, for example, it is proportional to $(1 + \cos^2\varphi)/2$. For natural light, however, there is no preferred direction perpendicular to the axis of propagation (x) of the incident radiation. Therefore, the scattered radiation for natural light must be rotationally symmetrical in relation to the plane perpendicular to this axis. This means that the factor $(1 + \cos^2\varphi)/2$ is valid not only for the scattered intensity in the xy-plane, but for any plane containing the direction x. The angular dependence of the scattered intensity can therefore be expressed because of this symmetry by a single ϑ_s, called the "scattering angle", given by the emergence direction of the primary beam (x) and an

---

[75] We may regard Fig. 32b as rotating around the y-axis.

arbitrary direction of observation. In the equatorial plane (xy) this is equivalent to φ, and not to the zenith distance ϑ.

If N is the number of molecules per cm^3, and we use again the assumption mentioned at the beginning of this chapter that the scattered radiation of the individual molecules can be regarded as incoherent, instead of (13) for the intensity ratio of the scattered radiation per cm^3 to the incident natural light, we have

$$\frac{\overline{I_\vartheta}}{I_0} = \left(\frac{2\pi}{\lambda_0}\right)^4 \frac{N\alpha^2}{R^2} \left(\frac{1+\cos^2\vartheta_s}{2}\right) \equiv q(\vartheta_s). \qquad (14)$$

Since the polarizability α of dilute gases is given by the well-known relation

$$\alpha = \frac{1}{N} \cdot \frac{\varepsilon - 1}{4\pi} \qquad (15)$$

which involves the dielectric constant ε of the gas, and, according to the Maxwell relation

$$\varepsilon = n^2, \qquad (16)$$

it follows that

$$\alpha = \frac{n^2 - 1}{4\pi N} \cong \frac{n-1}{2\pi N}, \qquad (17)$$

because for dilute gas the refractive index n lies close to unity, and $n^2 - 1 \cong 2(n-1)$. By inserting this in (14), we have:

$$\frac{\overline{I_\vartheta}}{I_0} = \frac{4\pi^2(n-1)^2}{N\lambda_0^4 R^2} \cdot \left(\frac{1+\cos^2\vartheta_s}{2}\right). \qquad (18)$$

This is the Rayleigh scattering formula for the dilute gas. The scattered intensity is inversely proportional to the fourth power of the wavelength λ_0, and depends on the angle ϑ_s between the incident and observed direction. The angular distribution in the polar diagram has the form given in Fig. 33, and is rotationally symmetrical in relation to the plane perpendicular to the propagation direction of the primary beam. For $\vartheta_s = 0$ and $\vartheta_s = 180°$, the radiation intensity is a maximum $(1 + \cos^2\vartheta_s = 2)$ and the scattered radiation is unpolarized. For $\vartheta_s = 90°$ and $\vartheta_s = 270°$, the scattered intensity is only half this value $(1 + \cos^2\vartheta_s = 1)$, and the scattered radiation is linearly polarized. In all other directions, we observe elliptically polarized scattered radiation. Eq. (18) is only valid for isotropic molecules (see p. 74). For anisotropic molecules, the induced moment μ does not oscillate in the direction of the exciting field vector $\vec{E}$, and the scattered radiation observed at 90° is partly "depolarized". We shall not discuss this phenomenon further.

If we integrate Eq. (14) over a spherical surface of radius $R=1$, we then obtain the intensity ratio of the total scattered radiation for N molecules per cm^3 in relation to that of the incident radiation:

$$\frac{I_{st}}{I_0} = \frac{1}{I_0} \int \overline{I_\vartheta}\, df = \frac{1}{I_0} \cdot \frac{8\pi^4 N \alpha^2}{\lambda_0^4} \int_0^\pi \int_0^{2\pi} (1 + \cos^2 \vartheta_s) \sin \vartheta_s\, d\vartheta_s\, d\varphi. \quad (14a)$$

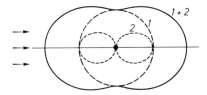

Fig. 33. Polar diagram of the dipole scattering of non-polarized light

The double integral has the value $16\pi/3$, so that, introducing (17),

$$\frac{I_{st}}{I_0} = \frac{128\pi^5 N \alpha^2}{3\lambda_0^4} = \frac{32\pi^3 (n-1)^2}{3\lambda_0^4 N} \equiv S'. \quad (19)$$

S' is identical with the apparent extinction coefficient, defined by (1), observed in transmission. This, which is also called the "turbidity" is thus an immediate measure of the total scattering power per cm^3 of the gas. Furthermore we have for the number of particles N per cm^3,

$$N = \frac{3G}{4\pi r^3 \cdot \varrho}, \quad (20)$$

where G is the weight of all the particles per cm^3, and ϱ is their actual density. Thus S' is proportional to the third power of the radius of the particles and therefore to their volume. S', as can be seen from both (19) and (1), has the dimension [cm^{-1}], although we might have expected that it would be dimensionless, since it is the ratio of two radiation fluxes. It must be remembered, however, that I_{st} refers to the unit path-length, thus in practice the observation distance R from the scattering volume must be so great that a distance of 1 cm can be regarded effectively as a point light source. In practical measurements, it is therefore necessary to introduce a correction to Eq. (19).

These considerations can also be applied to dilute solutions of small isotropic molecules in a homogeneous solvent also consisting of isotropic molecules [76]. For the polarizability α_i of the dissolved molecules, it is then no longer possible to use (17). From (2) and (17) it follows for the solvent that:

$$\alpha_0 = \frac{\mu_0}{|\vec{E}|} = \frac{n_0^2 - 1}{4\pi N},$$

while for the pure dissolved material:

$$\alpha = \frac{\mu}{|\vec{E}|} = \frac{n^2 - 1}{4\pi N}.$$

If a dissolved molecule is introduced into the solvent, its induced moment is different from that of the solvent molecule which it displaces by the quantity:

$$\frac{\Delta\mu}{|\vec{E}|} = \frac{n^2 - n_0^2}{4\pi N} = \Delta\alpha. \tag{21}$$

This difference is decisive for the scattering. Since $\Delta\mu \cdot N$ represents the so-called "dielectric polarization" per unit volume, the induced additional moment in individual particles of volume v in the dilute solution amounts to:

$$\Delta\mu = \frac{n^2 - n_0^2}{4\pi} \cdot v \cdot \vec{E}. \tag{22}$$

A further correction is, however, necessary for this equation, which involves replacing the external field $\vec{E}$ in the pure solvent by an "internal field" $\vec{F}$, since the different polarization within and outside the dissolved particle causes apparent charges to arise at the boundary, producing an additional field. From the laws of electrostatics we have:

$$\vec{F} = \frac{3n_0^2}{n^2 + 2n_0^2} \cdot \vec{E}. \tag{23}$$

From (22) and (23) we have for the additional induced moment in the particle:

$$\Delta\mu = \frac{3n_0^2}{4\pi} \left(\frac{n^2 - n_0^2}{n^2 + 2n_0^2} \right) v\vec{E}. \tag{24}$$

If this value of $\Delta\mu$ is inserted in (3) instead of that of μ, the same procedure as that used on p. 79 leads, after integration over all values of the angle

---

[76] Debye, P.: J. Phys. Coll. Chem. **51**, 18 (1947). In the pure solvent the waves scattered at the tightly packed molecules largely undergo extinction by mutual interference. The residual slight scattered radiation, which must be subtracted from that of the solution, depends on density fluctuations of the solvent because of the thermal movement of the molecules.

ϑ_s, to the total scattered radiation flux per unit volume:

$$\frac{I_{st}}{I_0} = 24\pi^3 \left(\frac{n^2 - n_0^2}{n^2 + 2n_0^2}\right)^2 \frac{\overset{*}{N}v^2}{\lambda^4}. \tag{25}$$

This corresponds to Eq. (19) for gases, λ being the wavelength in the pure solvent, and $I_{st}/I_0 = S'$ the "turbidity" measured in transmission.

It must be emphasized again that the calculation assumes that the molecules are isotropic and very small compared with the wavelength λ. If these conditions are not fulfilled, intramolecular interferences arise, since the scattering molecules can no longer be regarded as oscillating dipoles of vanishing size.

b) Theory of Scattering at Large Isotropic Spherical Particles

For molecules or particles of dimensions comparable with that of the incident radiation (the diameter being not less than, say, $\lambda/10$), the electrons in the various parts of the particle are excited with different phases, so that the secondary waves emitted from them are coherent and capable of interference. They will thus be almost completely extinguished in certain directions of scattering, which implies that the total scattered intensity diminishes compared with Rayleigh scattering, and that the dependence on the angle must deviate considerably from that shown in Fig. 33. In addition, the scattering will depend less strongly on the wavelength, and the polarization of the scattered radiation must also be altered.

We shall concern ourselves here only with the basic theory and its applications, particularly in respect to problems of multiple scattering to be considered later. For further information, the reviews referred to on p. 73 should be consulted.

We consider first the phase difference which arises with the scattering of the primary wave by two scattering centers A and B located a fixed distance apart which will be designated by the vector $\vec{r}$ (see Fig. 34). The path difference of two coherent waves is given by

$$\Delta \equiv AP - BQ = r\cos\alpha - r\cos\beta. \tag{26}$$

If we designate the direction of the incident wave with the unit vector $\vec{j}$, and that of the scattered wave with the unit vector $\vec{i}$, then this difference can also be expressed in the form of two scalar vector products:

$$\Delta = rj\cos\alpha - ri\cos\beta = (\vec{r}\vec{j}) - (\vec{r}\vec{i}) = (\vec{r}(\vec{j}-\vec{i})) \equiv (\vec{r}\vec{s}). \tag{27}$$

$\vec{s} \equiv \vec{j} - \vec{i}$ is also a vector, whose value, according to Fig. 35, depends on the angle ϑ_s between the incident and scattered wave:

$$|\vec{s}| = 2 \sin \frac{\vartheta_s}{2}. \tag{28}$$

From the path difference Δ we may obtain the phase difference δ as an angle by multiplying by $2\pi/\lambda_0$. The phase difference between the two waves scattered in the same direction is thus given by

$$\delta = \Delta \cdot \frac{2\pi}{\lambda_0} = (\vec{r}\vec{s}) \frac{2\pi v}{c}. \tag{29}$$

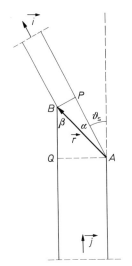
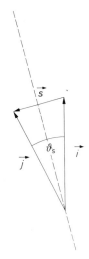

Fig. 34. The phase difference in case of scattering on two centers having a given distance

Fig. 35. Connection between direction of scattering and angle of scattering

Because of this phase difference, the amplitude A of the scattered wave at distance R from the scattering center, instead of being given by Eq. (6), is now given by

$$A = \left(\frac{2\pi v}{c}\right)^2 \frac{\mu_0}{R} \sin \vartheta \cdot \exp\left[i2\pi v\left(t - \frac{R}{c} + \frac{(\vec{r}\vec{s})}{c}\right)\right]$$

or

$$\frac{A}{A_{\text{Rayleigh}}} \equiv A' = \exp\left[i2\pi v\left(t + \frac{(\vec{r}\vec{s})}{c}\right)\right], \tag{30}$$

Theory of Scattering at Large Isotropic Spherical Particles

if the distance R is taken as an integral multiple of the wavelength. A' designates the amplitude *relative to that of the Rayleigh scattering*. If N scattering centers exist, whose positions are designated by $\vec{r}_n$ in relation to the origin A, then the total amplitude at the observation site [77] is given by

$$A' = \exp[i2\pi v t] \cdot \sum_n \exp[i2\pi v(\vec{r}_n \vec{s})/c]. \quad (31)$$

Interference occurs not only between the scattered radiation from the center A and that from all other scattering centers, but also between the scattered radiation from each of the N centers with that from all the other centers. We can write the relative complex conjugate amplitude for a second scattering center as origin with the sequence number m in the form

$$A'^* = \exp[-i2\pi v t] \cdot \sum_m \exp[-i2\pi v(\vec{r}_m \vec{s})/c]. \quad (32)$$

We obtain the total relative scattered intensity, apart from constant factors, by multiplication of the two amplitudes and get

$$I'_s = \sum_n \sum_m \exp[-i2\pi v(\vec{r}_{n,m} \vec{s})/c], \quad (33)$$

if we write

$$\vec{r}_n - \vec{r}_m \equiv \vec{r}_{n,m} \quad (34)$$

as the vector distance between the two scattering centers. The angular distribution of the scattered radiation is thus in this case not given only by $(1 + \cos^2 \vartheta_s)/2$ as in the case of Rayleigh scattering according to Eq. (14), but depends on the scalar product $(\vec{r}_{n,m} \vec{s})$, which is itself according to Eq. (28) a function of ϑ_s. If we consider the particle of dimensions comparable with or greater than the wavelength as made up of a large number of fixed point dipole radiators, we may in simple cases evaluate the relative amplitude of the scattered radiation given by Eqs. (31) and (32) due to interference of the scattered waves. Thus, if the scattering elements are uniformly distributed on a spherical surface, the origin of coordinates is located at the center, so that the vector $|\vec{r}|$ is constant, and the position of the individual scattering centers on the surface of the sphere is determined by particular values of the angle ψ between the vector $\vec{r}$ and the vector $\vec{s}$ and the corresponding azimuth φ. We may then express the sum in Eq. (31) by an integral over the spherical surface, so that

$$A' = \exp[i2\pi v t] \int_0^\pi \int_0^{2\pi} \exp[iwr\cos\psi r^2 \sin\psi \, d\psi \, d\varphi,$$

---

[77] When coherence exists between the waves, amplitudes not intensities, are added.

in which, according to (28) and (29),

$$w \equiv \frac{4\pi \sin(\vartheta_s/2)}{\lambda_0}. \tag{35}$$

Integration gives

$$A' = 4\pi r^2 \frac{\sin wr}{wr} \cdot \exp[i2\pi vt]. \tag{36}$$

We here have a further factor dependent on ϑ_s, namely $\frac{\sin wr}{wr}$, which in addition to $(1 + \cos^2 \vartheta_s)/2$ plays its part in determining the angular distribution of the scattered radiation. For particles that are very small compared with the wavelength ($\sin wr \sim wr$), this factor is equal to unity and Rayleigh angular distribution again occurs.

If scattering elements are present not only on the surface of the sphere but also throughout the entire sphere, then it is necessary to integrate over all distances r from the origin of the coordinates:

$$A' = \exp[i2\pi vt] \int_0^r 4\pi r^2 \frac{\sin wr}{wr} dr$$

or, after integration

$$A' = \frac{4\pi}{3} r^3 \cdot P(wr) \cdot \exp[i2\pi vt]. \tag{37}$$

The function

$$P(wr) \equiv \frac{3}{(wr)^3} (\sin wr - wr \cdot \cos wr) \tag{38}$$

also approaches unity for vanishingly small values of wr, and is the additional factor determining the angular distribution of the scattered radiation at a homogeneous sphere whose radius is comparable with the wavelength. Multiplying (37) by the complex conjugate amplitude, we then obtain the scattered intensity relative to that of the scattering dipole as given by the Rayleigh Law. Eq. (37) was also given by *Rayleigh*[78]. Fig. 36 shows both the scattering functions $\sin wr/wr$, and $P(wr)$ in dependence on wr. It is seen that it approaches unity in both cases for vanishingly small values of wr, though as wr approaches unity it diminishes very rapidly. That is, for

$$r \cong \frac{1}{w} = \frac{\lambda_0}{4\pi \sin(\vartheta_s/2)} \tag{39}$$

---

[78] *Rayleigh, J. W.*: Proc. Roy. Soc., **90**, 219 (1914). Eq. (37) was also compared with the results of the rigorous theory of *Mie*, and it was found that for sufficiently small particles (e.g. $2\pi r/\lambda \cong 1$ and $m = 1,3$) the deviations were small.

at large scattering angles ϑ_s we must expect interference of the scattered waves, although at small angles r must be considerably greater than λ_0 for interference to be appreciable. For large values of wr in both cases maxima and minima arise in the scattering functions. As has been shown [79], these simple scattering functions are also valid for solutions if the refractive index of the solvent is only slightly different from that of the dissolved material.

Fig. 37 shows the scattered intensity at various values of $2r/\lambda_0$ relative to that of the scattering dipoles according to the Rayleigh equation for spherical molecules. The values are given as a function of $\sin(\vartheta_s/2)$ as calculated from (37). As unit of intensity was chosen the scattered intensity when $\vartheta_s = 0$, which is the Rayleigh scattering in the direction of the primary radiation. It is seen that with increasing values of the scattering angle ϑ_s and increasing diameter of the sphere $2r$, the scattered intensity decreases as a result of interference. Hence, the scattered radiation becomes unsymmetrical, the backward scattering ($\vartheta_s = 180°$) becomes increasingly smaller than the forward scattering ($\vartheta_s = 0°$) as the ratio $2r/\lambda_0$ increases. Corresponding formulae have also been derived for other models of the scattering particles such as rods, coiled threads, discs, and ellipsoids [80].

The general rigorous theory of single scattering of a plane wave at spherical particles, both dielectric and absorbing, of any given size, has been developed by *Mie* [81]. While in the case of Rayleigh scattering the only scattered radiation considered is that due to the electric dipoles, the theory of *Mie* permits the evaluation of the angular distribution of intensity and of the polarization of the scattered radiation, in terms of the additive contributions of a series of electrical and magnetic multipoles, which are located at the center of the sphere.

The amplitudes and phases of the corresponding partial waves obtained by integration of the Maxwell equations are functions of the angle of scattering ϑ_s, the relative refractive index

$$m = \frac{n}{n_0}, \tag{40}$$

where n denotes the refractive index of the scattering particles, and n_0 that of the surrounding medium, and the variable

$$x \equiv \frac{2\pi r}{\lambda} = \frac{2\pi r n_0}{\lambda_0}, \tag{41}$$

---

[79] *Debye, P.*: J. phys. Coll. Chem. **51**, 18 (1947).

[80] See, for example, *Debye, P.*: J. Appl. Phys. **15**, 338 (1944). — *Kuhn, W.*: Helv. Chim. Acta **29**, 432 (1946). — *Neugebauer, F.*: Ann. Physik **42**, 509 (1943). — *Debye, P.*, and *E. W. Annacker*: J. Coll. Sci **5**, 644 (1955) and others.

[81] *Mie, G.*: Ann. Physik **25**, 377 (1908).

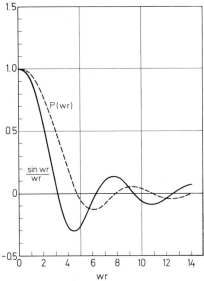

Fig. 36. Scattering functions $\sin wr/wr$ and $P(wr)$, dependent on wr

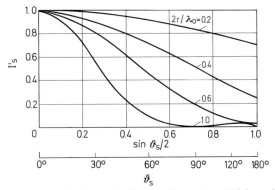

Fig. 37. Relative scattering intensity I'_s according to Eq. (37) for spherical particles in dependence on $\sin(\vartheta_s/2)$, and for various values $2r/\lambda_0$ as parameter. The ordinates give the respective factor which takes into account – in comparison with the Rayleigh scattering – the weakening of the scattering intensity due to interference

which gives the ratio of the particle circumference $2\pi r$ and wavelength λ. For a dielectric non-absorbing particle and an unpolarized primary radiation of intensity I_0, the relative scattered intensity at a distance R from the center of the sphere appears as

$$\frac{I_{\vartheta_s}}{I_0} = \frac{\lambda^2}{8\pi^2 R^2}(i_1 + i_2) \equiv q(\vartheta_s), \tag{42}$$

in which i_1 and i_2 are the intensities of two independent (incoherent) components of the scattered radiation whose electric vector oscillates perpendicular or parallel to the plane, determined by the directions of incidence and observation (vertical and horizontal components of the scattering). The scattered radiation in all directions is partially linearly polarized, and the degree of polarization is given by $(i_1 - i_2)/(i_1 + i_2)$. i_1 and i_2 can be represented by the following series:

$$i_1 = \left[\sum_{n=1}^{\infty} \frac{2n+1}{n(n+1)} (a_n \pi_n(\cos \vartheta_s) + b_n \tau_n(\cos \vartheta_s))\right]^2$$
$$i_2 = \left[\sum_{n=1}^{\infty} \frac{2n+1}{n(n+1)} (b_n \pi_n(\cos \vartheta_s) + a_n \tau_n(\cos \vartheta_s))\right]^2 . \quad (43)$$

a_n and b_n are complex functions of x and m, whose terms correspond to the electrical and magnetic partial waves referred to above. π_n and τ_n are spherical harmonics or their derivatives with respect to $\cos \vartheta_s$, which show maxima and minima for $n > 2$ when plotted against ϑ_s. The first four functions are:

$$\pi_1(\cos \vartheta_s) = 1 \, ; \quad \pi_2(\cos \vartheta_s) = 3 \cos \vartheta_s \, ;$$
$$\tau_1(\cos \vartheta_s) = \cos \vartheta_s \, ; \quad \tau_2(\cos \vartheta_s) = 3 \cos 2\vartheta_s \, . \quad (44)$$

If these are inserted in (43), we have

$$i_1 = [\tfrac{3}{2}(a_1 + b_1 \cos \vartheta_s) + \tfrac{5}{6}(3 a_2 \cos \vartheta_s + 3 b_2 \cos 2\vartheta_s) + \cdots]^2 ;$$
$$i_2 = [\tfrac{3}{2}(b_1 + a_1 \cos \vartheta_s) + \tfrac{5}{6}(3 b_2 \cos \vartheta_s + 3 a_2 \cos 2\vartheta_s) + \cdots]^2 . \quad (45)$$

For small particles (e.g. x and $mx < 0.8$) it is sufficient to use the contributions of the electric dipole and quadrupole moments (a_1 and a_2) and the magnetic dipole moment (b_1). The functions a_1, a_2, and b_1 may be expressed as series in increasing powers of x, which may often be neglected after the first term [82]. With these simplifications, approximate values for i_1 and i_2 are obtained, which for sufficiently small particles ($x < 0.8$) are sufficiently accurate. Values of i_1 and i_2 have been tabulated for a series of values of x and m [83].

---

[82] In this approximation we have

$$a_1 = \frac{2}{3}\left(\frac{m^2-1}{m^2+2}\right)x^3 \, ; \quad a_2 = \frac{-1}{15}\left(\frac{m^2-1}{2m^2+3}\right)x^5 \, ; \quad b_1 = \frac{-1}{45}(m^2-1)x^5 . \quad (46)$$

[83] *Lowan, A. N.:* Natl. Bur. Stand. Appl. Math. Ser. **4** (1948). — *Blumer, H.:* Z. Physik **32**, 119 (1925); **38**, 304, 920 (1926). — *Holl, H.:* Optik **2**, 213 (1947); **4**, 173 (1948/49).

If we neglect all terms beyond a_1, we obtain pure dipole radiation with symmetrical angular distribution shown in Fig. 33. We then have

$$i_1 + i_2 = \left(\frac{m^2 - 1}{m^2 + 2}\right) x^6 (1 + \cos^2 \vartheta_s), \tag{47}$$

introducing (42)

$$\frac{I_{\vartheta_s}}{I_0} = \frac{\lambda^2 x^6}{8\pi^2 R^2} \left(\frac{m^2 - 1}{m^2 + 2}\right)^2 (1 + \cos^2 \vartheta_s). \tag{48}$$

By integration over a spherical surface of radius $R = 1$, in a manner similar to (19) we have, making use of (41) for N particles per cm^3 [84],

$$\frac{I_{st}}{I_0} = \frac{8\pi^4 r^6 N}{\lambda^4} \left(\frac{m^2 - 1}{m^2 + 2}\right)^2 \cdot \frac{16\pi}{3} \equiv S'. \tag{49a}$$

Since also $4\pi r^3/3 = v$, the volume of the sphere, we obtain finally:

$$\frac{I_{st}}{I_0} \equiv S' = \frac{24\pi^3}{\lambda^4} \left(\frac{m^2 - 1}{m^2 + 2}\right)^2 N v^2, \tag{49b}$$

which is identical with (25) if the intermolecular interference between the particles is again neglected. Rayleigh scattering therefore constitutes a limiting case of the Mie theory for very small particles ($x < 0.8$), if only dipole radiation is taken into account. If terms of the second order (a_2, b_1, etc.) exert any influence, deviations arise from the angular distribution of the scattered intensity given by Fig. 33, for two reasons:

1. The intensity of forward scattering becomes greater than that of backward scattering, in a ratio given by

$$\frac{I_{st\ forward}}{I_{st\ backward}} \equiv \varphi = 1 + 4x^2 \left(\frac{m^2 + 2}{30} + \frac{1}{6} \frac{m^2 + 2}{2m^2 + 3}\right) + \cdots \tag{50}$$

neglecting terms in x^4 etc. This ratio is always greater than unity for dielectric particles, but Eq. (50) is correct only when the condition of sufficiently small values of x is fulfilled. Otherwise, higher terms must be taken into account.

2. The scattered radiation when $\vartheta_s = 90°$ is no longer completely, though still to a considerable extent, linearly polarized. When $\vartheta_s = 90°$, for Rayleigh scattering, $i_2 = 0$ and $i_1 \sim x^6$, as is seen from (45) ($a_2, b_1 = \cdots = 0$), so that $I_s \sim \lambda^2 x^6 \sim \lambda^{-4}$. In the case of Mie scattering,

---

[84] In this case also it is assumed that the particles are not "localized", that is, that no definite relation exists between the phases of the waves scattered from various particles.

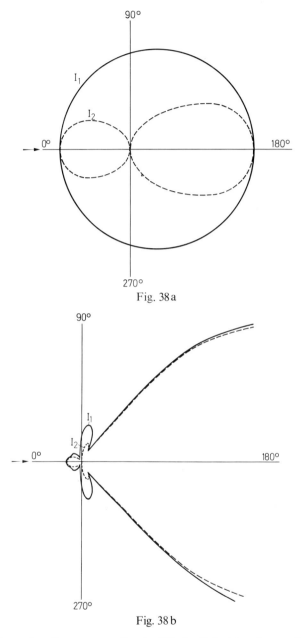

Fig. 38. Scattering diagram according to *Mie* for dielectric spheres with a) $x = 0.8$ and $m = 1.25$, b) $x = 4$ and $m = 1.25$. i_1 = vertical component, i_2 = horizontal component

however, in this approximation $i_2 = \frac{3}{2} b_1 \sim x^{10}$ and $I_{st} \sim \lambda^2 x^{10} \sim \lambda^{-8}$. This horizontally polarized fraction is smaller than i_1 by the factor x^4, but depends much more strongly on the wavelength. If $x > 1$, higher terms in (45) must be taken into account, so that for larger spheres the scattering distribution is very complicated, and maxima and minima arise (see Fig. 36). Because the numerical evaluation of i_1 and i_2 as functions of x and m is very tedious, these values have been provided only for $m = 1.25$,

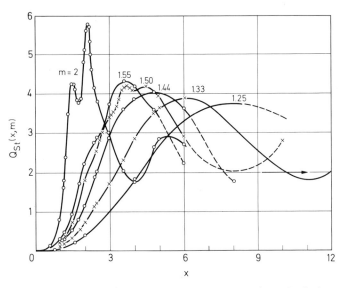

Fig. 39. Efficiency factor $Q_{st}(x, m)$ according to *Mie* for spherical particles as function of x for various values of m as parameter

1.333, 1.44, 1.50, 1.55, and 2.00, and for values of x between 0.1 and 8 (in one case also to $x = 40$ [85]). For example, Fig. 38 shows the scattering diagram (radiation characteristic) for dielectric spheres with a) $x = 0.8$ and $m = 1.25$; and b) $x = 4$ and $m = 1.25$ [86]. Here the vertical component i_1 and horizontal component i_2 are represented separately.

The tables often give both the Mie functions and the so-called *efficiency factor* $Q_{st}(x, m)$ [87]. This is defined as the ratio of the optically

---

[85] Values obtained by *A. N. Lowan* are given in Natl. Bur. Stand. Appl. Math. Ser. **4** (1948). — *Gumprecht, R. O., et al.*: J. Opt. Soc. Am. **42**, 226 (1952). New values at *Dettmar, H. K., W. Lode*, and *E. Marre*: Koll. Z. **188**, 28 (1963).

[86] *Blumer, H.*: Z. Physik **32**, 119 (1925); **38**, 304, 920 (1926).

[87] See, for example, *Sinclair, D.*, and *V. K. La Mer*: Chem. Rev. **44**, 245 (1949).

effective cross section to the geometrical cross section $N\pi r^2$ of the particles. According to this the total scattered radiation flux is

$$I_{st} = N\pi r^2 Q_{st}(x, m) I_0$$

or [88]

$$\frac{I_{st}}{I_0} \equiv S' = N\pi r^2 Q_{st}(x, m). \tag{51}$$

The "turbidity" defined on p. 73 is equal to the "scattering surface" of all particles per cm^3. If we measure the easily accessible turbidity S' as a function of λ and plot this against $1/\lambda$ and evaluate $Q_{st}(x, m)$ for a previously determined value of m as a function of x, both curves must have their maximum at the same value of x, so that r can be obtained. Once r is known, it is possible to use (51) to determine the particle number N per cm^3. For given values of N, particles whose x or r are maxima also show maximum turbidity.

Fig. 39 gives the values of $Q_{st}(x, m)$ for several values of m as a function of x [89]. The curves are simplified and actually show in general several secondary maxima [90].

A particle, for example, for which $r = 0.4$ μ, and $m = 1.44$ with $\lambda = 0.5$ μ giving $x \cong 5$, has a scattering surface of $Q_{st} = 4$, so that it scatters four times as much light as its geometrical cross-section of 0.50 [μ^2] would indicate. For Q_{st}, using the approximation of p. 87, taking account of terms in a_1, a_2, and b_1, we obtain

$$Q_{st} = \frac{8}{3} x^4 \left(\frac{m^2 - 1}{m^2 + 2}\right)^2 \left[1 + \frac{6}{5} \frac{m^2 - 1}{m^2 + 2} x^2 + \cdots \right]. \tag{52}$$

The first term corresponds to Rayleigh scattering, and follows directly from (49a). Eq. (52) is, of course, only valid for $x < 0.8$, whereas for larger values of x further terms in the series must be taken into account. It is found that exponent n for the λ^{-n} dependence of the scattering, which in the Rayleigh region is 4, diminishes to zero with further increase in x, so that the scattering then becomes independent of λ. In the scattering

---

[88] According to p. 79 S' has the dimension [cm^{-1}], and N the dimension [cm^{-3}]. Q_{st} is dimensionless.

[89] *Edsall, I. T.*, and *W. B. Dankliker*: Fortschr. Chem. Forsch. **2**, 1 (1951); see also *Sinclair, D.*: J. Opt. Soc. Am. **37**, 475 (1947). — *La Mer, V. K.*: J. Phys. Coll. Chem. **52**, 65 (1948). — *Jobst, G.*: Ann. Physik **76**, 863 (1925).

[90] For very large values of x, Q_{st} does not approach unity, as would have been expected, but the value 2. This results from the geometrical reflection being augmented by an equal energy contribution due to diffraction, which is scattered over such a small angular range that it can only be observed at very large distances. See also p. 79.

of white light, this may be observed as the gradual change of the color of the scattered light from blue to white. Fig. 40 shows the variation of n with the diameter, $2r$, of polystyrene spheres in aqueous suspension ($m = 1.24$), as determined by the measurements of *Heller*[91] in the visible region of the spectrum. The decrease of n with r for constant values of λ is so reproducible that a method for the determination of particle sizes can be based upon it.

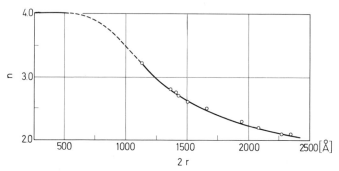

Fig. 40. Variation of the wavelength exponents n of the *Mie* scattered radiation with the diameter of dielectric spheres

For very large values of x, $Q_{st}(x, m)$ becomes independent of x, which has already been indicated in Fig. 38. Hence, according to Eq. (51) for a given wavelength

$$S' \sim N\pi r^2 \sim \frac{1}{r}. \tag{53}$$

The turbudity is then inversely proportional to the radius of the spheres.

For very large spheres, that is for the asymptotic case in which $x \to \infty$, the Mie formulas give a scattering distribution which can also be obtained from the law of geometrical optics with respect to intensity and phase by dealing separately with reflection, refraction, and diffraction[92]. According to the Mie formula there is no distinction in principle between reflection, refraction, and diffraction on the one hand, and scattering on the other hand. In the scope of this book no exhaustive treatment of this can be given, but comprehensive reviews are mentioned on p. 73 of which special mention should be made of that by H. C. van de Hulst.

---

[91] *Heller, W.*: J. Chem. Phys. **14**, 566 (1946); see also *la Mer, V. K.*: J. Phys. Coll. Chem. **52**, 65 (1948).

[92] Compare *Debye, P.*: Ann. Physik **30**, 59 (1909). — *van de Hulst, H. C.*: Rech. Astr. Obs. Utrecht **11**, part 1 (1946). — *Franz, W.*: Z. Naturforsch. **9a**, 705 (1954).

In applying the Mie theory to *absorbing spheres*, as in the case of Eq. (II, 43) we must introduce a complex refractive index, so that Eq. (40) must be replaced by

$$m' = \frac{n}{n_0} - \frac{ni\varkappa}{n_0} \equiv m - mi\varkappa. \tag{54}$$

In this, n_0 is the refractive index of the non-absorbing medium, in which the scattering center is embedded, and $m\varkappa$ is the absorption coefficient.

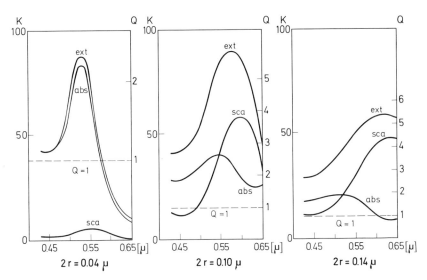

Fig. 41. Efficiency factors Q_{total}, Q_{st}, and Q_{abs} of gold sols (colloidal gold solutions) in water for various particle sizes in dependence on λ_0

As can be seen from Table 2, n and $n\varkappa$, and therefore m and $m\varkappa$, vary greatly as one changes materials or wavelength. There are therefore a great number of approximate formulas which can be derived from the rigorous Mie theory by the development of special series, depending on whether m or $m\varkappa$ or both are given particularly large or particularly small values. It is then possible, for example, to obtain Q as a function of x and m', but the total efficiency factor Q_{total} is made up of Q_{st} (real portion) and Q_{abs} (imaginary portion) of the measured apparent total absorbance:

$$Q_{\text{total}} = Q_{\text{st}} + Q_{\text{abs}}. \tag{55}$$

Q_{st}, for metals in the approximation already used, takes into account only a_1, a_2, and b_1, given by Eq. (52) whereas Q_{abs} is given in the same approximation by the imaginary portion:

$$Q_{abs} = -\left[4x\frac{m^2-1}{m^2+2} + \frac{4}{15}x^3\left(\frac{m^2-1}{m^2+2}\right)^2\frac{m^4+27m^2+38}{2m^2+3}\right]. \quad (56)$$

For small values of x (for metals for instance for x less than 0.3), the first term of the series again suffices to obtain absorption and scattering for particular values of m'. For detailed consideration the reviews given on p. 73 must be consulted [93]. For example, Fig. 41 gives the efficiency factors Q_{total}, Q_{st}, and Q_{abs} for gold sols in water of various particle sizes as a function of λ_0. It can be seen how strongly the values of Q_{st} and Q_{abs} depend upon the particle size.

c) Multiple Scattering

With diminishing distance between the scattering particles, single scattering increasingly gives way to multiple scattering. For high particle densities (as in clouds, opaque glass, concentrated sols, stellar atmospheres, crystal powders, pigments, etc.) multiple scattering predominates. The question now arises whether the angular distribution of the scattered intensity at dielectric spheres according to the Rayleigh or Mie theories also applies to these cases. Is, for example, in the case of a plane incident wave after multiple scattering, the forward scattering greater than backward scattering when $2\pi r \geq \lambda$? This problem has been investigated by *Theissing* [94]. It was assumed that the individual particles are distributed in a statistically random way, and are sufficiently far apart that no phase relations or interference between the scattered radiation from the various particles need be taken into account. The only distinction between this and Mie scattering is that the waves are scattered not once but twice, three times, or more, before leaving the medium.

We consider a plane non-absorbing layer of thickness d, on which, in the x-direction, a plane unpolarized wave impinges at $x=0$. The incident radiation flux will be taken as $I_0(0)=1$. Then the radiation flux at any distance x within the layer is made up of a fraction $I_0(x)$ of the parallel unscattered wave, a fraction $I_1(x)$ of a once scattered wave, a further fraction $I_2(x)$ of a twice scattered wave, etc., so that

$$I(x) = \sum_k I_k = I_0 + I_1 + I_2 + \cdots. \quad (57)$$

Each value of I_k has its own angular distribution $q_k(\vartheta_s)$ where ϑ_s is the scattering angle defined by (14). In the first place only $q_1(\vartheta_s)$ of the Mie

[93] See further Chromey, F. C.: J. Opt. Soc. Am. **50**, 730 (1960). — Brockes, A.: Optik **21**, 550 (1964).
[94] Theissing, H. H.: J. Opt. Soc. Am. **40**, 232 (1950).

scattering is known, being given by Eq. (42)[95]. We consider now the variation in I_k with x. For the unscattered (parallel) fraction, according to (1), we have

$$-\frac{dI_0}{dx} = S'I_0 \quad \text{or} \quad I_0(x) = e^{-S'x}; \quad I_0(0) = 1, \quad (58)$$

in which S' is the scattering coefficient according to the Mie theory, obtained by integration of (42), over a spherical surface of radius $R = 1$ [see, for example, (49a)]. For the scattered fractions it is convenient to divide the total space into forward (f) and backward (b) hemispheres. The fraction of the total scattering S'_k in the forward direction is then, in a similar way to (14a), given by

$$f_k \equiv \int_0^{\pi/2}\int_0^{2\pi} \left(\frac{q(\vartheta_s)}{S'}\right)_k \sin\vartheta_s \, d\vartheta_s \, d\varphi = 2\pi \int_0^{\pi/2} \left(\frac{q(\vartheta_s)}{S'}\right)_k \sin\vartheta_s \, d\vartheta_s, \quad (59)$$

while the fraction of total scattering in the back hemisphere is given similarly by

$$b_k \equiv 2\pi \int_{\pi/2}^{\pi} \left(\frac{q(\vartheta_s)}{S'}\right)_k \sin\vartheta_s \, d\vartheta_s. \quad (60)$$

Both f_k and b_k depend on the corresponding angular distribution of the fraction I_k, and we have

$$f_k + b_k = 1. \quad (61)$$

If we designate the fraction of the radiation scattered in the forward hemisphere by F_k, and that in the backward hemisphere by B_k, we have also

$$\Sigma I_k = \Sigma F_k + \Sigma B_k. \quad (62)$$

Using these expressions we have for the change in the *primary* forward scattered radiation $F_1(x)$ within the distance dx

$$dF_1(x) = [-F_1(x) + f_1 I_0(x)] S' dx. \quad (63)$$

In the layer thickness element dx, dF_1 is proportional to $-F_1$ itself, because the primary scattered radiation is converted into secondary scattered radiation. dF_1 is also proportional to the still present parallel radiation $I_0(x)$ the fraction f_1 of which is transformed into primary scattered radiation. Inserting Eq. (58) for $I_0(x)$ and rearranging, a linear differential equation of first order is obtained:

$$\frac{1}{S'}\frac{dF_1(x)}{dx} + F_1(x) - f_1 e^{-S'x} = 0 \quad (64)$$

---

[95] In (42) and (1) I_0 is the incident radiation flux, which is here set equal to 1.

The solution is
$$F_1 = f_1 S' x e^{-S'x}. \tag{65}$$

Similarly, we obtain for the alteration in the primary back scattered radiation $B_1(x)$ within the range dx

$$\frac{1}{S'} \frac{dB_1(x)}{dx} - B_1(x) + b_1 e^{-S'x} = 0 \tag{66}$$

whose solution is

$$B_1 = \frac{b_1}{2} e^{-S'x} - \frac{b_1}{2} e^{-2S'd} \cdot e^{S'x}. \tag{67}$$

For the *secondary* forward and back scattering we commence with differential equations corresponding to (63) and (66)[96]:

$$\frac{1}{S'} \frac{dF_2(x)}{dx} + F_2(x) - f_2 F_1(x) - b_2 B_1(x) = 0, \tag{68}$$

$$\frac{1}{S'} \frac{dB_2(x)}{dx} - B_2(x) + b_2 F_1(x) + f_2 B_1(x) = 0, \tag{69}$$

and obtain the solutions

$$F_2 = \frac{b_1 b_2}{2} \cdot S'x \cdot e^{-S'x} + f_1 f_2 \frac{(S'x)^2}{2!} \cdot e^{-S'x}$$
$$- \frac{b_1 b_2}{4} \cdot e^{-2S'd} e^{S'x} + \frac{b_1 b_2}{4} \cdot e^{-2S'd} e^{-S'x}, \tag{70}$$

$$B_2 = \frac{f_1 b_2 + f_2 b_1}{4} \cdot e^{-S'x} + \frac{f_1 b_2}{4} S'x \cdot e^{-S'x}$$
$$+ \frac{f_2 b_1}{2} S'x \cdot e^{-2S'd} e^{S'x} - \frac{f_1 b_2 + f_2 b_1}{4} \cdot (2S'd+1) \cdot e^{-2S'd} e^{S'x}. \tag{71}$$

It is seen that for very large particle densities, that is, for large values of $S'd$, the third and fourth terms in both equations vanish. Similarly, we obtain two equations each for F_3 and B_3, F_4 and B_4, etc. in which again the terms with $e^{-2S'd}$ may be neglected for large values of $S'd$. The sums

$$I_0(x), \quad I_1(x) = F_1(x) + B_1(x), \quad I_2(x) = F_2(x) + B_2(x) \quad \text{etc.}$$

give the contributions of unscattered, singly scattered, doubly scattered, etc., waves as a function of x.

In the limiting case for very small particle densities $S'd \to 0$, only single scattering needs to be considered. We then obtain for the radiation

---

[96] $b_2 B_1$ corresponds to a radiation flux in the forward direction, while $b_2 F_1$ and $f_2 B_1$ correspond to radiation flux in a backward direction.

emerging in the forward hemisphere when $x = d$, using (65)

$$F_1(d) = f_1 S'd \cdot e^{-S'd} \cong f_1 S'd(1 - S'd) \cong f_1 S'd, \tag{72}$$

while for the radiation emerging in the back hemisphere from (67),

$$B_1(0) = \frac{b_1}{2}(1 - e^{-2S'd}) \cong b_1 S'd. \tag{73}$$

The total scattering is then given by

$$F_1(d) + B_1(0) \cong (f_1 + b_1) S'd = S'd. \tag{74}$$

If we start, not from the two radiation fluxes F and B, but from the total fraction I_1 of the radiation flux once scattered in the entire space, we obtain, for small values of $S'd$, for the fraction scattered in the forward hemisphere

$$F_1(d) = I_1(d) f_1, \tag{75}$$

while for that scattered in the back hemisphere

$$B_1(0) = I_1(d) b_1. \tag{76}$$

Comparison with (72) and (73) shows that

$$I_1(d) = S'd \cdot e^{-S'd} \cong S'd \tag{77}$$

in agreement with (74). Thus, for the limiting case when $S'd \ll 1$, instead of considering both radiation fluxes F_1 and B_1, we can consider I_1 in dependence on x, by initially taking $f_1 = 1$, and $b_1 = 0$, and then multiply $I_1(d)$ by f_1 or b_1. If we then apply this to the twice scattered wave in a similar way, we obtain from (70), taking $f_1 = f_2 = 1$, and $b_1 = b_2 = 0$,

$$I_2(x) = \frac{(S'x)^2}{2!} \cdot e^{-S'x}. \tag{78}$$

From $I_2(d)$ we may then obtain an approximate value for the backward scattered radiation by multiplying by b_2. Similarly we obtain equations for higher orders of scattering k, e.g.

$$I_3(x) = \frac{(S'x)^3}{3!} \cdot e^{-S'x}, \tag{79}$$

$$I_k(x) = \frac{(S'x)^k}{k!} \cdot e^{-S'x}. \tag{80}$$

These I_k values are shown in Fig. 42 plotted against the scattering density $S'x$. It is seen that, for a given value of $S'd$, it is only necessary to take

7 Kortüm, Reflectance Spectroscopy

into account a certain number of orders of scattering, as both higher and smaller orders of scattering contribute negligibly to the total radiation flux $\sum_k I_k$.

Of particular interest is the angular distribution $q(\vartheta_s)/S'$ of the various orders of scattering. First only the angular distribution of the Mie scattering, that is, scattering of the first order, is known, being given by (42). According to Hartel[97] it is possible, with certain simplifying

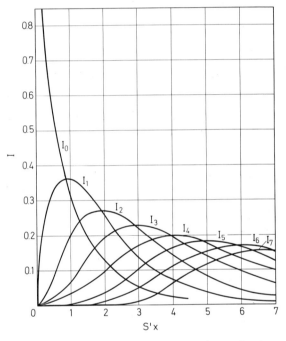

Fig. 42. Radiation intensities of different scattering order in case of multiple scattering (I_0 unscattered; I_1 scattered once; I_2 scattered twice; etc.) as function of the scattering density $S'x$

assumptions ($m \cong n/n_0 \cong 1$, and restriction to electrical multipole radiation) to express the angular factors $\left(\dfrac{q(\vartheta_s)}{S'}\right)_k$ by $\left(\dfrac{q(\vartheta_s)}{S'}\right)_1$:

$$\left(\frac{q(\vartheta_s)}{S'}\right)_k = \frac{1}{4\pi} \sum_{n=0}^{\infty} \frac{a_n^k}{(2n+1)^{k-1}} \cdot \pi_n(\cos \vartheta_s). \tag{81}$$

---

[97] Hartel, W.: Licht **10**, 141 (1940).

Thus we obtain the angular distribution of the second order (twofold scattering) from (81) as

$$4\pi \left(\frac{q(\vartheta_s)}{S'} \right)_2 = 1 + \frac{a_1^2}{3} \pi_1(\cos \vartheta_s) + \frac{a_2^2}{5} \pi_2(\cos \vartheta_s)$$

$$+ \frac{a_3^2}{7} \pi_3(\cos \vartheta_s) + \cdots , \qquad (82)$$

in which the values of π_n are the spherical harmonics of Eq. (43) and the values of a_n are the complex functions of x and m [see Eq. (46)].

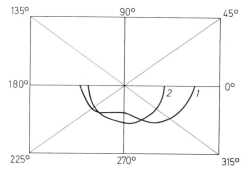

Fig. 43. Angular distribution of the scattering intensity after single (curve 1) and twofold (curve 2) scattering on dielectric spheres with $x = 0.6$ and $m = 1.25$

The evaluation of these angular factors according to (82) shows that with increase in the order k fewer terms in the sum need to be considered as these converge increasingly rapidly to zero. If we represent the angular distributions for the various orders given by (81) in a polar diagram, we find that with increasing value of k, the forward scattering according to the Mie theory steadily diminishes, and the total scattering approximates to an isotropic angular distribution. Fig. 43 shows this for dielectric spheres with $x = 0.6$ and $m = 1.25$ for the first order using the Mie theory ($k = 1$) and for the second order ($k = 2$). It is seen that under these conditions twofold scattering is sufficient to attain practically an isotropic scattering distribution. For $x = 5$ and $m = 1.25$ approximately eight-fold scattering is necessary.

The most important result of this semi-quantitative investigation is that for sufficiently large scattering density $S'd$, that is, for sufficiently great numbers of particles and layer thickness, an isotropic scattering distribution always arises ultimately inside the sample regardless of the

7*

scattering law used [98]. If the incident radiation is initially isotropic (diffuse irradiation), isotropic distribution can also be assumed in the boundary layers. With multiple scattering, the characteristic properties of single scattering disappear more or less rapidly according to the conditions given. We should thus expect that with multiple scattering in general both the remitted and the transmitted radiation (if any) would possess isotropic intensity distribution if $S'd$ is sufficiently large. This corresponds to the "diffusely reflected" radiation for densely packed coarse particles when the laws of reflection, refraction, and diffraction may still be used (see p. 26). The validity of the Lambert Cosine Law for very fine particles, ignoring any regularly scattered portions ($x \ll 1$), fully confirms this conclusion.

d) The Radiation-Transfer Equation

If $S'd$ is very large and $x \equiv \dfrac{2\pi r}{\lambda} \geq 1$, i.e., if the particles are large compared to the wavelength and so tightly packed that phase relations and interferences arise among the scattered beams, there is no general quantitative solution to the problem of multiple scattering. This is also the case if we assume an isotropic radiation field within the medium under investigation. In such cases we must resort to purely phenomenological theories. The practical usefulness of such theories depends upon how many experimentally determined constants are necessary to have a reasonably descriptive model for the properties of interest in such a system. Occasional attempts have been made to achieve this with single-constant theories, but the complexity of the processes in multiple scattering leads one to anticipate that they cannot suffice for the necessary characteristics of an simultaneously absorbing and scattering medium. The majority of authors, therefore, use *Two-constant Theories*, where the constants are intended to characterize the absorbance and scattering per centimeter of layer thickness of the medium under consideration. The constants are determined from transmission and reflectance measurements.

We start from a general *Radiation-Transfer Equation* which describes the change in intensity of a beam of radiation of given wavelength in a path length ds within the medium [99]:

$$-dI = \kappa \varrho I \, ds, \tag{83}$$

---

[98] Compare *Hamaker, H. C.*: Philips Res. Rep. **2**, 55 (1947). — *Rozenberg, G. V.*: Bull. Acad. Sci. U.S.S.R., Phys. Ser **21**, 1465 (1957); — *Blevin, W. R.*, and *W. J. Brown*: J. Opt. Soc. Am. **51**, 129, 975 (1961).

[99] A thorough representation of radiation transfer is given by *Chandrasekhar, S.*: Radiative Transfer. Oxford: Clarendon Press 1950; this was reissued by Dover Publications, Inc., New York 1960.

where κ is an attenuation coefficient corresponding to the total radiation loss due to both absorption and scattering, and ϱ is the density of the medium. Because the scattered fraction of the radiation is not really lost but reappears in another direction, (83) can be written in the form

$$-dI = \kappa \varrho I \, ds - j \varrho \, ds, \tag{84}$$

where j is a scattering function defined as

$$j(\vartheta, \varphi) = \frac{\kappa}{4\pi} \int_0^\pi \int_0^{2\pi} p(\vartheta, \varphi; \vartheta', \varphi') I(\vartheta', \varphi') \sin \vartheta' \, d\vartheta' \, d\varphi'. \tag{85}$$

$p(\vartheta, \varphi; \vartheta', \varphi')$ is designated as the *phase function*. It is a measure of the intensity of the scattered radiation in the solid angle $d\omega' = \sin \vartheta' \, d\vartheta' \, d\varphi'$, if a beam of radiation in the direction (ϑ, φ) strikes a mass element of the medium. By integrating over all space and dividing by 4π we obtain the scattering coefficient j. The *Radiation-Transfer Equation* is commonly written in the form

$$-\frac{dI}{\kappa \varrho \, ds} = I - \frac{j}{\kappa}. \tag{86}$$

The ratio j/κ of scattering to absorption coefficient is called the *source function*.

The problems which are of special interest here concern plane-parallel layers. Here we advantageously measure distances ds perpendicular to the plane of the layer and designate this normal as x, so that $ds = dx/\cos \vartheta$ where ϑ is the inclination to the outward normal. The equation of transfer then becomes

$$-\cos \vartheta \, \frac{dI(x, \vartheta, \varphi)}{\kappa \varrho \, dx} = I(x, \vartheta, \varphi) - \frac{j}{\kappa}(x, \vartheta, \varphi), \tag{87}$$

where φ is the azimuth referred to an arbitrarily chosen axis. If we finally employ the so called *optical thickness*

$$d\tau \equiv \kappa \varrho \, dx \tag{88}$$

and, in addition, set

$$\mu \equiv \cos \vartheta, \tag{89}$$

where ϑ represents the angle with respect to the *inward* surface normal, use of (85) leads to

$$\mu \frac{dI(\tau, \mu, \varphi)}{d\tau} = I(\tau, \mu, \varphi) - \frac{1}{4\pi} \int_{-1}^{+1} \int_0^{2\pi} p(\mu, \varphi; \mu', \varphi') I(\tau, \mu', \varphi') \, d\mu' \, d\varphi'. \tag{90}$$

This is the general form of the *Radiation-Transfer Equation* for plane parallel media. Because j/κ is a function of the intensity at each point, it is an integro-differential equation. We distinguish between two kinds of plane parallel media: 1. The semi-infinite medium has one planar boundary ($\tau = 0$) and extends perpendicular to this to infinity ($\tau \to \infty$); 2. The finite medium has two planar parallel boundaries at $\tau = 0$ and $\tau = T$.

Decisive for radiation transfer by scattering is the phase function. Let us designate with θ the angle between the incidence direction and the scattered direction of a radiation beam scattered by a mass element of a medium. Then the integral

$$\int_0^{4\pi} p(\cos\theta) \frac{d\omega'}{4\pi} \equiv \omega_0 \tag{91}$$

represents the total scattering in all directions. ω_0 is called the *albedo for single scatter*. If no "true" absorption occurs, $\omega_0 = 1$; otherwise, $1 - \omega_0$ is the fraction of the radiation absorbed. If every element scatters isotropically, $p(\cos\theta)$ is independent of θ and

$$p(\cos\theta) = \text{const} = \omega_0. \tag{92}$$

This is the most simple case of a phase function. The corresponding transfer equation from (90) is

$$\mu \frac{dI(\tau,\mu)}{d\tau} = I(\tau,\mu) - \frac{1}{2}\omega_0 \int_{-1}^{+1} I(\tau,\mu')\,d\mu', \tag{93}$$

because I as well as j/κ become independent of the azimuth (axial symmetry around the x-axis).

For *nonisotropic* scattering we frequently use the phase function

$$p(\cos\theta) = \omega_0(1 + x\cos\theta); \quad 0 \leq x \leq 1. \tag{94}$$

In the most general case we can represent $p(\cos\theta)$ as a series of Legendre polynomials [100]:

$$p(\cos\theta) = \sum_0^\infty \omega_l P_l(\cos\theta). \tag{95}$$

---

[100] The Legendre polynomials of the lth degree are given by

$$P_l(x) = \frac{1}{2^l l!} \frac{d^l}{dx^l}(x^2 - 1)^l.$$

Accordingly,

$P_0(x) = 1$

$P_1(x) = x = \cos\theta$

$P_2(x) = \frac{1}{2}(3x^2 - 1) = \frac{1}{4}(3\cos 2\theta + 1)$

$P_3(x) = \frac{1}{2}(5x^3 - 3x) = \frac{1}{8}(5\cos 3\theta + 3\cos\theta)$ etc.

Chapter IV. Phenomenological Theories of Absorption and Scattering of Tightly Packed Particles

We can proceed to study this matter in different ways. We can, for example, concieve of the coefficients of absorption and scattering as properties of the irradiated layer; i.e., consider the latter as a *Continuum*. On the other hand, we can assume the layer composed of a series of partial layers whose thicknesses are determined by the size of the scattering and absorbing particles. The scattering and absorption of these particles are then used as optical constants. We can call this procedure *Discontinuum Theory*. In what follows, we consider first of all the continuum theory [101].

a) The Schuster Equation for Isotropic Scattering

The first attempt which equals a simplified solution of the Radiation-Transfer Equation in its simplest form for isotropic scattering (III, 93) originated with *Schuster*[102]. He was interested in the absorption and emissivity of stellar atmospheres where the emissivity is partly the result of scattering and partly of luminescence induced by high temperatures. So far as we are concerned with radiation in the UV and visible parts of the spectrum, we can, at normal temperatures, naturally neglect the terms that relate to temperature-induced emission.

Although the layer is naturally penetrated in all directions by the radiation, the Schuster method consists of simplifying the radiation field into two oppositely directed radiation fluxes in the x and $-x$ directions, respectively. This corresponds to an analogous treatment of diffusion of the kinetic theory of gases where all the moving molecules in a right angled box are conceived of as moving in one or another of three equal pairs of currents parallel to the three axes of the box. It is easy to show that, in this manner, we can obtain a first approximation to the solution of the transfer equation (III, 93) for the case of isotropic scattering.

If we designate the radiation flux in the positive x-direction (perpendicular to the boundary plane of the layer) by I and that in the negative x-direction by J (resulting from scattering), we can separate

---

[101] A survey is given by *Kottler, F.:* Prog. Optics **3**, 3 (1964).
[102] *Schuster, A.:* Astrophys. J. **21**, 1 (1905).

Eq. (III, 93) into the following pair of equations:

$$-\frac{1}{2}\frac{dI}{d\tau} = I - \frac{\omega_0}{2}(I+J), \\ +\frac{1}{2}\frac{dJ}{d\tau} = J - \frac{\omega_0}{2}(I+J). \quad (1)$$

The factor $\pm\frac{1}{2}$ on the left side arises because of the averaging of μ over all possible angles to the x-direction (compare p. 107). If we express the albedo ω_0 in terms of the scattering coefficient σ and the true absorption coefficient a of single scattering, being constant in a continuum:

$$\omega_0 = p(\cos\theta) = \frac{\sigma}{a+\sigma}, \quad (2)$$

(1) becomes

$$-\frac{dI}{d\tau} = \frac{2a+\sigma}{a+\sigma}I - \frac{\sigma}{a+\sigma}J, \\ +\frac{dJ}{d\tau} = \frac{2a+\sigma}{a+\sigma}J - \frac{\sigma}{a+\sigma}I. \quad (3)$$

If we make the substitutions

$$\frac{2a}{a+\sigma} \equiv k \quad \text{und} \quad \frac{\sigma}{a+\sigma} \equiv s, \quad (4)$$

the two simultaneous differential equations can be written as

$$-\frac{dI}{d\tau} = (k+s)I - sJ, \\ +\frac{dJ}{d\tau} = (k+s)J - sI. \quad (5)$$

The undetermined integrals are

$$I = A(1-\beta)e^{\alpha\tau} + B(1+\beta)e^{-\alpha\tau}, \\ J = A(1+\beta)e^{\alpha\tau} + B(1-\beta)e^{-\alpha\tau}, \quad (6)$$

with

$$\alpha \equiv \sqrt{k(k+2s)} \quad \text{und} \quad \beta \equiv \sqrt{k/(k+2s)}. \quad (7)$$

Thereby we obtain

$$A = -\frac{(1-\beta)e^{-\alpha\tau}}{(1+\beta)^2 e^{\alpha\tau} - (1-\beta)^2 e^{-\alpha\tau}} I_0, \\ B = -\frac{(1+\beta)e^{\alpha\tau}}{(1+\beta)^2 e^{\alpha\tau} - (1-\beta)^2 e^{-\alpha\tau}} I_0, \quad (8)$$

if we allow for the boundary conditions

$$I = I_0 \quad \text{at} \quad \tau = 0,$$

$$I = I_{(\tau)}; \quad J = 0 \quad \text{at} \quad \tau = \tau.$$

For $\tau \to \infty$ (semi-infinite medium), the diffuse reflectance becomes

$$R_\infty = \frac{J_{(\tau=0)}}{I_0} = \frac{1-\beta}{1+\beta}. \tag{9}$$

Eq. (9) can be rewritten as

$$\frac{(1-R_\infty)^2}{2R_\infty} = \frac{k}{s}, \tag{10}$$

the so-called *reflectance function* for isotropic scattering assuming two oppositely directed radiation fluxes in the direction of the surface normal. Since this time, numerous other workers [103] have continued to concern themselves with this simplified theory of diffuse remission and transmission. The majority of these use differential equations similar to those of *Schuster*, and arrive, therefore, at similar expressions. This applies especially to the theory of *Kubelka* and *Munk*, which is regarded nowadays as the most general and usual of those available. As *Kubelka* has shown, the results obtained in the theories of *Gurevic* and *Judd*, developed independently of each other and of *Kubelka* and *Munk*, may be regarded as special cases of the Kubelka-Munk equations. In the same way, *Ingle* [104], has shown that the formulas of *Smith*, *Amy* and *Bruce*, may also be derived from the Kubelka-Munk equations. For these reasons, the Kubelka-Munk theory has received general acceptance, especially as

---

[103] *Channon, H. J., F. F. Renwick,* and *B. V. Storr:* Proc. Roy. Soc. London **94**, 222 (1918); — *Renwick, F. F.:* Photogr. J. **58**, 140 (1918); — *Bruce, H. D.:* Tech. Pap. 306, Natl. Bur. Std. (1926); — *Silberstein, L.:* Phil. Mag. **4**, 1291 (1927); — *Gurevič, M.:* Physik. Z. **31**, 753 (1930); — *Kubelka, P.,* u. *F. Munk:* Z. Tech. Physik **12**, 593 (1931); — *Smith, T.:* Trans. Opt. Soc. London **33**, 150 (1931); — *Ryde, J. W.,* and *B. S. Cooper:* Proc. Roy. Soc. London A **131**, 451 (1931); — *Judd, D. B.:* J. Res. Natl. Bur. Std. **12**, 354; **13**, 281 (1934); — *Amy, L.:* Rev. Optique **16**, 81 (1937); — *Neugebauer, H. E. I.:* Z. Tech. Physik **18**, 137 (1937); — *Duntley, S. Q.:* J. Opt. Soc. Am. **32**, 61 (1942); — *Hulbert, E. O.:* J. Opt. Soc. Am. **33**, 42 (1943); — *Mecke, R.:* Meteorol. Z. **61**, 195 (1944); — *Hamaker, H. C.:* Philips Res. Rep. **2**, 55, 103 (1947); — *Kubelka, P.:* J. Opt. Soc. Am. **38**, 448 (1948); — *Broser, J.:* Ann. Physik **5**, 401 (1950); — *Mühlschlegel, B.:* Ann. Physik **9**, 29 (1951); — *Bodo, Z.:* Acta Phys. Hung. **1**, 135 (1951); — *Bauer, G. T.:* Acta Phys. Acad. Sci. Hung. **14**, 209 (1962); — *Melamed, N. T.:* J. Appl. Phys. **34**, 560 (1963); — *ter Vrugt, J. W.:* Philips Res. Rep. **20**, 23 (1965).

[104] *Ingle, G. W.:* ASTM Bull. **116**, 32 (1942).

it is relatively simple and uses concepts such as "reflectance", "contrast ratio", "hiding power", "brigtening power", which were already widely used in practice.

As very complex processes are involved in scattering transmission and remission, it is necessary to make simplifying assumptions as a basis for the theory. These assumptions are: the Lambert Cosine Law is valid, involving isotropic distribution of scattering, so that any regular reflection (Chapter II) is ignored; the particles in the layer are regarded as being randomly distributed, and very much smaller than the thickness of the layer itself; the layer is subject only to diffuse irradiation[105]. Under these conditions, the two-constant theory of *Kubelka* and *Munk* leads to conclusions accessible to experimental test. In practice these are found to be at least qualitatively confirmed, and, under suitable conditions fulfilling the assumptions made, quantitatively as well.

b) The Kubelka-Munk Exponential Solution

Let a plane parallel layer of thickness d, capable of both scattering and absorbing radiation, be irradiated in the $-x$-direction with a diffuse, monochromatic radiation flux, $I_{(x=d)}$ (Fig. 44)[106]. Let the extension of the layer in the yz-plane be great compared with d, so that edge

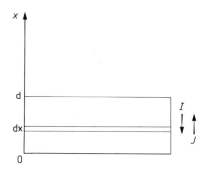

Fig. 44. Concerning the derivation of the simultaneous differential equations according to *Kubelka-Munk*

---

[105] A diffuse irradiation is the inverse concept of diffuse reflection, as defined on p. 26. It corresponds to an isotropic angular distribution of the incident radiation density, and can be verified by means of a so-called integrating sphere (see p. 219). Because with multiple scattering, isotropic distribution of the radiation within the sample is assumed (see p. 99), in the case of diffuse irradiation, isotropic distribution can be assumed to predominate throughout the whole sample.

[106] This somewhat unusual arrangement will simplify subsequent calculations; in this regard, Eq. (20) and (21) have signs opposite that of (5).

effects may be ignored. We consider an infinitesimal layer of thickness dx parallel to the surface. The radiation flux in the negative x-direction will be designated by I, and that in the positive x-direction due to the scattering with J. As the layer dx will be subject to radiation at all possible angles with respect to x, the *average* path length of the radiation within dx is not equal to dx but evidently larger than this (compare p. 104). For a particular direction ϑ, the path length is:

$$d\xi = \frac{dx}{\cos\vartheta}. \tag{11}$$

This is seen from Fig. 45. If we designate the angular distribution of the radiation flux falling on the layer of thickness dx by $\partial I/\partial \vartheta$, the relative intensity in the direction ϑ is given by $\dfrac{1}{I_0} \cdot \dfrac{\partial I}{\partial \vartheta} d\vartheta$, where I_0 is the total

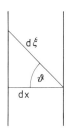

Fig. 45. Derivation of the average path length in case of isotropic scattering

flux falling in the hemisphere. To obtain the *mean value* of the path length of the radiation within the layer dx, we must integrate over all values of ϑ from 0 to $\pi/2$. Thus:

$$\overline{d\xi_I} = dx \int_0^{\pi/2} \frac{1}{I_0 \cos\vartheta} \cdot \frac{\partial I}{\partial \vartheta} d\vartheta \equiv u\, dx. \tag{12}$$

Similarly, for the radiation in the positive direction J:

$$\overline{d\xi_J} = dx \int_0^{\pi/2} \frac{1}{J_0 \cos\vartheta} \cdot \frac{dJ}{\partial \vartheta} d\vartheta \equiv v\, dx. \tag{13}$$

Isotropic angular distribution of the scattered radiation is characterized by its possessing the same intensity in all directions. For a plane upon which diffuse radiation impinges, the angular distribution is given by [107]:

$$\frac{\partial I}{\partial \vartheta} = 2I_0 \sin\vartheta \cos\vartheta \quad \text{or} \quad \frac{\partial J}{\partial \vartheta} = 2J_0 \sin\vartheta \cos\vartheta. \tag{14}$$

By inserting this in (12) or (13) we obtain:

$$u = \int_0^{\pi/2} 2\sin\vartheta \, d\vartheta = 2 \quad \text{or} \quad v = \int_0^{\pi/2} 2\sin\vartheta \, d\vartheta = 2. \tag{15}$$

For completely diffuse irradiation of the layer dx, we then have

$$\overline{d\xi_I} = \overline{d\xi_J} = 2\, dx. \tag{16}$$

The mean path length of the diffuse irradiation is thus twice the geometrical layer thickness.

If parallel radiation is used, according to Eq. (11) $d\xi = 2\,dx$, when $\cos\vartheta = \frac{1}{2}$ or $\vartheta = 60°$. If the layer is scattering diffusely (isotropic scattering), for the scattered portion of the radiation $u = 2$ while for the unscattered portion which retains the original direction of $\vartheta = 60°$, once again $u = 2$, so that the same relation applies for the total radiation. It is therefore possible to use parallel light at an angle of $60°$ instead of diffuse light without affecting the validity of Eq. (16).

If we designate the absorption coefficient of the material by k, and its scattering coefficient per centimeter by s, then within the layer dx under the irradiation conditions mentioned, the fraction $kI2\,dx$ will be absorbed, and the fraction $sI2\,dx$ will be lost by back scattering. The radiation flux J from below will itself give the fraction $sJ2\,dx$ through scattering in the negative x-direction, so that the intensity change from I in the layer element dx is made up of three parts:

$$-dI = -kI2\,dx - sI2\,dx + sJ2\,dx. \tag{17}$$

Similarly, we obtain the intensity reduction of J in the positive x-direction:

$$dJ = -kJ2\,dx - sJ2\,dx + sI2\,dx. \tag{18}$$

As J increases with increase in x, but I diminishes, dI and dJ are of opposite signs.

---

[107] See, for example, *Kortüm, G.*, and *M. Kortüm-Seiler*: Z. Naturforsch. **2a**, 652 (1947). — *Pohl, R. W.*: Optik. Berlin: Springer-Verlag 1959. According to the Lambert Cosine Law, the expression $dI = B\,df \cos\vartheta\,df'/R^2$ is valid for the radiation flux falling on a surface element df' from another surface element df at the direction ϑ to the normal. R is the distance separating the surface elements and B is the radiation density. We can interpret df' as the circular zone of a sphere so that $df' = 2\pi r\,dr = 2\pi R \sin\vartheta R\,d\vartheta$. From this follows $dI = 2\pi B\,df \cdot \sin\vartheta \cos\vartheta\,d\vartheta$ or $dI = \pi B\,df \sin 2\vartheta\,d\vartheta$ which, because integration over the entire forward hemisphere (i.e., $\vartheta = 0$ to $\vartheta = \pi/2$) gives $I_0 = \pi B\,df$, is identical with Eq. (14).

If we designate

$$2k \equiv K \quad \text{and} \quad 2s \equiv S, \tag{19}$$

we obtain the two fundamental simultaneous differential equations which describe the absorption- and scattering-process [108]:

$$-\frac{dI}{dx} = -(K+S)I + SJ, \tag{20}$$

$$\frac{dJ}{dx} = -(K+S)J + SI. \tag{21}$$

If we now write for brevity

$$\frac{S+K}{S} = 1 + \frac{K}{S} \equiv a, \tag{22}$$

the equations may be written in the form

$$-\frac{dI}{S\,dx} = -aI + J, \tag{23}$$

$$\frac{dJ}{S\,dx} = -aJ + I. \tag{24}$$

If we divide the first equation by I, and the second by J and add the two, we find, writing

$$\frac{J}{I} \equiv r, \tag{25}$$

that

$$\frac{dr}{S\,dx} = r^2 - 2ar + 1 \tag{26}$$

or

$$\int \frac{dr}{r^2 - 2ar + 1} = S \int dx. \tag{27}$$

Integrating over the whole thickness of the layer, we have for the two boundaries:

$$x = 0: \ (J/I)_{x=0} = R_g = \text{ reflectance of the background}, \tag{28}$$
$$x = d: \ (J/I)_{x=d} = R = \text{ reflectance of the sample}. \tag{29}$$

---

[108] The scattering coefficient S of the Kubelka-Munk Theory and the quantity S' of the Mie theory (see p. 91) are not identical. In the Mie theory any alteration in direction of the incident wave due to impact with an obstacle is designated as scattering, while in the Kubelka-Munk theory, the radiation is only regarded as scattered when it is backward reflected into the hemisphere whose bounding plane lies perpendicular to the x-direction. S thus corresponds to the integral given by (III, 60).

Integration of (27) by partial fractions gives

$$\ln \frac{(R-a-\sqrt{a^2-1})(R_g-a+\sqrt{a^2-1})}{(R_g-a-\sqrt{a^2-1})(R-a+\sqrt{a^2-1})} = 2Sd\sqrt{a^2-1}. \quad (30)$$

When $d = \infty$, that is when the layer is of infinite thickness, $R_g = 0$, and

$$(-a-\sqrt{a^2-1})(R_\infty - a + \sqrt{a^2-1}) = 0$$

or solved for R_∞

$$R_\infty = \frac{1}{a+\sqrt{a^2-1}} = a - \sqrt{a^2-1}$$

$$= 1 + \frac{K}{S} - \sqrt{\frac{K^2}{S^2} + 2\frac{K}{S}} = \frac{S}{S+K+\sqrt{K(K+2S)}}. \quad (31)$$

R_∞, the so-called *diffuse reflectance* of the sample is readily measurable, and is a function only of K/S, that is, it depends exclusively on the *ratio*

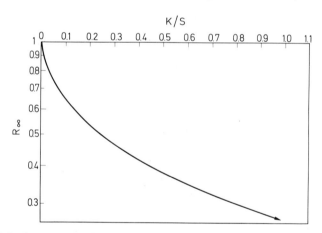

Fig. 46. Reflectance R_∞ of a thick layer of a white pigment (plotted logarithmically) in dependence on the increasing addition of soot

of the absorption and scattering coefficients, and not upon their absolute values. In Fig. 46, R_∞ is plotted logarithmically against K/S, the alteration in K/S being obtained by adding gradually increasing small quantities of carbon black to a white paint. Making the undoubtedly correct assumption that the scattering coefficient is not changed by this (see p. 209), K/S must be proportional to the addition of carbon black. The initially steep fall in the curve for small additions of carbon black shows how

The Kubelka-Munk Exponential Solution

sensitive the reflectance of an ideal white material is to very small absorbing impurities. This is the reason for the well-known fact that the value of $R_\infty = 1$ is practically never attained (see p. 146).

If we resolve (31) for K/S, we obtain analogously as in (10)

$$\frac{K}{S} = \frac{(1-R_\infty)^2}{2R_\infty} \equiv F(R_\infty), \tag{32}$$

an equation called the "*Kubelka-Munk function*".

Rearranging (31), we can represent both a and $\sqrt{a^2-1} \equiv b$ as a function of R_∞:

$$a = \frac{1}{2}\left(\frac{1}{R_\infty} + R_\infty\right), \tag{33}$$

$$\sqrt{a^2-1} \equiv b = \frac{1}{2}\left(\frac{1}{R_\infty} - R_\infty\right). \tag{34}$$

a and b are mutually related by the expressions:

$$b = \sqrt{a^2-1} \quad \text{und} \quad a = \sqrt{b^2+1} \tag{35}$$

and are only used separately for a convenient abbreviation. From (31) and (34) we have further

$$R_\infty = a - b = \frac{1}{a+b}. \tag{36}$$

If we insert (33) and (34) in (30), we obtain, for a layer of finite thickness d,

$$\ln\frac{(R-1/R_\infty)(R_g-R_\infty)}{(R_g-1/R_\infty)(R-R_\infty)} = Sd\left(\frac{1}{R_\infty} - R_\infty\right) \tag{37}$$

or, resolving for R, we have:

$$R = \frac{(1/R_\infty)(R_g-R_\infty) - R_\infty(R_g-1/R_\infty)\cdot\exp[Sd(1/R_\infty - R_\infty)]}{(R_g-R_\infty) - (R_g-1/R_\infty)\cdot\exp[Sd(1/R_\infty - R_\infty)]}. \tag{38}$$

Diffuse reflectance of such a layer therefore depends on the reflectance R_g of the backing, on the diffuse reflectance R_∞ of an analogous but infinitely thick layer, and on the product Sd, which is sometimes called the "scattering power".

If we select an ideal black nonreflecting background, so that $R_g = 0$, and $R \equiv R_0$, Eq. (38) becomes simplified to [109]:

$$R_0 = \frac{\exp[Sd(1/R_\infty - R_\infty)] - 1}{(1/R_\infty)\cdot\exp[Sd(1/R_\infty - R_\infty)] - R_\infty}. \tag{39}$$

---

[109] The index 0 will always be used to designate an ideal black background. This condition is obtained in practice if the layer is freely suspended so that, when $x = d$, no further radiation is reflected.

Solving for Sd, we obtain

$$\ln\frac{1-R_0R_\infty}{1-\dfrac{R_0}{R_\infty}} = Sd\left(\frac{1}{R_\infty}-R_\infty\right), \tag{40}$$

which when $R_g = 0$ can obviously also be obtained from (37). From (37) and (40) we obtain an expression relating the various reflectance values:

$$\frac{\left(R-\dfrac{1}{R_\infty}\right)(R_g-R_\infty)}{\left(R_g-\dfrac{1}{R_\infty}\right)(R-R_\infty)} = \frac{1-R_0R_\infty}{1-\dfrac{R_0}{R_\infty}} \tag{41}$$

which may be rearranged to give

$$R_0 = \frac{R_\infty(R_g-R)}{R_g - R_\infty(1-R_gR_\infty+R_gR)}. \tag{42}$$

By measurement of R_0 and R_∞ it is therefore possible by use of Eq. (40) to obtain both the "scattering power" Sd and, for known values of d, the scattering coefficient S:

$$S = \frac{2{,}303}{d}\frac{R_\infty}{1-R_\infty^2}\cdot\log\frac{R_\infty(1-R_0R_\infty)}{R_\infty-R_0}. \tag{43}$$

Using (31) it is now possible, finally, to obtain the absorption coefficient K:

$$K = \frac{2{,}303}{2d}\frac{1-R_\infty}{1+R_\infty}\cdot\log\frac{R_\infty(1-R_0R_\infty)}{R_\infty-R_0}. \tag{44}$$

In practice, the ratio $R_0/R_{(R_g)}$ is often used to characterise a diffusely reflecting layer which, depending on the reflectance R_g of the background, may have various values. When $R_g = 1$, for an ideal white background, the quantity $R_0/R_{(R_g=1)}$ is called the "ideal contrast ratio", which in practice cannot be measured, as $R_g = 1$ cannot be realised (see p. 146). We therefore replace this by the ratio $R_0/R_{(R_g=0.98)}$, using as white background a layer of freshly prepared magnesium oxide or titanium dioxide. If this contrast ratio is equal to 0.98, the layer in question is described as "completely hiding". The reciprocal of this is therefore called the "hiding power". The contrast ratio $R_0/R_{(R_g=0.98)}$ may be used as a measure of the "opacity" of the layer. In the paper industry the reflectance of the white background is fixed as $R_g = 0.89$ (Technical Association of the Pulp and Paper Industry), so that $C_{0.89} \equiv R_0/R_{(R_g=0.89)}$ is the so-called "Tappi-opacity". In testing white pigments, the "brightening power"

F is used. This is defined [110] as the weight-ratio W_e/W_x of a calibrating pigment (e) to that of the pigment to be measured (x) which imparts the same reflectance R_∞ to the same quantity of a coloured or grey suspension:

$$F \equiv \left(\frac{W_e}{W_x}\right)_{R_\infty e = R_\infty x} \cdot 100. \tag{45}$$

The calibrating pigment corresponds to the value $F = 100$.

If the value of R_0 obtained from (39) for constant S, and for R_∞ as parameter is plotted logarithmically against Sd, we obtain the curves shown in Fig. 47. These curves first rise very steeply, then become flatter

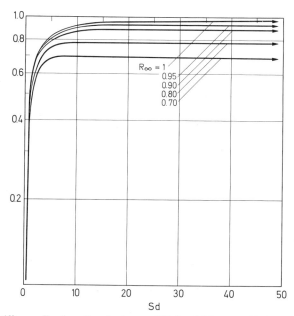

Fig. 47. Diffuse reflection R_0 of a layer of finite thickness with black background as a function of the scattering power Sd, with different values of R_∞ as parameter

until they approach the limiting value R_∞ asymptotically, the rate of approach being slower at larger values of R_∞. This general steep rise of the curves shows that Eq. (39) remains valid, not only for an absolutely black background ($R_g = 0$), but also for backgrounds deviating appreciably from this condition, provided that R and R_g are sufficiently different from each other. This also has been experimentally confirmed (see p. 240).

[110] *Grassmann, W.*, and *H. Clausen*: Deut. Farben-Z. **1953**, 211.

For the limiting case $K=0$, that is, for non-absorbing materials, Eq. (38) and (39) lead to indeterminate expressions. We then return to the differential Eqs. (20) and (21), which in this case can be written:

$$-\frac{dI}{dx} = -SI + SJ; \quad \frac{dJ}{dx} = -SJ + SI. \tag{46}$$

Integration gives

$$R = \frac{(1-R_g)Sd + R_g}{(1-R_g)Sd + 1}, \tag{47}$$

whereas for a black background ($R_g = 0$),

$$R_0 = \frac{Sd}{Sd+1} \quad \text{or} \quad Sd = \frac{R_0}{1-R_0}. \tag{48}$$

For the limiting case $S=0$, that is, for non-scattering materials, for which Eqs. (38) and (39) are also indeterminate, we obtain from the differential equations

$$-\frac{dI}{dx} = -KI \quad \text{and} \quad \frac{dJ}{dx} = -KJ \tag{49}$$

which gives, after integration,

$$R = R_g e^{-2Kd} \quad \text{and} \quad R_\infty = 0, \tag{50}$$

whereas for a black background,

$$R_0 = 0. \tag{51}$$

Eq. (50) corresponds to the Lambert Law.

Eq. (50) shows, as *Kubelka* and *Munk* themselves remarked, that the reflectance $R_\infty = 0$ is practically unattainable, since for this either $d = \infty$, or $K = \infty$, or, from (32) $S = 0$. Neither the absorption can be increased indefinitely nor the layer thickness, and the condition $S = 0$ can also in practice not be attained, for a very small scattering (at least molecular scattering) of the transmitted radiation is unavoidable. Hence it is impossible, for example, to prepare an absolutely black paint. How sensitive a black surface is to very small scattering is seen from Fig. 47, when we consider the abcsissa in terms, not of d at constant S, but of S at constant d. The initially sharp rise of R_0 for small S-values shows that an absolutely black surface is even more difficult to attain than an absolutely white one, which follows directly by comparing Figs. 46 and 47. Fortunately, as was shown on p. 113, this fact is of no practical importance in the measurement of R_0.

Eqs. (37) to (40) derived by *Kubelka* and *Munk* correlate the values of R, R_0, R_g, R_∞, and Sd. To use these, it is necessary to insert the measured quantities in one or other of these equations and solve for the unknown quantities. This very tedious procedure is practically of no use and it has been found necessary to set up the solutions of the equations graphically or in tabular form, from which the magnitudes required can be

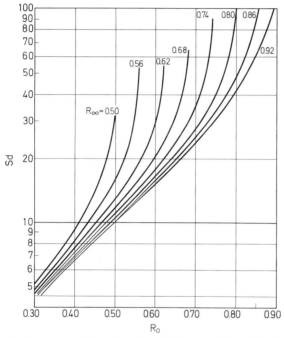

Fig. 48. Scattering power Sd as function of R_0, i.e., of the reflection in front of a black background, for different values of R_∞ as parameter

read off directly using the measured data. Diagrams of this kind have been developed, particularly by *Steele*[111] and by *Judd*[112, 113]. Simplification of the calculations is performed by the use of hyperbolic functions which will be described in greater detail later. An example of this kind of diagram is shown in Fig. 48, in which the "scattering power" Sd is given as a function of R_0, reflectance with a black background for various values of R_∞ as parameter, using a logarithmic scale. Using values

---
[111] *Steele, F. A.*: Paper Trade J. **100**, 37 (1935).

[112] *Judd, D. B.*: J. Res. Natl. Bur. Std. **19**, 287 (1931).

[113] *Judd, D. B.*: Colour in business, science and industry. 2nd. edition. New York: John Wiley & Sons 1963.

of R_0 and R_∞ known from measurement, it is possible to obtain quickly the value of the scattering power Sd. If the layer thickness is also known the scattering coefficient S is obtainable in addition.

c) The Hyperbolic Solution Obtained by Kubelka and Munk

As we indicated, the tedious graphical evaluation of the *Kubelka-Munk* equations can be avoided by the introduction of hyperbolic functions[114]. *Kubelka*[115] succeeded in obtaining explicit hyperbolic solutions for all the pertinent variables which greatly simplified their evaluation from the measured data. His procedure commences with differential equation (26).

The general solution of this equation is[116]:

$$\int \frac{dr}{r^2 - 2ar + 1} = \frac{-2}{\sqrt{4a^2 - 4}} \cdot \left(\begin{matrix}\tanh\\\coth\end{matrix}\right)^{-1} \frac{2r - 2a}{\sqrt{4a^2 - 4}}$$
$$= \frac{1}{b} \cdot \left(\begin{matrix}\tanh\\\coth\end{matrix}\right)^{-1} \frac{a - r}{b}, \qquad (52)$$

in which, from Eq. (35) $b \equiv \sqrt{a^2 - 1}$. Whether $\tanh$ or $\coth$ is to be used depends on the argument. $\tanh^{-1}$ should be used for $-1 \leq x \leq +1$, while $\coth^{-1}$ is appropriate for $-1 \geq x \geq +1$. In the present case, from (22) $a > 1$, while from (34) $b < 1$. The maximum value of r is R_∞, that is, $r \leq R_\infty$. Since from (36) $R_\infty = a - b$, then $a - r \geq b$, or $\frac{a - r}{b} \geq 1$. It is therefore necessary to use $\coth^{-1}$. If we integrate over the whole thickness d of the layer, using (28) and (29) we obtain

$$\int_0^R \frac{dr}{r^2 - 2ar + 1} = \frac{1}{b}\left(\coth^{-1}\frac{a - R}{b} - \coth^{-1}\frac{a - R_g}{b}\right). \qquad (53)$$

---

[114] E.g. *Steele, F. A.*: Paper Trade J. **100**, 37 (1935). — *Ryde, J. W.*: Proc. Roy. Soc. London A **131**, 451 (1931).

[115] *Kubelka, P.*: J. Opt. Soc. Am. **38**, 448 (1948).

[116] See, for example, Handbook of Mathematical Tables 1962. We have:

$$\sinh x = \frac{1}{2}(e^x - e^{-x}); \quad \cosh x = \frac{1}{2}(e^x + e^{-x});$$

$$\tanh x = \frac{\sinh x}{\cosh x} = \frac{e^x - e^{-x}}{e^x + e^{-x}}; \quad \coth x = \frac{\cosh x}{\sinh x} = \frac{e^x + e^{-x}}{e^x - e^{-x}}.$$

The inverse functions are:

$$\sinh^{-1} x = \ln(x + \sqrt{x^2 + 1}); \quad \cosh^{-1} x = \ln(x + \sqrt{x^2 - 1});$$

$$\tanh^{-1} x = \frac{1}{2} \ln \frac{1 + x}{1 - x}; \quad \coth^{-1} x = \frac{1}{2} \ln \frac{x + 1}{x - 1}.$$

Since in general $\coth^{-1} x - \coth^{-1} y = \coth^{-1} \dfrac{1-xy}{x-y}$, it follows that

$$Sd = \frac{1}{b} \coth^{-1} \frac{b^2 - (a-R)(a-R_g)}{b(R_g - R)} \qquad (54)$$

from which

$$R = \frac{1 - R_g(a - b \coth bSd)}{a + b \coth bSd - R_g}. \qquad (55)$$

Eq. (54) is identical with Eq. (37).

For an ideal black, non-reflecting background ($R_g = 0$, $R \equiv R_0$), Eq. (54) may be simplified to

$$Sd = \frac{1}{b} \coth^{-1} \frac{1 - aR_0}{bR_0} \qquad (56)$$

and Eq. (55) to

$$R_0 = \frac{1}{a + b \coth bSd}$$

$$= \frac{\sinh bSd}{a \sinh bSd + b \cosh bSd}. \qquad (57)$$

These two equations are identical with (40) and (39). For an infinitely thick layer ($d \to \infty$), $\coth bSd = 1$, and so

$$R_\infty = \frac{1}{a+b},$$

which agrees with (36).

To obtain a similar expression for the transmission of such a layer, we commence with Eq. (25) and, by means of (57), obtain for a layer of thickness x with a black background [117]

$$J = R_0 I = \frac{I}{a + b \coth bSx}. \qquad (58)$$

Inserting this in Eq. (23) we obtain

$$-\frac{dI}{S\,dx} = -aI + \frac{I}{a + b \coth bSx}, \qquad (59)$$

which now may be integrated. When $bSx \equiv u$, and $dx = \dfrac{du}{bS}$, we have

$$-\frac{1}{S} \int_{I(x=0)}^{I_0(x=d)} \frac{dI}{I} = -a \int_0^d dx + \frac{1}{bS} \int_0^d \frac{1}{a + b \coth u} du$$

---

[117] Alternately we can consider a freely suspended layer which also is unreflecting when $x = 0$.

and

$$-\frac{1}{S}\ln\frac{I_{0(x=d)}}{I_{(x=0)}}$$
$$= -ad + \frac{1}{bS} \cdot \frac{1}{a^2-b^2}\left(abSd - b\ln\frac{a\sinh bSd + b\cosh bSd}{b}\right).$$

Since according to (35) $a^2 - b^2 = 1$, we obtain finally

$$\frac{I_{(x=0)}}{I_{0(x=d)}} \equiv T = \frac{b}{a\sinh bSd + b\cosh bSd}, \qquad (60)$$

in which T is the transmittance of the layer and $I_{0(x=d)}$ the radiation flux incident on the layer. For an infinitely thick layer $(d \to \infty)$, of course $T = 0$. The inverse function to (60) is [118]

$$Sd = \frac{\sinh^{-1}\frac{b}{T} - \sinh^{-1}b}{b}. \qquad (61)$$

If we express a and b by (33) and (34), and insert in (60) we obtain by replacing the hyperbolic functions with exponential functions (see p. 116) for the transmittance of the layer the expression:

$$T = \frac{(1-R_\infty)^2 \, e^{-bSd}}{1 - R_\infty^2 \cdot e^{-2bSd}}, \qquad (62)$$

which corresponds to Eq. (39) for the reflectance of the same layer using a black background. In these two equations T and R_0 are represented as functions of R_∞ and Sd.

From the relations so far derived, further equations can be obtained which relate the several variables. Thus, from (57), (60), and (35), we arrive at

$$T^2 + b^2 = (a - R_0)^2, \qquad (63)$$

which is an equation which relates by means of (33), the three measured quantities T, R_0 and R_∞, before evaluating the "scattering power" Sd. Solving with respect to R_0 we obtain, using (35)

$$R_0 = a - \sqrt{T^2 + b^2}. \qquad (64)$$

---

[118] If we write $a\sinh bSd + b\cosh bSd = k \cdot \sinh(bSd + \varphi) = b/T$, then

$$k \cdot (\sinh bSd \cosh\varphi + \cosh bSd \sinh\varphi) = b/T.$$

From these equations it follows that

$$k \cdot \cosh\varphi = a, \quad k \cdot \sinh\varphi = b, \quad \text{and} \quad k^2 = a^2 - b^2 = 1.$$

Thus $\varphi = \sinh^{-1}b = \cosh^{-1}a$, and $bSd + \sinh^{-1}b = \sinh^{-1}(b/T)$, which is identical with (61).

Resolution with respect to T gives

$$T = [(a - R_0)^2 - b^2]^{1/2}, \tag{65}$$

while with respect to a, using (35) and (33) we have

$$a = \frac{1 + R_0^2 - T^2}{2R_0} = \frac{1}{2}\left(\frac{1 + R_\infty^2}{R_\infty}\right), \tag{66}$$

from which R_∞ can be obtained as a function of the measured quantities R_0 and T. Using (55) and (57), we can obtain a, and hence R_∞, as function of R, R_0, and R_g such that

$$a = \frac{1}{2}\left(R + \frac{R_0 - R + R_g}{R_0 R_g}\right). \tag{67}$$

Solving for R we have

$$R = \frac{R_0 - R_g(2aR_0 - 1)}{1 - R_0 R_g} \tag{68}$$

and solving for R_0 we have

$$R_0 = \frac{R - R_g}{1 - R_g(2a - R)} \tag{69}$$

which is identical with (42). These equations relate the various reflectance quantities to each other.

Again, using (55), (57), and (60), we have

$$R = R_0 + \frac{T^2 R_g}{1 - R_0 R_g}, \tag{70}$$

an equation which contains neither Sd nor R_∞. Measurements of R_0 and T make it possible to determine the ratio R_0/R (see p. 112) for any value of R_g. Resolution with respect to T gives

$$T = \left[(R - R_0)\left(\frac{1}{R_g} - R_0\right)\right]^{1/2}. \tag{71}$$

A method has been given shortly whereby the Kubelka-Munk coefficients R_∞, R_0, K, S, a, and b can be easily determined by transmission measurements on two layers with a thickness ratio of 1:2 [119].

In the following table the equations developed by *Kubelka* are repeated in a composite list in a form suitable for practical use.

---

[119] Caldwell, B. P.: J. Opt. Soc. Am. **58**, 755 (1968).

Table 4. *Relations between the quantities R, R_g, R_0, R_∞, Sd, and T of the Kubelka-Munk theory*

$$R = f(Sd, R_g, R_\infty) = \frac{1 - R_g(a - b \coth bSd)}{a + b \coth bSd - R_g} \tag{55}$$

$$R_0 = f(Sd, R_\infty) = \frac{1}{a + b \coth bSd} \tag{57}$$

$$= \frac{\sinh bSd}{a \sinh bSd + b \cosh bSd}$$

$$T = f(Sd, R_\infty) = \frac{b}{a \sinh bSd + b \cosh bSd} \tag{60}$$

$$Sd = f(R, R_g, R_\infty) = \frac{1}{b} \cdot \coth^{-1} \frac{b^2 - (a - R)(a - R_g)}{b(R_g - R)} \tag{54}$$

$$Sd = f(R_0, R_\infty) = \frac{1}{b} \cdot \coth^{-1} \frac{1 - aR_0}{bR_0} \tag{56}$$

$$Sd = f(T, R_\infty) = \frac{1}{b} \left(\sinh^{-1} \frac{b}{T} - \sinh^{-1} b \right) \tag{61}$$

$$= \frac{1}{b} \left(\sinh^{-1} \frac{b}{T} + \ln R_\infty \right)$$

$$R_0 = f(T, R_\infty) = a - \sqrt{T^2 + b^2} \tag{64}$$

$$T = f(R_0, R_\infty) = \sqrt{(a - R_0)^2 - b^2} \tag{65}$$

$$R_\infty = f(R_0, T) : a = \frac{1 + R_0^2 - T^2}{2R_0} \tag{66}$$

$$R = f(R_0, R_g, R_\infty) = \frac{R_0 - R_g(2aR_0 - 1)}{1 - R_0 R_g} \tag{68}$$

$$R_0 = f(R, R_g, R_\infty) = \frac{R - R_g}{1 - R_g(2a - R)} \tag{69}$$

$$R_\infty = f(R, R_0, R_g) : a = \frac{1}{2} \left(R + \frac{R_0 - R + R_g}{R_0 R_g} \right) \tag{67}$$

$$R = f(R_0, R_g, T) = R_0 + \frac{T^2 R_g}{1 - R_0 R_g} \tag{70}$$

$$T = f(R, R_0, R_g) = \sqrt{(R - R_0) \left(\frac{1}{R_g} - R_0 \right)} \tag{71}$$

It is often possible to replace the exact equations derived here by approximate ones, if the deviations between the two lie within the limits of error of the measurements. This is especially true for small values of bSd, which may arise either through low "scattering power" ($Sd \to 0$) or low absorption ($a \to 1$, or $b \equiv \sqrt{a^2 - 1} \to 0$). This type of approximate

The Hyperbolic Solution Obtained by Kubelka and Munk

equations is obtained, for example, by expressing the hyperbolic functions in series [120], and neglecting the higher terms. For the equations of most significance in reflectance spectroscopy, (57), (60), (56), and (61), we thus obtain by neglecting all but the first term in the series:

$$R_0 \underset{(sd\to 0)}{\overset{a\to 1}{\cong}} \frac{Sd}{aSd+1} \cong \frac{Sd}{Sd+1}, \qquad (57\,a)$$

$$T \underset{(sd\to 0)}{\overset{a\to 1}{\cong}} \frac{1}{aSd+1} \cong \frac{1}{Sd+1}, \qquad (60\,a)$$

$$Sd \underset{(a\to 1)}{\cong} \frac{R_0}{1-aR_0} \cong \frac{R_0}{1-R_0}, \qquad (56\,a)$$

$$Sd \underset{(a\to 1)}{\cong} \frac{1-T}{T} . \qquad (61\,a)$$

(56a) and (57a) are identical with (48).

Other approximate forms are obtained for low scattering power $(S \ll K)$ as follows: From (22) we have $a \to \frac{K}{S}$, and from (34) $b = \frac{1}{S}\sqrt{K(K+2S)} \to \frac{K}{S}$, so that

$$R_0 \underset{(S\ll K)}{\cong} \frac{S}{K(1+\coth Kd)}, \qquad (57\,b)$$

$$T \underset{(S\ll K)}{\cong} \frac{1}{\sinh Kd + \cosh Kd} \cong e^{-Kd}. \qquad (60\,b)$$

When $S=0$, that is for non-scattering materials, $R_0 = R_\infty = 0$, and the transmittance T follows the Bouguer-Lambert Law.

To apply the equations, a number of conditions must be satisfied, based on the assumptions involved in their development. These may be briefly summarized as:

a) The host medium (matrix) containing the scattering particles must be the same as that from which the radiation impinges on the surface

---

[120] Thus:
$$\sinh x = x + \frac{x^3}{3!} + \frac{x^5}{5!} + \cdots$$

$$\cosh x = 1 + \frac{x^2}{2!} + \frac{x^4}{4!} + \cdots$$

$$\sinh^{-1} x = x - \frac{x^3}{6} + \frac{3x^5}{40} - + \cdots$$

$$\coth x = \frac{1}{x} + \frac{x}{3} - \frac{x^3}{45} + - \cdots$$

$$\coth^{-1} x = \frac{1}{x} + \frac{1}{3x^3} + \frac{1}{5x^5} + \cdots$$

of the layer, so that no additional reflection loss shall occur at the surface because of the refractive index difference. In practice then, only particles embedded in air, such as powders, paper, textiles, etc., fulfill these conditions. Suspensions in liquids, colloidal dispersions, etc., will only be capable of treatment by these equations when additional corrections are introduced for the surface reflection loss (see p. 131).

b) Larger scattering particles must possess a distribution function for their inclination angles to the macroscopic surface such that the surface approaches ideal diffuse (isotropic) reflection. Thus, laminar particles, which are preferentially oriented parallel to each other, will produce deviations from these equations.

c) The scattering particles must be distributed homogeneously over the entire layer, so that S and K are constant throughout the thickness of the layer. This is, for example, not the case if the packing density of a powder or of a paper varies within the thickness of the layer, which frequently occurs in practice.

The last case has been investigated by *Kubelka*[121]. We may again commence with the two simultaneous differential Eqs. (20) and (21), though now S and K must be represented as functions of x. In general, the functions $S(x)$ and $K(x)$ are not known in advance, but in special cases the reflectance and transmittance of such inhomogeneous layers can nevertheless be determined.

A simple case is found when the ratio K/S is constant, that is, S and K depend upon x in the same way. This can happen in the case of powders of variable packing density. The scattering power of such a layer of thickness d is then no longer given by Sd (see p. 112) but by the integral

$$P = \int_0^d S(x)\,dx. \tag{72}$$

We may then use all the equations already derived provided that the scattering power Sd is replaced by P. When the packing density varies with x, both S and K should be proportional to the density ϱ of the scattering powder so that S/K is constant, while

$$S(x) = S' \cdot \varrho(x) \tag{73}$$

in which S' is constant and characterises the scattering capacity of the medium independently of its density. Then from (72):

$$P = S' \cdot \int_0^d \varrho(x)\,dx = S'G. \tag{74}$$

---

[121] *Kubelka, P.*: J. Opt. Soc. Am. **44**, 330 (1954).

G is the weight of the layer of thickness d and 1 cm^2 cross-section, and therefore easily measurable, so that as in the earlier discussion, the quantity S' may be determined from reflectance- and transmittance-measurements.

There is often special interest in the problem of diffuse reflectance and transmittance of *non-homogeneous*, absorbing and diffusely scattering layers ($K = K(x)$, and $S = S(x)$), in dependence on the direction of the diffuse irradiation: that is, the question of whether these properties are different at irradiation from the forward or reverse side; and also the question of how the reflectance and transmittance of two or more layers located one behind the other corresponds to the individual values of the single layers. We consider here, of course, freely suspended layers without background, so that the reflectance values are always R_0 values (see note on p. 111).

Whereas for a homogeneous layer (K and S constant) the reflectances R_1 and R_l for irradiation from above and below respectively, and similarly T_1 and T_l must clearly be equal, this is only true for T, and not for R in the case of non-homogeneous layers. R_1 and R_l are clearly different, as can be seen immediately if the inhomogeneous layer is regarded as consisting of two homogeneous layers with different values of K or S. Thus, for the same value of S, on irradiating the side with the larger value of K, the reflectance R will be smaller than when the side with the smaller value of K is irradiated.

If two such inhomogeneous layers are combined, and their total reflectance $R_{1,2}$ and total transmittance $T_{1,2}$ are measured, then the path of the diffuse radiation may be represented schematically by Fig. 49. The radiation flux strikes the sample 1 from above, is partially reflected (R_1), and partially transmitted (T_1). The fraction T_1 falls on sample 2,

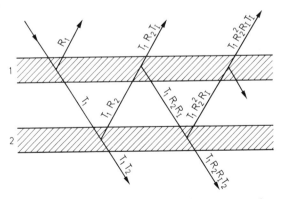

Fig. 49. Reflectance and transmittance of two inhomogeneous layers according to *Kubelka*

the fraction $T_1 R_2$ is reflected, and the fraction $T_1 T_2$ transmitted. The fraction $T_1 R_2$ falls from beneath on sample 1, the new fraction $T_1 R_2 T_1$ being transmitted and the fraction $T_1 R_2 R_1$ reflected, etc., as indicated in Fig. 49. By summing up all the transmitted and reflected fractions, we obtain, as on p.10 two geometrical series:

$$T_{1,2} = T_1 T_2 (1 + R_1 R_2 + R_1^2 R_2^2 + \cdots) = \frac{T_1 T_2}{1 - R_1 R_2} \tag{75}$$

and

$$R_{1,2} = R_1 + T_1 T_1 R_2 (1 + R_1 R_2 + R_1^2 R_2^2 + \cdots) = R_1 + \frac{T_1 T_1 R_2}{1 - R_1 R_2}. \tag{76}$$

Since $T_1 = T_I$, we have

$$R_{1,2\,(\text{inhomogeneous})} = R_1 + \frac{T_1^2 R_2}{1 - R_1 R_2}. \tag{77}$$

If the two layers are homogeneous, then in both formulae the products $R_1 R_2$ may be replaced by $R_1 R_2$ giving

$$T_{1,2\,(\text{homogeneous})} = \frac{T_1 T_2}{1 - R_1 R_2}, \tag{75a}$$

$$R_{1,2\,(\text{homogeneous})} = R_1 + \frac{T_1^2 R_2}{1 - R_1 R_2}. \tag{77a}$$

These equations have been obtained by a number of workers [122]. Eq. (77a) has the same form as Eq. (II, 18) for the total reflectance of a plane parallel plate.

To obtain similar equations for three non-homogeneous layers, we may consider two of these (for example, 2 and 3) as a single layer, which is then combined with the third (1), giving by Eq. (75), the transmittance

$$T_{1,2,3} = \frac{T_1 \cdot T_{2,3}}{1 - R_1 R_{2,3}}. \tag{78}$$

If $T_{2,3}$ and $R_{2,3}$ are represented by the corresponding expressions in (75) and (77) we obtain the desired transmittance. Similarly, for the reflectance of three non-homogeneous layers we obtain from (77) the expression:

$$R_{1,2,3} = R_1 + \frac{T_1^2 R_{2,3}}{1 - R_1 R_{2,3}}. \tag{79}$$

---

[122] See, for example, *Stokes, G. G.*: Proc. Roy. Soc. (London) **11**, 545 (1860/62); — *Gurevich, M. M.*: Physik Z. **31**, 753 (1930). — *Benford, F.*: J. Opt. Soc. Am. **36**, 524 (1946). — *Kubelka, P.*: J. Opt. Soc. Am. **44**, 330 (1954).

The Hyperbolic Solution Obtained by Kubelka and Munk

The calculations can obviously be extended to cover any number of layers. If the individual layers are themselves homogeneous, there is no need to distinguish between R_1 and R_I.

In deriving Eqs. (75) to (79), as already indicated, it has been assumed that the layers are freely suspended, so that all the R values are really R_0 values, corresponding to reflectance with a black background. Cases often arise in practice, however, in which the reflectance of a non-homogeneous layer with a background of reflectance R_g must be determined. If then in Eq. (77) we consider the second layer as constituting a reflecting background, so that $R_2 = R_g$, we obtain for the reflectance of an inhomogeneous layer with background quite simply the expression

$$R_{1,R_g} = R_1 + \frac{T_1^2 R_g}{1 - R_I R_g}. \tag{80}$$

Solving this equation for T, we obtain

$$T_1 = \sqrt{(R_{1,R_g} - R_1)\left(\frac{1}{R_g} - R_I\right)}, \tag{81}$$

an equation from which the transmittance of a non-homogeneous layer can be determined by reflectance measurements only. Eq. (80) and (81) are identical with (70) and (71) if $R_1 = R_I$, which is the case for homogeneous layers.

Since the transmittance is not changed when the direction of irradiation is reversed, as will be shown below, Eq. (81) can also be written in the form:

$$T_1 = \sqrt{(R_{I,R_g} - R_I)\left(\frac{1}{R_g} - R_1\right)}. \tag{82}$$

If T_1 is eliminated from the two equations, we obtain a relation between the different types of reflectance of a non-homogeneous layer:

$$\frac{R_{1,R_g} - R_1}{1 - R_1 R_g} = \frac{R_{I,R_g} - R_I}{1 - R_I R_g}. \tag{83}$$

R_1 and R_I denote the reflectances of the first layer by irradiation at the front or back side respectively.

For special cases dealt with on p. 122, when K/S is constant, and $S = S(x)$, not only the transmittance, but also the reflectance are independent of the direction of irradiation of the inhomogeneous layer, so that $R_1 = R_I$. This follows immediately from (64) which can be written by means of (35) in the form

$$R = a - \sqrt{T^2 + a^2 - 1}. \tag{84}$$

Since $a \equiv (S+K)/S$ is constant according to the assumptions made, and independent of the direction of irradiation, this must also be true for R, so that $R_1 = R_I$. A further special case in which this arises is the one in which a, although not constant, varies symmetrically with respect to the center of the layer, that is, $a(x) = a(d-x)$. There is then no difference between the two directions of irradiation, and again $R_1 = R_I$. Such a centrosymmetrical layer therefore behaves like a homogeneous one. This has been confirmed by *Stenius*[123] by measurements on folded papers. Papers generally show an unsymmetrical distribution of absorption and scattering power in individual layers, but when they are folded these properties become symmetrical with respect to the center plane, so that the Kubelka-Munk theory may be applied (see p. 199).

It remains to be shown that for two non-homogeneous layers also, $T_{1,2} = T_{II,I}$, and that this is valid for the transmittance of two layers of any kind situated the one behind the other, whether they are irradiated from above or below. If the two non-homogeneous layers of Fig. 49 are irradiated from below, then, similarly to (75)

$$T_{II,I} = \frac{T_{II} T_I}{1 - R_2 R_1}. \tag{85}$$

Dividing (85) by (75) we obtain:

$$\frac{T_{II,I}}{T_{1,2}} = \frac{T_{II} T_I}{T_1 T_2}. \tag{86}$$

If we regard the two non-homogeneous samples as consisting each of two homogeneous layers a and b, with constant but different values of S and K, then, as in the case of (86),

$$\frac{T_{II,I}}{T_{1,2}} = \frac{T_{IIa} T_{IIb} T_{Ia} T_{Ib}}{T_{1a} T_{1b} T_{2a} T_{2b}}. \tag{87}$$

Since, however,

$$T_{Ia} = T_{1a}; \quad T_{Ib} = T_{1b}; \\ T_{IIa} = T_{2a}; \quad T_{IIb} = T_{2b}, \tag{88}$$

it follows that

$$T_{II,I} = T_{1,2},$$

that is, the transmittance of the two layers together is also independent of the direction of irradiation. This result may be generalised, so that for n non-homogeneous layers in series we may write,

$$T_{N,...,III,II,I} = T_{1,2,3,...,n}. \tag{89}$$

---

[123] S: son Stenius, A.: Svensk Papperstidning **54**, 663 (1951).

This indicates that the direction of irradiation may be reversed, but that the given sequence of layers: 1, 2, 3, ... may not be arbitrarily altered, without modifying the transmittance. This is easy to demonstrate [124].

d) Use of Directed Instead of Diffuse Irradiation

An essential assumption of the Kubelka-Munk theory was that the irradiation of the sample was diffuse (see p. 106). To be shure, also in the case of directed irradiation, however, within the sample, because of multiple scattering, isotropic angular distribution of the radiation is quickly established (see p. 99). For small scattering densities, however, as in the case of thin samples with low scattering coefficients, it is possible that a fraction, however small, of the radiation may penetrate the sample unscattered, and in general the number of multiple scatterings may be too small to give an isotropic scattering distribution. This problem was first discussed by *Silberstein* [125]. It has been treated for the case of spherical particles by *Ryde* [126] and *Duntley* [127], and carried through to a general solution which is converted again into the Kubelka-Munk solution when the irradiation is diffuse.

Ryde starts with differential equations similar to (20) and (21):

$$-\frac{dI}{dx} = -(K+B)I + BJ + \overset{*}{F}\overset{*}{I}_x, \tag{90}$$

$$\frac{dJ}{dx} = -(K+B)J + BI + \overset{*}{B}\overset{*}{I}_x. \tag{91}$$

I and J represent the diffuse radiation flux in the $-x$ and $+x$ directions respectively, as in Fig. 44. Instead of the scattering coefficient S in Eq. (20) and (21) we here distinguish between a scattering coefficient F in a forward direction and a scattering coefficient B in a backward direction. This is based on the theory of Mie, according to which the scattering in the forward direction predominates over that in the reverse direction the larger the ratio $x \equiv 2\pi r/\lambda$ (relation of particle size and wavelength), given by Eq. (III, 41) (see p. 85). B thus corresponds to all scatterings within the region $\pi/2 \leq \vartheta_s \leq \pi$, and F to all scatterings in the region $0 \leq \vartheta_s \leq \pi/2$, when ϑ_s denotes the angle of scattering defined on p. 77 [128].

---

[124] See *Kubelka, P.*: J. Opt. Soc. Am. **44**, 330 (1954).

[125] *Silberstein, L.*: Phil. Mag. **4**, 129 (1927).

[126] *Ryde, J. W.*: Proc. Roy. Soc. London A **131**, 451 (1931). — *Ryde, J. W.*, and *B. S. Cooper*: Proc. Roy. Soc. London A **131**, 464 (1931).

[127] *Duntley, S. Q.*: J. Opt. Soc. Am. **32**, 61 (1942).

[128] F and B thus correspond to the double integrals in (III, 59) or (III, 60); further $F + B = S$.

$\overset{*}{I}_x$ in the two last terms of (90) and (91) is the fraction of parallel radiation, $\overset{*}{I}_{(x=d)}$, falling perpendicularly on the layer, at the position x, being reduced to the value $\overset{*}{I}_x$ through absorption and scattering. $\overset{*}{I}_x$ is thus the fraction of the parallel radiation flux still present. $\overset{*}{F}$ and $\overset{*}{B}$ are the corresponding scattering coefficients in the forward and reverse directions for parallel and perpendicular irradiation [129]. For the parallel radiation flux we have the third differential equation:

$$\frac{d\overset{*}{I}_x}{dx} = -(\overset{*}{K}+\overset{*}{F}+\overset{*}{B})\,\overset{*}{I}_x \equiv \overset{*}{q}\overset{*}{I}_x. \tag{92}$$

The indefinite integration of these equations gives:

$$I = k_1\left(1-\frac{K}{bB}\right)e^{bBx} + k_2\left(1+\frac{K}{bB}\right)e^{-bBx} - \overset{*}{Q}e^{-\overset{*}{q}x}, \tag{93}$$

$$J = k_1\left(1+\frac{K}{bB}\right)e^{bBx} + k_2\left(1-\frac{K}{bB}\right)e^{-bBx} - \overset{*}{P}e^{-\overset{*}{q}x}, \tag{94}$$

$$\overset{*}{I}_x = e^{-\overset{*}{q}x}. \tag{95}$$

In these equations k_1 and k_2 are integration constants, while

$$b \equiv \frac{1}{B}\sqrt{K(K+2B)} \tag{96}$$

in the same way as (31) and

$$\overset{*}{Q} \equiv \frac{(K+\overset{*}{K})\overset{*}{F}+(B+\overset{*}{F})(\overset{*}{B}+\overset{*}{F})}{(\overset{*}{K}^2-K^2)+2\overset{*}{K}(\overset{*}{F}+\overset{*}{B})-2KB+(\overset{*}{F}+\overset{*}{B})^2}, \tag{97}$$

$$\overset{*}{P} \equiv \frac{(K-\overset{*}{K})\overset{*}{B}+(B-\overset{*}{B})(\overset{*}{B}+\overset{*}{F})}{(\overset{*}{K}^2-K^2)+2\overset{*}{K}(\overset{*}{F}+\overset{*}{B})-2KB+(\overset{*}{F}+\overset{*}{B})^2}. \tag{98}$$

For a freely suspended layer or one with a black background (see note on p. 111) the boundary conditions given by Fig. 44 apply:

$$x = d: \quad \overset{*}{I}_{x=d} = 1; \qquad I = 0; \qquad J = \overset{*}{R}_0;$$

$$x = 0: \quad \overset{*}{I} = e^{-\overset{*}{q}d}; \qquad I + \overset{*}{I} = \overset{*}{T}; \qquad J = 0.$$

---

[129] All the quantities designated with * refer to perpendicular parallel irradiation. The absorption coefficient $\overset{*}{K}$ for the incident radiation will be different from that for diffuse radiation K, because the path length within the layer is different (see p. 108), and the particles will not always be completely randomly distributed. In the work of *Ryde*, on the other hand, K is set equal to $\overset{*}{K}$, and $F+B = \overset{*}{F}+\overset{*}{B}$.

Thus integration gives the following expressions, written in hyperbolic form:

$$\overset{*}{T} = \frac{\overset{*}{Q}bB + \overset{*}{P}e^{-\mathring{q}d}B\sinh bBd}{(K+B)\sinh bBd + bB\cosh bBd} - (\overset{*}{Q}-1)e^{-\mathring{q}d}, \tag{99}$$

$$\overset{*}{R}_0 = \frac{\overset{*}{P}bB e^{-\mathring{q}d} + \overset{*}{Q}B\sinh bBd}{(K+B)\sinh bBd + bB\cosh bBd} - \overset{*}{P}. \tag{100}$$

With diffuse irradiation, the difference between B and $\overset{*}{B}$, and F and $\overset{*}{F}$ vanishes, while it follows from (97) and (98) that $\overset{*}{Q}=1$ and $\overset{*}{P}=1$. Thus, Eq. (99) and (100) become simplified to

$$T = \frac{b}{a\sinh bBd + b\cosh bBd} \tag{99a}$$

and

$$R_0 = \frac{1}{a + b\coth bBd}, \tag{100a}$$

which are identical with (60) and (57) if B is put equal to S. For diffuse irradiation, the equations derived thus pass over into those of the Kubelka-Munk theory.

For infinite layer thickness $(d \to \infty)$ (100) is converted to

$$\overset{*}{R}_\infty = \frac{\overset{*}{Q}B}{a+b} - \overset{*}{P} \tag{101}$$

and (99) is converted into

$$\overset{*}{T} = 0. \tag{101a}$$

This corresponds to Eq. (36) for diffuse irradiation. For a layer thickness d so great that $e^{-\mathring{q}d} \equiv \overset{*}{I}_d$ may be neglected in practice, that is, when practically none of the parallel irradiation is transmitted unscattered, we have from (99) and (99a)

$$\overset{*}{T} = \overset{*}{Q}T, \tag{102}$$

while from (100) and (100a) we have

$$\overset{*}{R}_0 = \overset{*}{Q}R_0 - \overset{*}{P}. \tag{103}$$

With respect to the scattering coefficients the following can be said: if unit radiation flux falls on a single spherical particle, the same fraction will always be scattered independently of whether the radiation is parallel or diffuse. It follows from this that

$$F + B = \overset{*}{F} + \overset{*}{B}. \tag{104}$$

This is also true for many particles if and only if these are completely randomly distributed. If the particles are very small compared with the wavelength λ, the angular distribution of the scattered radiation will be symmetrical (Fig. 33, Rayleigh Scattering), so that $\overset{*}{F}=\overset{*}{B}$ and $F=B$, and also $B=\overset{*}{B}$. It follows then that for $2\pi r \ll \lambda$ Eq. (99) and (100) also are transformed into those of the Kubelka-Munk theory. Hence the magnitude of the four scattering coefficients must depend on the ratio $x=2\pi r/\lambda$, which is the case in the theory of Mie. Similarly, borrowing the result from the Mie theory, the expression $\dfrac{m^2-1}{m^2+2}$ may be used in the evaluation of the four scattering coefficients, where $m=\dfrac{n_1}{n_0}$, the relative refractive index of the particles and the surrounding medium. *Ryde* and *Cooper* have given these coefficients as a function of $x=2\pi r/\lambda$. Whereas in the case of diffuse irradiation the theory of Kubelka and Munk only uses two constants, K and S, to express the experimental data, it is necessary in the case of directed irradiation to use six constants, B, $\overset{*}{B}$, F, $\overset{*}{F}$, K, $\overset{*}{K}$, in order to evaluate the experimental results.

Further expressions were given by *Stenius*[130] for a layer with a reflecting background using perpendicular irradiation, including the general condition that the reflectance is the same independently of whether diffuse or perpendicular and parallel radiation is used. This condition is:

$$\frac{K}{B} = \frac{[\overset{*}{P}-(\overset{*}{Q}-1)]^2}{2\overset{*}{P}(\overset{*}{Q}-1)}. \tag{105}$$

If the expressions for $\overset{*}{P}$ and $\overset{*}{Q}$ are inserted, we obtain a complex equation whose individual terms always contain K or $\overset{*}{K}$, and one or more of the scattering coefficients B, $\overset{*}{B}$, F, or $\overset{*}{F}$. Two solutions of (105) are, therefore, $K=\overset{*}{K}=0$ and $B=\overset{*}{B}=F=\overset{*}{F}=0$, which correspond to a scattering medium without absorption or to a medium without scattering with any given absorption, each giving the same diffuse reflectance for perpendicular and for diffuse irradiation. Whereas the latter is of no physical significance, the first result is important in practice, as we shall show later (see p. 170).

e) Consideration of Regular Reflection at Phase-Boundaries

A situation often met in practice is that of scattering particles not surrounded by air (as they are in the case of aerosols, crystal powders, paper, etc.), but embedded in a transparent medium whose refractive index is considerably greater than that of air. Examples include opal glasses, pigments in plastic, colloids in a suspension, etc. It therefore

---

[130] S: son Stenius, Å.: Svensk Papperstidning **56**, 607 (1953).

follows that in these cases, the measured reflectance and transmittance must also depend on the refractive index of the medium, because additional regular reflection will take place at the air-medium phase boundaries. As, in addition, the regular reflectance depends on the kind of irradiation, whether directed or diffuse, as we have shown on p. 12 in detail, the reflectance and transmittance in such cases for directed and diffuse irradiation will show additional differences. The corresponding equations have been given by *Ryde*[131] and by *Duntley*[132].

For example, let us consider the reflectance and transmittance of an opal glass plate[133] for parallel and perpendicular, or for diffuse irradiation. The regular reflectance is given, according to Eq. (II, 16a) for perpendicularly incident parallel radiation by

$$\overset{*}{r}_1 = \left(\frac{n-1}{n+1}\right)^2, \tag{106}$$

if the refractive index of air is taken as equal to unity[134]. The corresponding value of r_1 for diffuse irradiation is obtained from Eq. (I, 25) or from Table 1. For $n = 1.5$ we obtain the two values $\overset{*}{r}_1 = 0.040$ and $r_1 = 0.092$. The radiation scattered within the sample emerges from the more dense medium ($n = 1.5$) at both sides into the optically less dense medium ($n = 1$), and is therefore in part totally reflected (cf. p. 13). This regular and partially total reflection at the inner side will be designated r_2. Although $r_2 > r_1$, this total reflection reduces the regular reflection for the overall process, since the total reflected fraction will, because of scattering, be only partially totally reflected a second time, and will partially pass through the surface. This process occurs continuously, so that as shown on p. 10, it will lead to a geometrical series, whose sum gives the reflection from the inner side of the glass surface. For the *measured* reflectance, including the external and the successive internal reflections, $\overset{*}{\varrho}$, and for the measured transmittance $\overset{*}{\tau}$ of such an opal glass sheet, using parallel and perpendicularly incident radiation, *Ryde* gives the following equations:

$$\overset{*}{\tau} = (1-\overset{*}{r}_1)(1-r_2) \cdot \frac{\overset{*}{T}(1-r_2 R_0) + r_2 \overset{*}{R}_0 T}{(1-r_2 R_0)^2 - r_2^2 T^2}, \tag{107}$$

$$\overset{*}{\varrho} = \overset{*}{r}_1 + (1-\overset{*}{r}_1)(1-r_2) \cdot \frac{\overset{*}{R}_0(1-r_2 R_0) + r_2 \overset{*}{T} T}{(1-r_2 R_0)^2 - r_2^2 T^2}. \tag{108}$$

---

[131] *Ryde, J. W.*: Proc. Roy. Soc. London A **131**, 451 (1931); – *Ryde, J. W.*, and *B. J. Cooper*: Proc. Roy. Soc. London A **131**, 464 (1931).

[132] *Duntley, S. O.*: J. Opt. Soc. Am. **32**, 61 (1942).

[133] The scattering properties of opal glass arise from spherical particles of calcium and sodium fluorides embedded in the glass.

[134] The quantities designated with * deal with parallel, perpendicularly incident radiation.

132   IV. Theories of Absorption and Scattering of Tightly Packed Particles

In this we assume that the parallel radiation flux $\overset{*}{I}$ in the inside of the sample (see p. 127) is small compared with the diffuse radiation flux. $\overset{*}{T}$ and $\overset{*}{R}_0$ are given by (99) and (100). For diffuse irradiation we obtain, replacing $\overset{*}{r}_1$ by r_1, $\overset{*}{T}$ by T, and $\overset{*}{R}_0$ by R_0:

$$\tau = \frac{(1-r_1)(1-r_2)T}{(1-r_2 R_0)^2 - r_2^2 T^2} \tag{109}$$

and

$$\varrho = r_1 + (1-r_1)(1-r_2)\frac{R_0(1-r_2 R_0) + r_2 T^2}{(1-r_2 R_0)^2 - r_2^2 T^2}. \tag{110}$$

Although r_1 and $\overset{*}{r}_1$ are easily accessible, experimental determination of r_2 involves great difficulties, because for small scattering and small layer thickness, the radiation reflected at the inner side is no longer uniformly diffuse. Hence r_2 must vary with d and B, especially if $\overset{*}{I}$ is relatively large. We must therefore choose the layer thickness in such a way that these sources of error are kept as small as possible [135]. r_2 may be determined according to *Ryde* in the following way: if we suppress the regular reflection at the exit phase boundary by placing a completely absorbing layer behind the opal glass plate, the terms $r_2 T^2$ and $r_2^2 T^2$ in Eq. (110) vanish, since here $r_2 = 0$, and we obtain for diffuse irradiation

$$\varrho_0 = r_1 + (1-r_1)(1-r_2)\frac{R_0(1-r_2 R_0)}{(1-r_2 R_0)^2}. \tag{111}$$

We have further from (109) and (110) for $T = R_0$

$$\tau = \varrho - r_1.$$

If we choose the layer thickness so that the condition $T = R_0$, or $\tau = \varrho - r_1$ is fulfilled, then by measurement of τ, ϱ and ϱ_0 we have, from Eq. (109), (110) and (111)

$$\frac{1}{r_2} = 1 + \tau \frac{(\varrho_0 - r_1)}{(1-r_1)(\varrho - \varrho_0)}, \tag{112}$$

so that r_2 can be evaluated. *Ryde* gives a value $r_2 \cong 0.4$ for opal glass.

By this means all the effects of regular reflection at the external and internal phase boundaries on the transmittance and reflectance of a simultaneously scattering and absorbing layer can be taken into account. In addition to the six constants of the Ryde theory, two further constants

---

[135] r_2 can be determined by means of the Fresnel equations if we assume that the radiation leaving the surface is completely diffuse. For a glass matrix with $n = 1.5$, this leads to $r_2 = 59.6\%$. Nevertheless, small deviations of the isotropic scattering distribution can lead to large errors [compare *Giovanelli, R. G.*: Opt. Acta **3**, 127 (1956)]. See also Eq. (113).

Consideration of Regular Reflection at Phase-Boundaries 133

r_1 and r_2 are necessary which must be experimentally determined, making a total of eight constants if the theory is to be fully evaluated. Obviously, at least eight experimental results are necessary to evaluate these constants. Since the equations are very intractable, the parameters occurring at powers of 2 and more, it is obviously desirable that the measuring technique should be so modified that some of these constants are eliminated. Methods for this have been developed by *Duntley*.

One of these consists in using a sample (such as opal glass or a plastic material) placed between two equivalent integrating spheres, so that the sample is diffusely irradiated, and the radiation can simultaneously be diffusely registrated. By this means, all complications are removed which depend upon the use of directed irradiation, so that $\overset{*}{B}$, $\overset{*}{F}$, and $\overset{*}{K}$ disappear. The integrating spheres are filled with a colorless oil with the same refractive index for the monochromatic radiation used as that of the medium in which the particles are embedded. Thus, $r_1 - r_2 = 0$. Then by means of simple photometric measurements using a double beam instrument, the diffuse reflectance R_0 and the diffuse transmittance T of the sample can be determined, as in the case of the simple Kubelka-Munk theory [136].

The second method given by *Duntley* for parallel, perpendicularly incident radiation, also uses an integrating sphere filled with oil, with two oil reservoirs facing each other, in which the sample can be immersed, depending on whether the transmittance or the reflectance is to be measured. The latter is measured against a block of magnesium carbonate as a white standard, whose absolute reflectance is regarded as known in advance (relative measurement, see p. 59). In this way the regular reflections r_1 and r_2 are eliminated, and $\overset{*}{T}$ or $\overset{*}{R}_0$ are measured directly. Because of the complexity of Eqs. (99) and (100), or (102) and (103), for the case when the sample is made so thick that $e^{-\overset{*}{q}d} = 1$ (see p. 129), the desired magnitudes K and B cannot be obtained directly from the measured values of $\overset{*}{T}$ and $\overset{*}{R}_0$. *Duntley* therefore describes a logarithmic diagram on which, in a similar way to that used in Fig. 48, the desired absorption- and scattering-coefficients can be obtained by means of an approximation procedure, which depends on the fact that $\overset{*}{Q}$ generally differs only slightly from unity, while $\overset{*}{P}$ differs only slightly from zero. As was shown on p. 129, for $\overset{*}{Q} = 1$ and $\overset{*}{P} = 0$, the theory is converted into the two-constant theory of *Kubelka-Munk*.

---

[136] The method requires very precise measurements, so that it is unsuitable for routine work. It is also presupposed that the sample is resistant to the oil, and that the dispersion curves of the oil and of the medium surrounding the scattering centers are very similar.

Using the Fresnel equations, it is easier to determine the reflectance $\mathring{R}_{\infty,n}$ of a semi-infinite scattering medium embedded in a matrix of refractive index n and perpendicularly irradiated. We must again assume, however, that the radiation originating in the *interior* that strikes the phase boundary is ideally diffuse. For this case, the total reflectance is given by

$$\mathring{R}_{\infty,n} = \mathring{r}_1 + \frac{\dfrac{T}{n^2} R_{\infty 1}(1-\mathring{r}_1)}{1 - R_{\infty 1}\left(1 - \dfrac{T}{n^2}\right)}. \tag{113}$$

$\mathring{r}_1$ is given by (106) and is the regularly reflected fraction. The second term of the sum is the diffusely reflected fraction and is the ratio of the radiation flux emerging from the surface to that falling on the scattering medium. The diffusely reflected fraction is equal to $R_{\infty,1}(1-\mathring{r}_1)$, corresponding to the reflectance of a medium with a matrix of $n=1$, of which, according to p. 15, only the fraction T/n^2 will be transmitted, passing from a more dense into a less dense medium. The flux incident on the scattering medium is not simply equal to 1 since the fraction $\left(1 - \dfrac{T}{n^2}\right) R_{\infty,1}$ must be subtracted from it. This is the fraction that again does not emerge from the surface but is reflected back into the more dense medium. We will return to this equation on p. 301.

In practice, the reflectance measurement frequently involves the following problem: either the sample in many commercial measuring instruments does not lie in a horizontal position, or else it is necessary to protect it from air or moisture. In both cases, a covering glass is necessary in order to secure a definite and reproducible surface. We then have the system of two different homogeneous layers (a non-scattering layer R_{12} with phase boundaries 1 and 2 and subsequently to that, a diffusely scattering layer 3), for which Eq. (77a) may be used [136a]. From this

$$R_{12,3} = R_{12} + \frac{T_{12}^2 R_3}{1 - R_{12} R_3}. \tag{114}$$

In this equation R_{12} denotes the reflectance and T_{12} the transmittance of the covering glass, R_3 the diffuse reflectance of the sample. If directed radiation is used at an angle of α, then R_{12} in the first term of the sum is given by (II, 20), but in the second term by (II, 26), since in multiple reflection between the covering glass and the sample, the covering glass is diffusely irradiated from beneath, even when directed irradiation is used from above. Correspondingly, we must replace T_{12}^2 by the value

---

[136a] See *Judd, D. B.*, and *K. S. Gibson*: J. Res. Natl. Bur. Std. **16**, 261 (1936).

of $T_{12\,\text{direct}} \cdot T_{12\,\text{diffuse}}$. If, on the other hand, the irradiation from above is diffuse, then R_{12} in both terms is given by (II, 26), and T_{12}^2 by $T_{12\,\text{diffuse}}^2$.

It is particularly recommended that the measurements should be made relative to a standard, which if possible should possess the same scattering coefficients and the same scattering distribution as the sample (see p. 177). If we take measures that the regularly reflected radiation from the covering glass cannot fall on the detector, then in measuring the sample and the standard the first term of the sum in (114) vanishes, and for the *relative* reflectance of the sample and the standard we obtain

$$R'_{\text{exp}} \equiv \frac{R_{12,3\,\text{sample}}}{R_{12,3\,\text{standard}}} = \frac{R_{3\,\text{sample}}}{R_{3\,\text{standard}}} \cdot \frac{1 - R_{12} R_{3\,\text{standard}}}{1 - R_{12} R_{3\,\text{sample}}}. \quad (115)$$

If $R_{3\,\text{standard}}$ can be regarded as approximately equal to unity, we have

$$R'_{\text{exp}} \simeq \frac{R_3(1 - R_{12})}{1 - R_{12} R_3}. \quad (115a)$$

Thus, R'_{exp} is always smaller than R_3, and will be relatively reduced the more the larger the absorption of the sample, that is, the smaller R_3. Eq. (114) can also be used, of course, to determine scattering coefficients if R_{30} is measured instead of $R_{3\infty}$, when the substance in question is sensitive to moisture or air.

For measurement of the transmittance T, except in the case of freely suspended layers such as paper, quartz cells must be used, consisting of quartz plates with welded quartz ring (layer thickness lying between 0.1 and 0.3 mm), which if necessary may be covered with a second quartz plate. In the first case the transmittance $T_{1,2}$ of the sample/quartz plate combination is measured, while in the second case, the transmittance $T_{1,2,3}$ of the combination quartz plate/ sample/quartz plate is determined. In both cases the effect of the quartz plates must be eliminated, which can be done as follows [137]:

First Case: From (75a) for the transmittance of two homogeneous layers located one behind the other, where P = sample, and Q = quartz, we have

$$T_{1,2} = \frac{T_P T_Q}{1 - R_P R_Q}. \quad (116)$$

Even when directed irradiation is used, R_Q and T_Q for diffuse irradiation must be inserted, as the radiation within the sample is scattered diffusely, that is, from Eq. (II, 26) $R_Q = 0.115$ for both surfaces of the quartz plate, and $T_Q = 0.845$, if the self-absorption of the quartz plates is neglected. R_P is the reflectance of the freely suspended layer alone, and therefore

---
[137] Kortüm, G., and D. Oelkrug: Z. Naturforsch. **19a**, 28 (1964).

equal to R_0. For R_P, therefore, Eq. (56) solved for R_0, and for T_P Eq. (61) solved for T, must be inserted in (116). By rearrangement, we then obtain directly the desired scattering coefficient from

$$Sd = \frac{1}{b}\left(\sinh^{-1}\frac{T_Q b}{T_{1,2}\sqrt{(a-R_Q)^2 - b^2}} - \sinh^{-1}\frac{b}{\sqrt{(a-R_Q)^2 - b^2}}\right). \quad (117)$$

If the quartz plate is eliminated, so that $T_Q = 1$, and $R_Q = 0$, Eq. (117) becomes identical with Eq. (61).

Second Case: For the transmittance of the combination quartz plate/sample/quartz plate we obtain from Eq. (78), assuming that the layers are homogeneous

$$T_{1,2,3} = \frac{\overset{*}{T_Q} T_P T_Q}{1 - 2R_P R_Q + R_P^2 R_Q^2 - T_P^2 R_Q^2}. \quad (118)$$

In this, it has been assumed that layers 1 and 3 are identical, that is, that identical quartz plates are used, so that $R_1 = R_3$. On the other hand, $T_1 \neq T_3$, since directed irradiation is used for transmission measurements, so that $\overset{*}{T_Q}$ concerns directed and T_Q diffuse radiation. If a single quartz plate is introduced also into the path of the reference beam, then $\overset{*}{T_Q} = 1$.

If the absorption of the sample is very small so that $T_P + R_P \cong 1$, then from (118) the desired transmittance of the sample is obtained in the form

$$T_P = \frac{T_{1,2,3} T_Q^2}{T_Q - 2T_{1,2,3} R_Q + 2T_{1,2,3} R_Q^2}. \quad (119)$$

In the case of absorbing samples T_P is not so easy to calculate. Nevertheless, if the relatively small quadratic terms in (118), which in any case appear with different signs, are neglected, then we obtain approximately

$$T_{1,2,3} \cong \frac{T_P T_Q}{1 - 2R_P R_Q}. \quad (120)$$

From this, as described above in the case of Eq. (116) we may obtain the scattering coefficient as

$$Sd = \frac{1}{b}\left(\sinh^{-1}\frac{T_Q b}{T_{1,2,3}\sqrt{(a-2R_Q)^2 - b^2}} - \sinh^{-1}\frac{b}{\sqrt{(a-2R_Q)^2 - b^2}}\right). \quad (121)$$

If the root is imaginary, we may write:

$$Sd = \frac{1}{b}\left[\cosh^{-1}\frac{T_Q b}{T_{1,2,3}\sqrt{b^2 - (a-2R_Q)^2}} - \cosh^{-1}\frac{b}{\sqrt{b^2 - (a-2R_Q)^2}}\right]. \quad (121\text{a})$$

Because of the approximations used, the scattering coefficients evaluated from transmittance measurements are inferior to those obtained in reflection.

f) Absolute and Relative Measurements

All the quantities arising in the Kubelka-Munk theory for a simultaneously scattering and absorbing layer are defined as the ratios of reflected or transmitted to incident radiation, and are thus absolute quantities. Experimentally, however, only the transmittance of a layer may be determined absolutely, while diffuse reflectance with apparatus at present available can usually only be obtained in reference to a comparison standard, and therefore as a relative quantity [138]. If these relative quantities are designated by R', we have the equation [139]

$$R' = \frac{R_{\text{sample}}}{R_{\text{standard}}}. \tag{122}$$

If $\varrho \equiv R_{\infty \text{standard}}$ were equal to unity, then the absolute and relative reflectance values would of course be identical, but we have already shown on p.110 that no white standard is at present known which possesses this property over the entire accessible range of wavelengths. To determine R_{sample}, therefore, it is necessary to know the absolute reflectance R_∞ of the standard.

In pratice, magnesium oxide, prepared freshly by burning metallic magnesium, has been found hitherto to be the best available white standard. In the visible region of the spectrum no substance is known whose reflectance is better than this. Moreover this standard can easily be prepared under somewhat defined conditions, and can be replaced when its reflectance diminishes. For this reason, the absolute reflectance ϱ of freshly prepared magnesium oxide has often been determined as a function of the wavelength. A number of methods are available for this.

The simplest way would of course be to determine ϱ directly from relative measurements. This method was proposed, for example, by *Stenius* [140]. If the relative values corresponding to the different reflectance values are inserted in Eq. (42), then writing $R = R' \varrho$, we have

$$R'_0 = \frac{R'_\infty (R'_g - R')}{R'_g - R'_\infty (1 - \varrho^2 R'_g R'_\infty + \varrho^2 R'_g R')}. \tag{123}$$

From this equation, the absolute reflectance ϱ of the standard can be determined from the relative quantities, R', R'_g, R'_0 and R'_∞, where R'

---

[138] For methods for the direct measurement of absolute reflectance values, see p. 142.

[139] Relative reflectance values will generally be designated with a prime (') in the following pages.

[140] S: son Stenius, Å.: Svensk Papperstidning **54**, 663 (1951).

and R'_0 refer to the same layer

$$\varrho = \left[\frac{R'_\infty(R'_g - R') + R'_0(R'_\infty - R'_g)}{R'_0 R'_\infty R'_g(R'_\infty + R')} \right]^{1/2}. \qquad (124)$$

Similarly, using Eq. (66),

$$\frac{1 + R_\infty^2}{R_\infty} = \frac{1 + R_0^2 - T^2}{R_0},$$

by substituting $\varrho R'_0$ for R_0 and $\varrho R'_\infty$ for R_∞ we obtain the relation [141]

$$\varrho = \left[\frac{1 - T^2 - R'_0/R'_\infty}{R'_0 R'_\infty - R'^2_0} \right]^{1/2}, \qquad (125)$$

in which R'_0 and T again refer to the same layer. To determine ϱ from Eqs. (124) and (125) four or three experimental quantities respectively are required, some of which appear in the form of differences, so that the accuracy of ϱ cannot be regarded as very great.

Eq. (125) has been tested by means of a thin scattering layer giving only slight absorption. The layer consisted of paper used for chromatography [142], for which R'_0 and R'_∞ were sufficiently different. The transmittance of fifteen equivalent papers was determined, and found to differ by not more than 2%. For measurement of R'_∞ the papers giving the best agreement in their T values were piled one on top of another, and additional papers were used as background. R'_0 and T were in each case measured for the same papers. All the reflectance measurements were carried out using freshly prepared magnesium oxide, backed by the purest commercial product, as standard, the absolute reflectance of which was to be determined. The measurements could only be carried out with light whose $\lambda \geq 400$ mµ, since for shorter wavelengths the paper is too strongly absorbing. The results are given in Fig. 51 as average values of determinations on four different magnesium oxide samples, for which the error in individual measurements of ϱ was approximately ± 0.006.

The majority of the measurements given in the literature for the absolute diffuse reflectance of magnesium oxide in the visible and ultraviolet region of the spectrum depend on the theory of the so-called Taylor sphere [143], which is represented schematically in Fig. 50. Parallel incident light falls on a sample P, whereby it is scattered in all directions. A screen S prevents the light scattered by P from impinging directly on

---

[141] Oelkrug, D.: Dissertation, Tübingen 1963.
[142] Oelkrug, D.: Dissertation, Tübingen 1963.
[143] Taylor, A. H.: J. Opt. Soc. Am. **4**, 9 (1920); **21**, 776 (1931); Sci. Pap. Natl. Bur. Std. **16**, 42 (1920); **17**, 1 (1920); **18**, 281 (1923); — Benford, F.: Gen. Elec. Rev. **23**, 72 (1920). — Mäder, F.: Bull. Schweiz. Elektrotech. Verein Nr. 20 (1947).

the measuring surface f_m imaged on the measuring window M. The entire sphere may be rotated about the axis I–II, so that the light incident from L through the opening shown dotted in the Figure may fall on to the wall of the sphere instead of on to the sample. Assuming that the

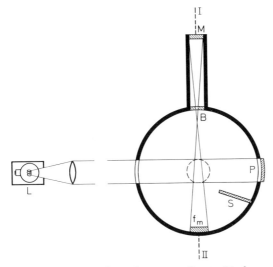

Fig. 50. Integrating sphere according to *Taylor*

inner wall of the integrating sphere produces ideal diffuse reflection, the radiation fluxes measured in both cases may be evaluated in the following way:

1. Irradiation I_0 on the Wall of the Sphere

The radiation flux is distributed as follows:

Diffuse reflection in the entire sphere is:	the radiation flux on the measuring surface is:	the radiation flux on the measuring window is:
$I_0 \varrho_W$,	$I_0 \varrho_W \dfrac{f_m}{F}$	$I_0 \varrho_W^2 \dfrac{f_m}{F} \cdot \dfrac{f_M}{F}.$

In this, ϱ_W is the absolute reflectance of the wall, F the total internal surface of the sphere, so that f_m/F or f_M/F, are the relative fractions of the measuring surface or the observation aperture. If we designate the mean reflectance of the entire sphere which, because of the presence of the sample and the different openings, must be somewhat different from

ϱ_W, as ϱ_K, then after single reflection at the entire sphere we have

of which we find on f_m and on the measuring window

$$I_0 \varrho_W \varrho_K, \quad I_0 \varrho_W \varrho_K \frac{f_m}{F}, \quad I_0 \varrho_W^2 \varrho_K \frac{f_m}{F} \cdot \frac{f_M}{F},$$

after double reflection at the entire sphere we have

of which on f_m and on the measuring window

$$I_0 \varrho_W \varrho_K^2, \quad I_0 \varrho_W \varrho_K^2 \frac{f_m}{F}, \quad I_0 \varrho_W^2 \varrho_K^2 \frac{f_m}{F} \cdot \frac{f_M}{F} \quad \text{etc.}$$

The measured radiation flux is thus, when the light is incident on the sphere wall, given by

$$I_W = I_0 \varrho_W^2 \frac{f_m f_M}{F^2} [1 + \varrho_K + \varrho_K^2 + \varrho_K^3 + \cdots]$$

$$= I_0 \varrho_W^2 \frac{f_m f_M}{F^2} \cdot \frac{1}{1 - \varrho_K}.$$

(126)

2. Irradiation I_0 on the Sample

If the sample is irradiated, instead of the sphere wall, the measuring surface f_m receives none of the direct reflection from the sample because of the screen, and therefore the first term of the geometrical series vanishes. Otherwise, one value of ϱ_W should be replaced by ϱ_P, the absolute reflectance of the sample, so that the measured radiation flux is now given by

$$I_P = I_0 \varrho_W \varrho_P \frac{f_m f_M}{F^2} [\varrho_K + \varrho_K^2 + \varrho_K^3 + \cdots]$$

$$= I_0 \varrho_W \varrho_P \frac{f_m f_M}{F^2} \varrho_K [1 + \varrho_K + \varrho_K^2 + \cdots] \quad (127)$$

$$= I_0 \varrho_W \varrho_P \frac{f_m f_M}{F^2} \varrho_K \cdot \frac{1}{1 - \varrho_K}.$$

In the series for the irradiated sphere wall there is one additional term not present in the series for the irradiated sample because of the screen in the latter case. It follows that in fact the absolute reflectance of the sample is measured, and not that relative to the wall. From (126) and (127) we have

$$\frac{I_P}{I_W} = \frac{\varrho_P \varrho_K}{\varrho_W} \quad \text{or} \quad \varrho_P = \frac{I_P}{I_W} \cdot \frac{\varrho_W}{\varrho_K}. \quad (128)$$

I_P and I_W are measured; the factor ϱ_W/ϱ_K is often set equal to unity by neglecting the small fractions of the sphere surface f_M and f_E corresponding to sample and measurement openings, compared with the total surface F. ϱ_K is given more accurately from the contributions of the wall and the sample, multiplied by the appropriate surfaces:

$$\varrho_K = \left(1 - \frac{f_M + f_E + f_P}{F}\right)\varrho_W + \frac{f_P}{F}\varrho_P. \tag{129}$$

If the sample is made from the same material as the wall of the integrating sphere, such as freshly prepared magnesium oxide, the equation obtained is

$$\varrho_K = \frac{I_P}{I_W} = \left(1 - \frac{f_M + f_E}{F}\right)\varrho_W \quad \text{or} \quad \varrho_W = \frac{I_P}{I_W} \cdot \frac{F}{F - (f_M + f_E)}. \tag{130}$$

This is one of the methods available for measuring the absolute reflectance of magnesium oxide for different wavelengths.

Using a procedure given by *Preston* [144], it is also possible to determine the absolute reflectance of magnesium oxide using an integrating sphere coated with this substance, and dispensing with the screen. A beam of light falls on the wall, and the radiation flux I_1 is measured at an exit window. The value of I_1 is given by (126). Then a segment of the sphere is removed, the assumption being made that all the openings of the sphere possess zero reflectance, and a second radiation flux, I_2 is measured. Then from (126) we have

$$\frac{I_1}{I_2} = \frac{1 - \varrho_{K_2}}{1 - \varrho_{K_1}} \quad \text{or} \quad I_1(1 - \varrho_{K_1}) = I_2(1 - \varrho_{K_2}). \tag{131}$$

If, using (130), we write

$$\varrho_K = \left(1 - \frac{\Sigma f}{F}\right)\varrho_W \equiv a\varrho_W,$$

then

$$I_1(1 - a_1\varrho_W) = I_2(1 - a_2\varrho_W)$$

or

$$\varrho_W = \frac{I_1 - I_2}{I_1 a_1 - I_2 a_2}. \tag{132}$$

If the fractions of the sphere surface a_1 and a_2 are accurately known, and I_1 and I_2 are measured, ϱ can be calculated.

This procedure for measurement can be modified in very diverse ways, such as irradiating the sphere with diffuse light instead of directed light, replacing the segment of sphere by a plain surface, using a dis-

---

[144] Preston, J. S.: Trans. Opt. Soc. London **31**, 15 (1929/30).

mountable sphere from which separate portions with accurately known fractions of the total internal surface may be removed, etc., making it possible to refine the theory still further [145]. Fig. 51 gives the results obtained by various authors [146-150] as a function of the wavelength used. As we can see, the values given differ from each other much more than the limits of error given by the individual authors (0.3 to 0.4%). This may arise partly from the method of measurement, as a "deflection" method presupposes proportionality between photo current and the strength of irradiation as well as the linearity of the amplifier, and partly from the dependence of the reflectance of magnesium oxide on the method of preparation [151], purity, layer thickness, and the background. Thus, *Teller* and *Waldron* [148] have shown that even a layer 8 mm thick of freshly prepared magnesium oxide still possesses some transmittance. In addition, a layer of this kind undergoes relatively rapid modification, so that in course of time adsorbed impurities and the decomposition of any magnesium nitride, formed during the combustion of magnesium, in the presence of ultraviolet radiation, may affect the behaviour of the material [152].

A very elegant method which determines absolute reflectances from transmittance values alone, uses Eq. (75 a) for two identical homogeneous layers ($T_1 = T_2 \equiv T_a$, and $R_1 = R_2 \equiv R_a$). We may rearrange this equation to give

$$R_a = \left(1 - \frac{T_a^2}{T_{2a}}\right)^{1/2}. \tag{133}$$

If the two layers are different, then correspondingly we have

$$R_1 R_2 = 1 - \frac{T_1 T_2}{T_{12}} \tag{134}$$

---

[145] See *Jacquez, J. A.,* and others: J. Opt. Soc. Am. **45**, 460, 781, 971 (1955); **46**, 428 (1956); J. Appl. Physiol. **8**, 212, 297 (1955). — *Budde, W.,* and *G. Wyszecki:* Farbe **4**, 15 (1955). — *Miller, O. E.,* and *A. J. Sant:* J. Opt. Soc. Am. **48**, 828 (1958). — *van den Akker, J. A.,* and others: J. Opt. Soc. Am. **56**, 250 (1966).

[146] *Benford, F., G. P. Lloyd,* and *S. Schwarz:* J. Opt. Soc. Am. **38**, 445, 964 (1948).

[147] *Middleton, W. E. K.,* and *C. L. Sanders:* J. Opt. Soc. Am. **41**, 419 (1951); **43**, 58 (1952).

[148] *Tellex, P. A.,* and *J. R. Waldron:* J. Opt. Soc. Am. **45**, 19 (1955).

[149] *Höfert, H. J.,* and *H. Loof:* Farbe **13**, 53 (1964).

[150] *Goebel, D. G.,* and others: J. Opt. Soc. Am. **56**, 783 (1966).

[151] With reference to technic of preparation see [147, 148]; *Dimitroff, J. M.,* and *D. W. Swanson:* J. Opt. Soc. Am. **45**, 19 (1955); Beckman Instruments Inc. Instruction Manual Nr. 24 500.

[152] See *Middleton, W. E. K.,* and *C. L. Sanders:* J. Opt. Soc. Am. **41**, 419 (1951); **43**, 58 (1953) and *Priest, I. G.:* J. Opt. Soc. Am. **20**, 157 (1930).

T_a and T_{2a} are easy to determine experimentally, so that R_a can be evaluated. Measurements of this kind have been carried out by *Launer*[153] who has compared the values of R_a calculated from (133) by the use of two transmittance measurements with that obtained directly in reflection using a magnesium oxide standard and known values of ϱ. The agreement is satisfactory, as can be seen from Table 5, deviations amounting in most cases to 1–2%.

In measuring transmittance the radiation emerges in a diffuse state. In order to collect this as completely as possible the cathode of the photocell is brought as close as possible to the sample being measured. As a result of this, a small fraction of the incident radiation is reflected from the cathode, and a correction for this is necessary. When the series for reflection between sample and cathode is summed up (see p. 124), we obtain, for a single layer

$$T_a = t_a(1 - R_a R_C), \qquad (135)$$

while for two layers in succession

$$T_{2a} = t_{2a}(1 - R_a R_C)/(1 + t_{2a} R_a R_C). \qquad (136)$$

t_a and t_{2a} are the measured transmittances, R_a and R_C are separately determined reflections from a layer and the cathode respectively. If the corrected values of T_a and T_{2a} are inserted in (133) we obtain an equation for evaluating R_a from transmittance measurements, which, when very small terms are neglected, gives

$$R_a = \left(1 - \frac{t_a^2}{t_{2a}}\right)^{1/2} + 0.002 + \frac{1}{2} R_C \left(\frac{t_a^2}{t_{2a}}\right)(1 - t_{2a}). \qquad (137)$$

In this $R_C \cong 0.1$; the number 0.002 applies to papers. The calculated values of R_a obtained from this formula are given in Table 5.

A similar method for the direct measurement of absolute reflectances has been proposed by *Shibata*[154] and tested on freshly prepared magnesium oxide. The method depends on using an opal glass plate instead of an integrating sphere in order to irradiate the sample diffusely. The opal glass sheet a is irradiated by directed radiation, and the sample b is placed behind it. The sample thus receives diffuse radiation because of the scattering effect of the opal glass. If we wish to use for transmittance and reflectance of two homogeneous layers, located one behind the other, the Eqs. (75a) and (77a) developed by *Kubelka* (see p. 124), it must be remembered that the opal glass plate is irradiated with directed and not diffuse radiation. The transmittance and reflectance for directed radiation

---

[153] *Launer, H. F.*: J. Opt. Soc. Am. **32**, 84 (1942).
[154] *Shibata, K.*: J. Opt. Soc. Am. **47**, 172 (1957).

Table 5. *Comparison between the reflectance of transparent papers measured, and calculated from transmittance measurements by Eq.* (133)

	623 mµ			546 mµ			436 mµ			405 mµ			365 mµ			"white" light	
	measured P %	calculated from transmittance measurements P %	D %	measured P %	calculated P %	D %	measured P %	calculated P %	D %	measured P %	calculated P %	D %	measured P %	calculated P %	D %	measured P %	calculated P %
Rag paper with 14% CaCO₃ sizing	81.2	80.7	81.8	81.2	81.4	82.4	78.6	78.3	80.5	76.9	77.3	79.9	73.2	73.9	72.6	80.6	81.4
Rag paper without sizing	73.8	72.6	74.7	73.6	73.1	74.7	69.5	68.1	69.8	66.5	65.6	67.1	60.8	61.0	57.8	73.9	73.3
Yellow wood pulp without sizing	68.2	66.6	70.3	65.9	63.9	68.1	56.2	52.1	56.7	50.5	51.5	53.4	41.6	38.0	37.4	66.5	65.8
Cigarette paper with 19% CaCO₃ sizing	60.9	59.8	63.6	61.9	61.5	64.2	62.5	62.2	65.2	62.6	61.6	64.6	60.1	60.1	62.0	61.5	62.4

P = perpendicular incidence; D = diffuse incidence.

will be designated by $\overset{*}{T}_a$ and $\overset{*}{R}_a$. Instead of (75a) and (77a) we then have

$$T_{a,b} = \frac{\overset{*}{T}_a T_b}{1 - R_a R_b}, \tag{138}$$

$$R_{a,b} = \overset{*}{R}_a + \frac{\overset{*}{T}_a T_a R_b}{1 - R_a R_b}. \tag{139}$$

If b also consists of a similar opal glass plate, we have

$$R_{a,a} = \overset{*}{R}_a + \frac{\overset{*}{T}_a T_a R_a}{1 - R_a^2}. \tag{140}$$

Dividing (140) by (139) we obtain

$$\frac{R_{a,a} - \overset{*}{R}_a}{R_{a,b} - \overset{*}{R}_a} = \frac{R_a(1 - R_a R_b)}{(1 - R_a^2) R_b}. \tag{141}$$

Solving now for $1/R_b$, we have finally

$$\frac{1}{R_b} = R_a + \frac{R_{a,a} - \overset{*}{R}_a}{R_{a,b} - \overset{*}{R}_a} \cdot \left(\frac{1}{R_a} - R_a\right). \tag{142}$$

Shibata has shown that R_a for diffuse irradiation, as well as the ratio included in the first parenthesis can be measured by simple methods on the additional assumption that the first opal glass plate causes the radiation to emerge completely diffuse. Then the absolute diffuse reflectance of the sample b may be determined from (142). The ϱ-values measured by *Shibata* for magnesium oxide are also given in Fig. 51.

In the *infrared* the absolute reflectance of magnesium oxide in the region from 1 to 15 μ has been measured [155] by introducing it into a hohlraum type radiator, so that it would serve as a radiation source in a commercial infrared spectrometer (see also p. 246). Above 2 μ numerous strong absorption bands arise, so that magnesium oxide cannot be used as a comparison standard in this range.

As Fig. 51 shows, the absolute ϱ-values of freshly prepared magnesium oxide, not only when determined by different methods, but also by the same methods by different observers, are rather uncertain. As we have indicated already, this may be partly attributed to systematic errors in the methods used, partly to the purity of the magnesium employed, partly to the ineffective reproducibility of a sufficiently thick coherent layer of the freshly prepared magnesium oxide, and, finally, partly also to the relatively rapid ageing of such a layer.

---

[155] Gier, J. T., R. V. Dunkle, and J. T. Bevans: J. Opt. Soc. Am. **44**, 558 (1954).

For these reasons, it has often been suggested [156] that barium sulphate, perhaps with a suitable binder, should be substituted for magnesium oxide as reflectance standard; it has been found [157], however, that in spite of the very good reproducibility of individual measurements, the absolute reflectance of barium sulphate also varies greatly from sample to sample, and also depends on impurities, packing density, particle size, and perhaps other factors. These defects are probably common to all white standards.

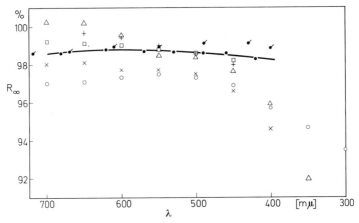

Fig. 51. Absolute, diffuse reflectance of freshly prepared MgO according to various authors: + *Benford, Lloyd, Schwarz;* ⊙ *Middleton, Sanders;* ● *Höfert, Loof;* × *Oelkrug;* △ *Shibata;* □ *Tellex, Waldron;* ● *Goebel*

It would therefore be desirable for practical purposes to use a more suitable white standard, preferably one which is not age dependent. This would of course need to be calibrated against freshly prepared magnesium oxide in the spectroscopic range used. With this in view, for example, for the photometer "Elrepho" produced by the firm Zeiss (see p. 222) a standard has been produced, consisting of a surface-treated milk-glass whose reflectivity using suitable care and cleanliness is constant over a long period. The same is true of a standard prepared by Heraeus in Hanau. This consists of a quartz cell with ground windows of Suprasil,

---

[156] *DIN-standards* 5033, Color measurement, 2 edition 1944. — *Miescher, K.,* and *R. Rometsch:* Experientia **6**, 302 (1950). — *Middleton, W. E. H.,* and *C. L. Sanders:* Illum. Eng. **48**, 254 (1953). — *Kortüm, G.,* and *P. Haug:* Z. Naturforsch. **8a**, 372 (1953). — *Grum, F.,* and *G. W. Luckey:* Appl. Optics **7**, 2289 (1968).

[157] *Budde, W.:* J. Opt. Soc. Am. **50**, 217 (1960). — *Laufer, J. S.:* J. Opt. Soc. Am. **49**, 1135 (1959). — *Goebel, D. G.,* and others: J. Opt. Soc. Am. **56**, 783 (1966).

filled with Suprasil powder of about 1 µ particle diameter and fused under nitrogen. This standard is also satisfactory in the ultraviolet. Several authors [158] have suggested Vitrolite Glass supplied by the National Bureau of Standards as a white standard. Its reflectance has been measured against MgO and lies between 0.922 and 0.885 in the spectral region 400–750 mµ.

To check the calibration values of these standards from time to time in a simple way, the firm of Carl Zeiss has developed a "magnesium oxide primary standard" [159] which avoids the undesirable properties of freshly prepared magnesium oxide referred to above. This primary standard is prepared from the purest magnesium oxide powder (Merck) by means of a powder press [160] in the form of a tablet of diameter ca. 45 mm and thickness ca. 5 mm. These tablets, prepared freshly under necessary conditions of cleanliness, deviate only by a maximum of $\pm 0.1\%$ in reflectance from each other, but may only be used after preparation for a maximum of 24 hours for comparison and calibration purposes. Their relative reflectance [161] with respect to magnesium oxide as standard has the values given in the Table 6, which also contains the absolute ϱ values of freshly prepared magnesium oxide measured by means of the Taylor sphere. These values represent the best available at this time.

Table 6. *Absolute reflectance of freshly prepared MgO and relative reflectance of the primary standards supplied by C. Zeiss*

λ [mµ]	420	460	490	530	570	620	680
ϱ absolute	0.983	0.986	0.986	0.987	0.987	0.988	0.986
R'_∞ primary standard	0.992	0.992	0.993	0.994	0.995	0.995	0.996

For the investigation of chemical problems by reflection measurements (see p. 256 ff.) it is often necessary to choose such an adsorbent as comparison standard in which definite interaction between adsorbent and adsorbate occurs, such as acid-base reactions, complex formation, ligand exchange, etc.

For this reason, it would be desirable to measure the reflectance of as many standards as possible throughout the entire accessible spectro-

---

[158] Gabel, J. W., and E. J. Stearns: J. Opt. Soc. Am. **39**, 48 (1949).

[159] Pamphlet 50-660/MgO-d, Carl Zeiss Oberkochen; see also S: son Stenius, Å.: J. Opt. Soc. Am. **45**, 727 (1955). — Gilmore, E. H., and R. H. Knipe: J. Opt. Soc. Am. **42**, 481 (1957).

[160] Available from Carl Zeiss, Oberkochen.

[161] Höfert, H.-J., and H. Loof: Farbe **13**, 53 (1964).

scopic range, in so far as these might be used as adsorbents. The absolute reflectance of the most important materials is given in Fig. 52 as a function of wavenumber[162]. The purest materials available commercially were used without further purification. These were dehydrated at the temperatures given in Fig. 52, cooled in a vacuum desiccator over phosphorus pentoxide, ground for 2 hours with the exclusion of moisture, and then immediately measured against the Suprasil standard referred

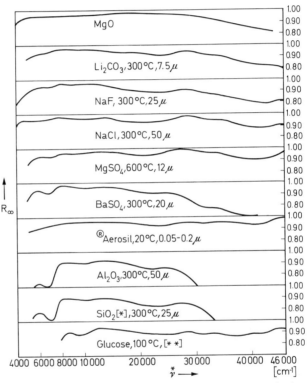

Fig. 52 Absolute reflectance of a series of white standards

to above, which itself had been calibrated against freshly prepared magnesium oxide. The absolute reflectance of the materials investigated may then be obtained from

$$R_{\infty \text{ sample}} = \frac{R_{\infty \text{ sample}}}{R_{\infty \text{ Suprasil}}} \cdot \frac{R_{\infty \text{ Suprasil}}}{R_{\infty \text{ MgO}}} \cdot \varrho_{\text{MgO}} . \qquad (143)$$

The mean values given in Fig. 51 were used for ϱ.

---

[162] *Kortüm, G., W. Braun,* and *G. Herzog:* Angew. Chem. **75**, 653 (1963); Angew. Chem. internat. Edit. **2**, 333 (1963).

Since the curves measured depend on moisture content, origin and purity of samples, and the correctness of the ϱ-values of the magnesium oxide, they are not reproducible to better than $\pm 2\%$. It is seen that, in addition to magnesium oxide, sodium chloride, and particularly Aerosil[163] possesss very good reflectance in the entire visible and ultraviolet region of the spectrum. Potassium chloride, potassium bromide and calcium fluoride behave in a similar way to sodium chloride. Potassium iodide often contains traces of elementary iodine, and must therefore always be recrystallised with the exclusion of air. The reflectance of barium sulphate in the short wave ultraviolet region is considerably better than that given in Fig. 52 if it is freshly precipitated. The same is true for other sparingly soluble substances. In general it is concluded from these observations that the apparent absorption of such standards which, from the standpoint of molecular theory, may be regarded as completely colorless (white), most probably will always depend upon the presence of impurities which it is very difficult to remove completely. The rapid "ageing" of these standards which is always observed, and which results in the reduction of their original reflectance, supports this conclusion.

If the standard or the diluting agent itself possesses appreciable absorption, as may occur if the substance to be investigated is adsorbed on particular adsorbents or catalysts, then in the measurement of the relative reflectance $R'_{\infty\,(\text{sample}+\text{standard}/\text{standard})}$, the self-absorption of the standard is not eliminated as it is in the case of transmittance measurements if the solution is measured against a correspondingly absorbing solvent. This follows because in transmittance measurements the absorption coefficient k is connected logarithmically with the measured quantity (transmittance), whereas in reflectance measurements this is not so. For transmittance measurements with an absorbing solvent we have:

$$E_1 = k_{st} d = \log \frac{I_0}{I_{st}}; \quad E_2 = (k_{st} + k_{sample}) d = \log \frac{I_0}{I_{st+sample}}$$

where the subscribt "st" designates the solvent. From this it follows that

$$E_2 - E_1 = k_{sample} d = \log \frac{I_{st}}{I_{st+sample}};$$

that is, we obtain the absorbance of the sample by means of a single measurement. For reflectance measurements, on the other hand, using (122), if the standard is measured against magnesium oxide, we have

$$R_{\infty\,\text{standard}} = \varrho R'_{\infty\,\text{standard}} \equiv R_{\infty 1},$$

---

[163] Aerosil consists of chemically pure silicic acid prepared by Degussa, Rheinfelden.

while if standard + sample are measured against the standard we have

$$R_{\infty \text{ standard} + \text{sample}} = R'_{\infty \text{ standard} + \text{sample}} \cdot \varrho R'_{\infty \text{ standard}} \equiv R_{\infty 2},$$

and if standard + sample are measured against magnesium oxide:

$$R_{\infty \text{ standard} + \text{sample}} = \varrho R''_{\infty \text{ standard} + \text{sample}} \equiv R_{\infty 2}.$$

It follows from this that:

$R''_{\infty \text{ standard} + \text{sample against magnesium oxide}}$
$= R'_{\infty \text{ standard} + \text{sample against standard}} \cdot R'_{\infty \text{ standard against magnesium oxide}} \cdot$

It is seen that two measurements are used, since from the Kubelka-Munk theory $F(R_\infty) = K/S$, so that

$$F(R_{\infty 2}) - F(R_{\infty 1}) = \frac{K_{\text{standard}} + K_{\text{sample}}}{S} - \frac{K_{\text{standard}}}{S} = \frac{K_{\text{sample}}}{S}, \quad (144)$$

because the scattering coefficient of the standard can be regarded as equal to that of the dilute sample (see p. 209). We must thus either measure standard + sample, and standard alone, against magnesium oxide, or measure standard + sample against standard, and standard against magnesium oxide in order to determine the absorption coefficient of the sample on the Kubelka-Munk basis. The absolute reflectance ϱ of magnesium oxide must be regarded as known in advance. The second method is the better, for the reasons already discussed.

g) Consideration of Self-Emission or Luminescence

In this section we compliment the equations derived for isotropic scattering and absorption in a homogeneous medium with $n_{\text{matrix}} = 1$ to include the case where radiation energy originates within the medium itself. The flux of this radiation in the x- and $-x$-directions is superimposed upon that entering the medium from without. Three characteristic cases of practical importance will be briefly outlined: 1. Self-emission of the medium as a result of high temperatures (black body radiation) or because the medium contains radioactive materials which excite luminescence within the medium. 2. Luminescence of the medium as a consequence of absorption of radiation of high energy from an external source (e.g., a phosphore). 3. Quantum yield and reabsorption of luminescence outside of the overlapping region of absorption and luminescence bands.

1. *Temperature induced emission* of a scattering and absorbing medium has already been considered by *Schuster* in his investigation of stellar atmospheres with isotropic scattering distribution (compare

p. 103). If the radiation flux $kI\,dx$ is absorbed in a path length dx, according to the Kirchhoff Law, the layer of thickness dx must render a contribution $k'E\,dx$ to the black radiation E in the x- or $-x$-direction. E depends on the temperature given and relates to the wavelength considered. The simultaneous differential Eqs. (5) in this case are

$$\left.\begin{array}{l}\dfrac{dI}{dx} = -(k+s)\,I + sJ + k'E \\[6pt] \dfrac{dJ}{dx} = (k+s)\,J - sI - k'E.\end{array}\right\} \tag{145}$$

k and k' are proportional to one another so that

$$E' \equiv \frac{k'}{k}\,E = \text{konst.} \tag{146}$$

at given temperature and wavelength. The indefinite integrals analogous to (6) are

$$\left.\begin{array}{l}I = A'(1-\beta)\,e^{\alpha x} + B'(1+\beta)\,e^{-\alpha x} + E' \\ J = A'(1+\beta)\,e^{\alpha x} + B'(1-\beta)\,e^{-\alpha x} + E',\end{array}\right\} \tag{147}$$

where α and β are again defined by (7). The constants A' and B' arise from the respective boundary conditions. If we again set

$$I = I_0 \quad \text{at} \quad x = 0 \quad \text{and} \quad J = 0 \quad \text{at} \quad x = d,$$

i.e., assume a freely suspended layer of thickness d, we obtain analogously to Eq. (8)

$$\left.\begin{array}{l}A' = \dfrac{-I_0(1-\beta)\,e^{-\alpha d} - (1+\beta)\,E' + (1-\beta)\,e^{-\alpha d}E'}{(1+\beta)^2\,e^{\alpha d} - (1-\beta)^2\,e^{-\alpha d}} \\[10pt] B' = \dfrac{I_0(1+\beta)\,e^{\alpha d} + (1-\beta)\,E' - (1+\beta)\,e^{\alpha d}E'}{(1+\beta)^2\,e^{\alpha d} - (1-\beta)^2\,e^{-\alpha d}}.\end{array}\right\} \tag{148}$$

Substituting this into (147) we obtain for a semi-infinite medium $(x \to \infty)$

$$\left.\begin{array}{l}\dfrac{J_{(x=0)}}{I_0} = \dfrac{1-\beta}{1+\beta} + \dfrac{E'}{I_0}\left(1 - \dfrac{1-\beta}{1+\beta}\right) \\[10pt] \phantom{\dfrac{J_{(x=0)}}{I_0}} = R_\infty + \dfrac{E'}{I_0}(1 - R_\infty),\end{array}\right\} \tag{149}$$

where R_∞ is the reflectance in the absence of temperature-induced emission. For $E' = 0$, (149) reduces again to (9).

2. The *luminescence excitation* in a fluorescent screen by X-rays or cathode rays, that is sometimes practiced in X-ray photography to sharpen the photographic density of the emulsion, has been discussed

by several authors [164]. To do this, a layer capable of luminescence and of thickness d is pressed to the back side of the film ($x=0$). If the number of X-ray quanta at $x=0$ is given by N_0, it decreases exponentially on penetration of the luminescent layer according to

$$N = N_0 e^{-\mu x}, \qquad (150)$$

where μ is the coefficient of absorption for the X-rays. In a layer element dx the decrease amounts to

$$-dN = \mu N_0 e^{-\mu x} dx \qquad (151)$$

and the luminescent radiation flux is

$$dI = -dJ = q\mu N_0 e^{-\mu x} dx, \qquad (152)$$

where q is the quantum yield. Analogously as in (145) we assume that an equivalent radiation flux of luminescence dissipates in the x- and $-x$-direction and is thereby partly absorbed and partly isotropically scattered. The appropriate differential equations for this case are

$$\left.\begin{aligned}\frac{dI}{dx} &= -(k+s)I + sJ + Ce^{-\mu x} \\ \frac{dJ}{dx} &= (k+s)J - sI - Ce^{-\mu x},\end{aligned}\right\} \qquad (153)$$

where C represents a composite of the constants in (152). The general solutions are

$$\left.\begin{aligned}I &= A(1-\beta)e^{\alpha x} + B(1+\beta)e^{-\alpha x} - \frac{C}{\mu-\alpha}e^{-\mu x} \\ J &= A(1+\beta)e^{\alpha x} + B(1-\beta)e^{-\alpha x} + \frac{C}{\mu+\alpha}e^{-\mu x}.\end{aligned}\right\} \qquad (154)$$

α and β again have the same designations as in (7). The constants A and B must be calculated from the suitable boundary conditions, into which, for example, the diffuse reflectance R_g of the film and perhaps a covering layer present at $x=d$ enter.

3. We can obtain the *absolute energy* or *luminescent quantum efficiency* by measuring the luminescence and *absolute* diffuse reflectance R_∞ of a powder layer. In this procedure, all measurements must be made at exactly the same optical geometry and we must assume that the layer reflects diffusely, i.e. the reflected primary radiation and secondary luminescent radiation are both isotropically scattered. The comparison of the reflected or emitted radiation flux in a given solid angular region

---

[164] *Hamaker, H. C.*: Philips Res. Rep. **2**, 55 (1947). — *Coltman, J. W.*, et al.: J. Appl. Phys. **18**, 530 (1947). — *Broser, I.*: Ann. Physik **5**, 401 (1950).

yields directly the quantum yield[165]. The best reflectance standard is MgO whose absolute reflectance is already known (see p. 146).

First of all we calculate the opposing radiation fluxes I and J of the Kubelka-Munk theory for the primary radiation as a function of distance x from the surface. We start with Eq. (52) and integrate first over a finite freely suspended layer of thickness d (Fig. 44). This means that at $x = 0$, also $r = 0$ as is the case when the layer has a black background. Instead of (54), integration between d and any arbitrary x gives

$$S(x - d) = \frac{1}{b} \coth^{-1} \frac{ar - 1}{rb}. \tag{155}$$

Here, $r = J/I$ according to (25) and a and b are defined by (22) and (35) respectively. Written another way, (155) becomes

$$\frac{ar - 1}{rb} = \frac{\coth bSx - \coth bSd}{1 - \coth bSx \cdot \coth bSd}. \tag{156}$$

For infinite layer thickness[166], $\lim_{d \to \infty} \coth bSd = 1$ so that

$$r = \frac{1}{a + b} \tag{157}$$

for an arbitrary x. At $x = d$, or at the surface, $r = J/I_0 = R_\infty$, in agreement with (36). If we substitute

$$r = \frac{J}{I} = \frac{1}{a + b}$$

in the differential Eqs. (20 and 21), they can be integrated directly and we obtain for the radiation fluxes I and J as functions of x at infinite layer thickness

$$I = I_0 e^{-bSx} \quad \text{and} \quad J = R_\infty I_0 e^{-bSx}. \tag{158}$$

Thus we are able to calculate the luminescent radiation flux at every position x of the sample. In the layer element between x and $x + dx$, the total radiation flux absorbed, according to (158) is given by

$$dA = K_A(I + J)\,dx = I_0 K_A(1 + R_{\infty A})\,e^{-b_A S_A x}\,dx, \tag{159}$$

where the index A relates to the wavelength λ_A of the primary radiation. The corresponding luminescent radiation flux dE at the (greater) wave-

---

[165] Compare Bril, A.: Luminescence of organic and inorganic materials, p. 479. New York: John Wiley & Sons 1962.

[166] Equations for intensity calculations of Raman spectra in finite layers of polycrystaline materials are derived by Schrader, B., and G. Bergmann: Z. Anal. Chem. **225**, 230 (1967).

length λ_E is then

$$dE = q\,dA. \qquad (160)$$

q is thus the luminescent quantum efficiency at λ_E.

Experience shows (Fig. 52) that the absolute reflectance of so-called white standards lies between 0.98 and 0.8 and still lower in many cases. This behavior can be generally traced to surface contamination of the crystallites. This means that the luminescent radiation will be reabsorbed

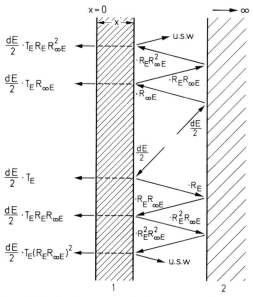

Fig. 53 Derivation of the luminescent radiation generated at x, at the surface of a semi-infinite medium. Actually, the layers 1 and 2 border directly on one another

not only in regions where luminescence and absorption overlap ($K_E \neq 0$), and that the externally measurable luminescence dE' will be less than dE. This will be especially true when the primary radiation penetrates deep within the sample and the luminescent radiation must then also cover a large distance, on the average, within the sample. dE' can be calculated in the following manner [167]:

Let us visualize the powder layer as formally divided into two layers. The first extends from $x=0$ to the position x under consideration and the second extends beyond this to infinity (Fig. 53). At x, luminescent radiation $dE/2$ is emitted into each hemisphere. Layer 1 may have the transmittance

---
[167] Oelkrug, D., and G. Kortüm: Z. Physik. Chem. N. F. **58**, 181 (1968).

Consideration of Self-Emission or Luminescence

T_E and reflectance R_E, given by Eqs. (60) and (57) respectively, while layer 2 has the reflectance $R_{\infty E}$. Referring again to Fig. 53, we thus obtain for the luminescent radiation flux at the surface

$$dE' = \frac{q\,dA}{2} \cdot T_E[(1 + R_E R_{\infty E} + R_E^2 R_{\infty E}^2 + R_E^3 R_{\infty E}^3 + \cdots)$$

$$+ R_{\infty E}(1 + R_E R_{\infty E} + R_E^2 R_{\infty E}^2 + \cdots)] \qquad (161)$$

$$= \frac{q\,dA}{2} \cdot \frac{T_E(1 + R_{\infty E})}{1 - R_E R_{\infty E}}$$

which gives, by integration,

$$E' = \frac{q}{2}(1 + R_{\infty E}) \int_0^\infty \frac{T_E}{1 - R_E R_{\infty E}} dA. \qquad (162)$$

Using (157) and (159), this can be rearranged to

$$E' = \frac{q}{2} I_0 K_A (1 + R_{\infty A})(1 + R_{\infty E}) \int_0^\infty \exp(-b_A S_A + b_E S_E) x\,dx$$

$$= \frac{q I_0 K_A (1 + R_{\infty A})(1 + R_{\infty E})}{2(b_A S_A + b_E S_E)}. \qquad (163)$$

If $K_E = 0$ and thereby $R_{\infty E} = 1$, we obtain

$$E' = E = q I_0 (1 - R_{\infty A}), \qquad (164)$$

which is the luminescent yield without reabsorption. Let us assume that $S_A = S_E$, or that the scattering coefficient in the wavelength region λ_E to λ_A is practically constant (compare p. 206). Then we obtain a simplified equation

$$E'_{(S_E = S_A)} = \frac{q I_0 \cdot F(R_{\infty A})(1 + R_{\infty A})(1 + R_{\infty E})}{R_{\infty A}^{-1} + R_{\infty E}^{-1} - R_{\infty A} - R_{\infty E}}, \qquad (165)$$

which allows the evaluation of E' by means of two absolute reflectance measurements at λ_A and λ_E. The ratio E'/E according to (165) and (164) as a function of $R_{\infty E}$ is given in Fig. 54 for different values of $R_{\infty A}$. We see that at a given $R_{\infty E}$, the adulteration of the luminescent yield becomes greater with increasing $R_{\infty A}$, i.e., with decreasing absorption of the substance at λ_A.

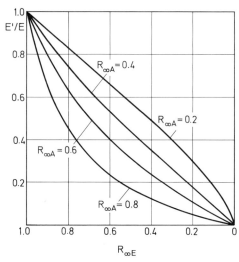

Fig. 54. Relation of measurable to true luminescence as a function of the reflectance at λ_E of the emission for various $R_{\infty A}$ at λ_A of the incident irradiation

h) Attempts at a Rigorous Solution of the Radiation-Transfer Equation

As was shown on p. 103 ff., the Schuster equation for isotropic scattering represents a first approximation of the general radiation-transfer equation. The success of the Schuster method, in which for simplification the radiation field is divided into two opposing radiation fluxes in the direction of the normal to the surface of a medium bounded by parallel planes, has encouraged attempts to generalize or refine the method. These attempts have consisted of dividing the radiation field into a large arbitrary number of radiation fluxes going in different directions and, in this manner, obtaining rigorous solutions for the radiation-transfer equation which could also be applicable for the case of non-isotropic scattering[168]. For this purpose, it is at first specified that the scattering medium is embedded in a matrix of refractive index 1 (air).

If we want to divide the radiation field into $2n$ radiation fluxes in the directions $\mu_i \equiv \cos\vartheta_i (i = \pm 1, \pm 2, \ldots, \pm n)$, we can replace the radia-

---

[168] See the detailed presentation by S. Chandrasekhar, l.c. We restrict ourselves here to limited information regarding the train of thought and results of the calculation. Because the approximation of isotropic scattering in tightly packed media is almost completely fulfilled, the differences between the results from exact solutions and the approximate solution according to Schuster or Kubelka-Munk fall within the limits of error of the experimental methods.

tion-transfer Eq. (III, 93) by $2n$ linear differential equations with constant coefficients a_j:

$$\mu_i \frac{dI(\tau, \mu_i)}{d\tau} = I(\tau, \mu_i) - \frac{1}{2}\omega_0 \sum_{j=-n}^{j=+n} a_j I_j. \quad (166)$$

The procedure generally depends on replacing the integrands in the integral $\int_{-1}^{+1} f(\mu)\,d\mu$ approximately by a polynomial that is integrated in place of the original inexplicit function (the so-called approximate quadrature)[169].

For this procedure to be practically useful, it is important that the lower orders of the approximation ($n = 3$ or 4) already sufficiently approach the true value of the integral. In the formula for the numerical quadrature of such an integral, given by Gauss, the interval from -1 to $+1$ is subdivided according to the zeros of the Legendre polynomial $P_l(\mu)$. The integral of the function $f(\mu)$ over this interval is expressed as a sum in the form

$$\int_{-1}^{+1} f(\mu)\,d\mu \cong \sum_1^m a_j f(\mu_j), \quad (167)$$

where μ_j represent the zeros of $P_l(\mu)$ and

$$a_j = \frac{1}{P_l'(\mu_j)} \int_{-1}^{+1} \frac{P_l(\mu)}{\mu - \mu_j}\,d\mu \quad (168)$$

are the Gaussian weighting coefficients. The μ_j- and a_j-values have been tabulated[170].

Chandrasekhar has given rigorous solutions of the equation system (166) from which the optical characteristics of isotropically scattering media ($p(\cos\vartheta) = \omega_0$) or for the phase function (III, 94) $p(\cos\vartheta) = \omega_0(1 + x\cos\vartheta)$ of simultaneously absorbing media can be derived. The solutions can be expressed as non-linear integral equations of the form

$$H(\mu) = 1 + \mu \cdot H(\mu) \int_0^1 \frac{\Psi(\mu')}{\mu + \mu'} H(\mu')\,d\mu', \quad (169)$$

---

[169] Compare in this regard, for example, *Margenau, H.*, and *G. M. Murphy*: The Mathematics of Physics and Chemistry, Vol. I. D. van Norstrand Comp., Inc. 1943.

[170] *Lowan, A. N., N. Davids*, and *A. Levenson*: Bull. Am. Math. Soc. **48**, 739 (1942).

where the so-called *characteristic function* Ψ is equal to $\omega_0/2$ for the case of isotropic scattering and equal to $\dfrac{\omega_0}{2}[1+x(1-\omega_0)\mu^2]$ for the phase function (III, 94) $p(\cos\theta) = \omega_0(1 + x\cos\theta)$, and through the moments of $H(\mu)$ defined by

$$\alpha_n = \int_0^1 H(\mu)\,\mu^n\,d\mu \qquad (170)$$

of various order $n = 0, 1, 2, \ldots$ of $H(\mu)$. *Chandrasekhar* has tabulated these so-called H- and α-integrals [171].

The reflectance of a semi-infinite medium with *isotropic scattering* irradiated in the direction μ_0 is given by

$$R_\infty(\mu_0) = 1 - H(\mu_0)(1-\omega_0)^{1/2}, \qquad (171)$$

or by

$$R_{\infty\,\text{diff}} = 1 - 2(1-\omega_0)^{1/2}\,\alpha_1 = 1 - 2(1-\omega_0)\int_0^1 \mu\cdot H(\mu)\,d\mu, \qquad (172)$$

if it is diffusely irradiated. Correspondingly, when the medium scatters according to the phase function $\omega_0(1 + x\cos\theta)$,

$$R_\infty(\mu_0) = 1 - H(\mu_0)\left[1 - \frac{\omega_0(\alpha_0 - c\alpha_1)}{2}\right] \qquad (173)$$

with

$$c \equiv \frac{x(1-\omega_0)\,\omega_0\alpha_1}{2 - \omega_0\alpha_0} \qquad (174)$$

or, again with diffuse irradiation,

$$R_{\infty\,\text{diff}} = 1 - 2\alpha_1\left[1 - \frac{\omega_0(\alpha_0 - c\alpha_1)}{2}\right]. \qquad (175)$$

With the help of the tables given by *Chandrasekhar*, these expressions for various ω_0-values can be easily calculated, but only for $x = 1$ and $x = 0$. Table 7 gives the reflectance for different ω_0-values calculated in this way [172]. The data consider isotropic scattering and the phase function $\omega_0(1 + \cos\theta)$ for both perpendicularly incident $(\mu_0 = 1)$ and diffuse irradiation.

If the scattering particles are embedded in a medium whose refractive index n is greater than 1, additional regular reflections occur at the phase boundaries. Of special meaning is the partly total reflection at the inner

---

[171] l.c. S. 125. 139, 141, 328.
[172] Giovanelli, R. G.: Opt. Acta **2**, 153 (1955).

side of the phase boundary (compare Chap. IV e). *Giovanelli* has specified the reflectance for two limiting cases: 1. the phase boundary reflects regularly, and 2. the phase boundary is a hypothetical one with the property that, in spite of $n > 1$, all reflected and transmitted radiation is ideally and diffusely scattered independently of the angle of incidence. The behavior of real media must then be such as to lie between these

Table 7. *Reflectance R_∞ ($\mu_0 = 1$) and $R_{\infty\,\text{diff.}}$ for isotropic scattering ($p(\cos\theta) = \omega_0$) and for the phase function $\omega_0(1 + \cos\theta)$ as a function of ω_0. The medium is considered semi-infinite in a matrix with $n = 1$*

ω_0	$\mu_0 = 1$		Diffuse irradiation	
	$p(\cos\theta) = \omega_0$ (isotropic)	$p(\cos\theta) = \omega_0(1 + \cos\theta)$	$p(\cos\theta) = \omega_0$ (isotropic)	$p(\cos\theta) = \omega_0(1 + \cos\theta)$
1.000	1.00000	1.00000	1.00000	1.00000
0.999	0.91285	0.89367	0.92971	0.91446
0.995	0.81705	0.77877	0.84985	0.81945
0.990	0.75275	0.70270	0.79457	0.75482
0.975	0.64092	0.57344	0.69501	0.64140
0.950	0.53555	0.45552	0.59667	0.53311
0.925	0.46655	0.38104	0.52965	0.46172
0.900	0.41495	0.32712	0.47802	0.40825
0.85	0.33966	—	0.40017	—
0.80	0.28526	0.20015	0.34187	0.27406
0.7	0.20867	0.13286	0.25655	0.19626
0.6	0.15541	0.09065	0.19471	0.14318
0.5	0.11521	0.06192	0.14653	0.10411
0.4	0.08336	0.04147	0.10934	0.07394
0.3	0.05721	0.02638	0.07445	0.04986
0.2	0.03524	0.01513	0.04626	0.03018
0.1	0.01639	0.00649	0.02170	0.01382
0	0.00000	0.00000	0.00000	0.00000

two limiting cases. Table 8 gives examples of the results of such calculations which were carried through for a great number of various experimental possibilities. Table 8 lists the theoretical values of the reflectance of semi-infinite media whose matrices have refractive indices lying between 1 and 1.5 for isotropic scattering. The values include the regular portion for perpendicular incidence ($\mu_0 = 1$). In addition, the intensity distribution of the radiation at the surface of an isotropically scattering semi-infinite or plane parallel medium with matrix $n \geq 1$ has been calculated for the case where a linear infinite light source is placed parallel to the phase boundary on the surface or within the

Table 8. *Total reflectance according to Giovanelli for semi-infinite media specifying isotropic scattering and perpendicularly incident irradiation. The refractive indices of the surrounding media are equal or greater than 1*

ω_0	a/σ	Refractive index of the matrix			
		$n=1$	$n=1.333$	$n=1.46$	$n=1.50$
1.00000	0.00000	1.0000	1.0000	1.0000	1.0000
0.99999	0.00001	0.991	0.986	0.983	0.982
0.99997	0.00003	0.984	0.975	0.971	0.970
0.99990	0.00010	0.972	0.956	0.984	0.945
0.99970	0.00030	0.951	0.925	0.913	0.909
0.99900	0.00100	0.9128	0.868	0.848	0.8414
0.9975	0.00251	0.866	0.802	0.774	0.765
0.9950	0.00503	0.8170	0.736	0.701	0.6910
0.9900	0.01010	0.7528	0.6520	0.613	0.6020
0.9750	0.02564	0.6409	0.520	0.479	0.4685
0.9500	0.05263	0.5356	0.4073	0.370	0.3612
0.9	0.1111	0.4150	0.295	0.266	0.2600
0.8	0.2500	0.2853	0.1898	0.174	0.1720
0.7	0.4286	0.2087	0.137	0.129	0.1289
0.6	0.6667	0.1554	0.1030	0.101	0.1024
0.5	1.0000	0.1152	0.079	0.082	0.0841
0.4	1.500	0.0834	0.061	0.068	0.0708
0.3	2.333	0.0572	0.047	0.057	0.0605
0.2	4.000	0.0352	0.036	0.048	0.0523
0.1	9.000	0.0164	0.028	0.041	0.0456
0.0	∞	0.0000	0.0204	0.0350	0.0400

scattering medium [173]. The process can also be applied to diffusely reflecting media of arbitrary geometric form.

A comparison between the results of the approximate Kubelka-Munk equation and those of the rigorous theory is of special interest for the later proof of theory through experiment. For the simplest case of the reflectance of a semi-infinite medium with a matrix with $n=1$ and an isotropic scattering distribution under diffuse irradiation (the assumptions in the Kubelka-Munk treatment), Eq. (31) or (172) is valid. R_∞ is a function of K/S or a/σ; i.e., a function of the ratio of the absorption coefficient to the scattering coefficient. The semi-logarithmic plot of $R_{\infty\,\text{diff}}$ against K/S or a/σ with a common abscissa scale is given in Fig. 55. The figure shows that the curves are very similar but only come

---

[173] *Giovanelli, R. G.*: Opt. Acta **3**, 24, 49 (1956). — *Jefferies, J. T.*: Opt. Acta **2**, 109 (1955).

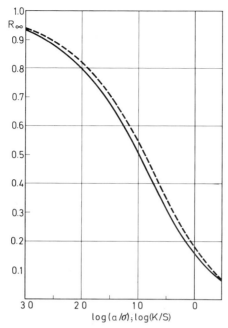

Fig. 55. Reflectance R_∞ of a medium ($n_{matrix} = 1$) with diffuse irradiation and isotropic scattering distribution calculated according to *Giovanelli* (———) and *Kubelka-Munk* (– – –) as functions of K/S or a/σ respectively with common abscissa

together at $R_\infty = 0$ and $R_\infty = 1$ and deviate from one another by as much as 8% in between[174]. This occurs mainly because of the various definitions of K and a or S and σ. *Blevin* and *Brown*[175] have shown that we can compare the equations more meaningfully if we multiply K/S by a normalization factor so that the curves come together about at the point $R_\infty = 0.5$. This causes the Kubelka-Munk curve to be shifted along the abscissa parallel to itself. The deviations in the curves now amount to a maximum of about ± 0.01 in R_∞ which corresponds to the limit of error in the experimental measurement. This is valid even for media whose refractive indices exceed 1 and where regular reflection thus additionally occurs. Fig. 56 shows this comparison of curves calculated according to (113) or (171) for perpendicular irradiation, isotropic scattering, and a matrix with $n = 1.5$. The abscissa scales are displaced with regard to one another as a result of the normalization so that the

---

[174] *Hecht, H. G.*: Modern aspects of reflectance spectroscopy, p. 1 ff. New York: Plenum Press 1968.

[175] *Blevin, W. R.*, and *W. J. Brown*: J. Opt. Soc. Am. **52**, 1250 (1962).

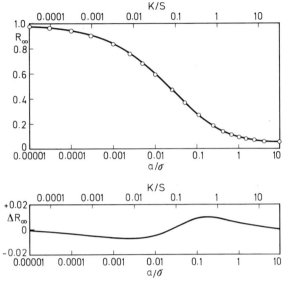

Fig. 56. Reflectance of a medium ($n_{\text{matrix}} = 1.5$) with perpendicular irradiation and isotropic scattering distribution calculated according to *Giovanelli* (———) and *Kubelka-Munk* (○ ○). K/S and a/σ abscissas shifted with respect to one another. Lower curve: Deviations between the two curves

curves fall together at $R_\infty = 0.469$. Beneath the graphical representation are the differences of both curves as a function of K/S or a/σ and they do not exceed the value 0.01 here either. This means that the Kubelka-Munk equation represents an extraordinarily good approximation for the rigorous equations, if we can assume an isotropic scattering distribution which may be the case as a rule in multiple scattering (compare p. 99). Even in cases where the scattering is angularly dependent, as perhaps in thin layers or where the concentration of scattering centers is too low, we can regard isotropic scattering as a good approximation if we replace the actual scattering coefficient σ by an effective scattering coefficient of the single scattering process [176]

$$\sigma_e = \sigma(1 - \bar{\mu}) \qquad (176)$$

where, according to the mean-value theorem,

$$\bar{\mu} = \frac{\int\limits_{-1}^{+1} I(\mu)\,\mu\,d\mu}{\int\limits_{-1}^{+1} I(\mu)\,d\mu} \qquad (177)$$

---

[176] *Blevin, W. R.*, and *W. J. Brown*: J. Opt. Soc. Am. **51**, 975 (1961).

and $I(\mu)$ again indicates the scattered intensity in the direction μ (for isotropic scattering, $\bar{\mu}=0$). The scattering distribution calculated for various simple phase functions gives a reflectance R_∞ as a function of a/σ that agrees very well with that for isotropic scattering (Fig. 55).

The problem of inhomogeneous media treated already by *Kubelka* (compare p. 123) has been thoroughly investigated by *Giovanelli*[177] because in practice it frequently plays a roll. Examples of this are unevenly distributed pigments in a matrix or in the solar photosphere and chromosphere. For the special case already mentioned by *Kubelka*, where the inhomogeneities extend only in the direction x perpendicular to the surface (e.g., paper), we can keep the equations for homogeneous media if we use the "optical thickness" as defined by (III, 88) as a variable. If, however, in spite of a constant ratio $a/\sigma = \omega_0$, the scattering particles ball up into statistically irregularly distributed aggregates, the problem becomes very difficult and lends itself only to approximate solutions for definite model media in which the inhomogeneities are treated as negligeable perturbations of the homogeneous distribution. One such case could perhaps be that where the attenuation coefficient $\kappa = a + \sigma$ and the scattering parameter $\lambda = a/\kappa$ are, to be sure, constant in the x-direction but vary as a sine-function in the y-direction parallel to the surface so that

$$\kappa = \kappa_0 + \kappa_1 \cos ly,$$
$$\lambda = \lambda_0 + \lambda_1 \cos ly,$$

where κ_1/κ_0 and λ_1/λ_0 are very small and l represents an inverse measure for the structure parameter of the medium. The calculation reveals that, compared with a homogeneous medium of the same kind, the reflectance of such a medium with gross inhomogeneities decreases considerably. The reason for this is, that the particle aggregation reduces the number of scattering centers and the penetration depth of the incident radiation is increased.

i) Discontinuum Theories

The continuum theories of absorption and scattering of powder layers are not completely satisfactory, since the scattering coefficient and absorption coefficient refer to units of thickness of homogeneous layers treated as a continuum, so that there is no direct connection between the optical properties of the particles making up the layer and the measured quantities. Of course the absorption coefficient K_R measured in remission should, for the reasons given on p. 108, be twice as

---

[177] *Giovanelli*, R. G.: Progress in optics, Vol. 2, p. 111 ff. Amsterdam: North Holland Publishing Co. 1963; Australian J. Phys. **10**, 337 (1957); **12**, 164 (1959).

large as the true absorption coefficient K_T determined from transmittance measurements, but as we shall show later (see Table 11, p. 195) this expectation is not confirmed by experiment. The reason is probably that when the radiation emerges from the crystal-air interface, some total reflection occurs, so that the light path in the crystal is extended. Thus, if we limit ourselves to the case of "diffuse reflection", taking the particle diameter to be considerably greater than λ, for which the interaction of the electromagnetic wave with a particle may still be conceived in terms of reflection, refraction, and diffraction (cf. p. 26), it would be desirable to correlate the measured quantities K_R and S with the properties in the crystals (particle size, shape, mean external and internal refractive index, surface reflection, absorption coefficient, K_T), so as to avoid being limited to purely phenomenological theories. The same applies also to particles of diameter comparable with λ, or smaller than λ, since in the theory of scattering the quantities characterising the particle, $2\pi r/\lambda$ and n/n_0 are involved in the case of single scattering (see p. 85).

To find this correlation, the layer of powder previously regarded as homogeneous can be regarded as composed of a large number of plane-parallel thin layers, of thickness equal to the mean particle diameter. This method has been used by a number of authors[178] to derive relations between diffuse reflectance of the layer and the properties of the particles. For this, we commence with Eqs. (78) and (79) for the transmittance and reflectance of several homogeneous parallel layers placed in series, as derived in connection with the continuum theory. It is now assumed that the layer consists of n individual thin layers of thickness equal to the mean diameter of the particles, then, in the case of the total layer, we have

$$T = \frac{T_1 T_2, \ldots, _n}{1 - R_1 R_2, \ldots, _n}, \tag{178}$$

$$R = R_1 + \frac{T_1^2 R_2, \ldots, _n}{1 - R_1 R_2, \ldots, _n}. \tag{179}$$

For $n \to \infty$, that is, for a layer of infinite thickness, we obtain $T_\infty = 0$, and

$$R_\infty = R_1 + \frac{T_1^2 R_\infty}{1 - R_1 R_\infty}, \tag{180}$$

---

[178] See, for example, *Stokes, G. G.*: Proc. Roy. Soc. London **11**, 545 (1860/62).— *Bodó, Z.*: Acta Phys. Acad. Sci. Hung. **1**, 135 (1951); — *Broser, I.*: Ann. Physik **5**, 401 (1950); — Z. Naturforsch. **6a**, 466 (1951); — *Johnson, P. D.*: J. Opt. Soc. Am. **42**, 978 (1952); — *Companion, A.*, and *G. H. Winslow*: J. Opt. Soc. Am. **50**, 1043 (1960); — *Bauer, G. T.*: Acta Phys. Acad. Sci. Hung. **14**, 209 (1962).

or, solving with respect to R_∞,

$$R_\infty = \frac{1 + R_1^2 - T_1^2}{2R_1} - \sqrt{\left(\frac{1 + R_1^2 - T_1^2}{2R_1}\right)^2 - 1}. \qquad (181)$$

To determine R_1 and T_1, *Bodó* assumes that the fraction α of the incident radiation is reflected from the individual particles in a regular way, while the fraction $1 - \alpha$ penetrates the crystal, and is attenuated by absorption to the fraction $(1 - \alpha)e^{-kd}$, if k represents the absorption coefficient of the material, and d the particle size. At the back of the crystal the fraction $1 - \alpha$ of the radiation still present is again regularly reflected, so that fraction $(1 - \alpha)^2 e^{-kd}$ passes through, etc. We thus obtain a geometrical series both for the total regular reflection and the transmitted radiation flux:

$$R_1 = \alpha + (1 - \alpha)^2 \alpha e^{-2kd} + (1 - \alpha)^2 \alpha^3 e^{-4kd} + (1 - \alpha)^2 \alpha^5 e^{-6kd} + \cdots$$
$$T_1 = (1 - \alpha)^2 e^{-kd} + (1 - \alpha)^2 \alpha^2 e^{-3kd} + (1 - \alpha)^2 \alpha^4 e^{-5kd} + \cdots$$

or

$$R_1 = \frac{\alpha \cdot \exp(kd) + (1 - 2\alpha) \exp(-kd)}{\exp(kd) - \alpha^2 \exp(-kd)}, \qquad (182)$$

$$T_1 = \frac{(1 - \alpha)^2}{\exp(kd) - \alpha^2 \exp(-kd)}. \qquad (183)$$

If these two values are inserted in (181), we obtain the measured quantity R_∞ of the layer as a function of α, k and d, that is of the properties of the individual particles. Similar expressions have been derived by other authors.

These equations can be tested by transmittance and reflectance measurements at various wavelengths of a colored glass which is subsequently pulverized and (for example by repeated sedimentation) divided into fractions of various particle sizes, the size within any fraction being as uniform as possible. We then obtain the quantities k and d, while α is extracted from Table 1 for diffuse irradiation. If the measured R_∞ of the powder is plotted as a function of $\log d$, and then evaluated from Eqs. (181), (182), and (183) as a function of $\log kd$, we obtain curves which, according to *Bodó*, can be approximately superimposed. Deviations of calculated and measured values don't seem to exceed $\pm 20\%$[179]. It must be remembered that the evaluation of R_1 and T_1 represents a very crude approximation, since neither the alteration in the effective path length through crystals of different shapes, nor the spaces between the particles, nor the partial total reflection at the back of the crystals, has been taken into consideration. It is possible that some of the neglected quantities may partially compensate each other.

---

[179] See also *ter Vrugt, J. W.*: Philips Res. Rep. **20**, 23 (1965).

A more recent attempt to correlate the measured quantity R_∞ for a powder with the properties of the particles making up the powder has been made by *Melamed* [180], who abandoned the idea of plane parallel layers and substituted them by a statistical summation with respect to the individual particles. The procedure for evaluation of R_∞ is as follows:

It is first assumed that the particles dealt with are identical spherical particles of diameter $d \gg \lambda$. Let their transmission [181] for a plane electromagnetic wave be t. It is assumed that diffuse reflection takes place at the surface of the sample, so that the Lambert Cosine Law applies. Let r_1 be the regular reflectance of the individual particle for diffuse irradiation from outside, and r_2 the "internal" regular reflectance of the particle, which takes account of the total reflection at the boundary surface between the particle and air (see p. 131) [182]. Let x be the fraction of the radiation which is scattered from the inside of the particle in a backward direction to the surface of the sample, and so contributes to the value of R_∞. In the case of isotropic scattering this represents simply the solid angle in the backward direction expressed as a fraction of 4π. If we resolve the sideways scattered radiation into its component x_b for backward scattering and x_f for forward scattering (compare p. 95), so that only two oppositely directed radiation fluxes are considered, we can evaluate x for weak absorption, for which $x_b \cong x_f$ as

$$x = \frac{x_b}{1-(1-2x_b)t}. \tag{184}$$

For particles arranged in the closest spherical packing, $x_b = 0.284$, corresponding to a solid angle $4\pi - \frac{\sqrt{3\pi}}{2}$. x may be regarded as the probability that the radiation emerging from a particle will be scattered in the direction of another particle lying one diameter nearer the surface of the sample.

For a particle in the surface of the sample, the radiation incident from outside will be reflected to the extent r_1 within the solid angle 2π instead of 4π, that is, $x_f = 0$, and x_b is approximately twice its value for particles inside the sample. For unit incident radiation, the initially reflected fraction is therefore $2xr_1$, which represents the contribution to R_∞, while the remainder, $1 - 2xr_1$, penetrates within the particle. Of this

---

[180] *Melamed, N. T.*: J. appl. Physics **6**, 34, 560 (1963).

[181] This transmission is valid for the case in which the particle is present in a non-absorbing medium of the same refractive index, so that there is no radiation loss through reflection at the phase boundaries (see p. 59).

[182] In the case of spherical particles no total reflection occurs as the radiation emerges from the inside, but the formulas obtained will later be extended to cover particles of any shape.

fraction, the subfraction $x(1-2xr_1)t$ contributes to R_∞, whereas the fraction $(1-x)(1-2xr_1)t$ penetrates further into the inside of the sample of reflectance R_∞, so that of this, the fraction

$$(1-x)(1-2xr_1)t R_\infty$$

is again reflected back to the surface particle, which will transmit the fraction $x(1-x)(1-2xr_1)(1-r_1)t^2 R_\infty$ on the direction of the surface. With the next reflection between the surface particle and the body of the sample, the fraction $x(1-x)(1-2xr_1)(1-r_1)t r_1^2 R_\infty^2$ is contributed to the value of R_∞, with the following reflection the fraction $x(1-x)(1-2xr_1)(1-r_1)t^2 r_1^2 R_\infty^3$, etc. We then have as the sum of these repeated reflections:

$$2xr_1 + x(1-2xr_1)t + x(1-x)(1-2xr_1)(1-r_1)t^2 R_\infty$$
$$+ x(1-x)(1-2xr_1)(1-r_1)t^2 r_1 R_\infty^2 + \cdots$$

which is a geometrical series giving

$$2xr_1 + x(1-2xr_1)t$$
$$+ x(1-x)(1-2xr_1)(1-r_1)t^2 R_\infty [1 + r_1 R_\infty + r_1^2 R_\infty^2 + r_1^3 R_\infty^3 + \cdots]$$
$$= 2xr_1 + x(1-2xr_1)t \qquad (185)$$
$$+ x(1-x)(1-2xr_1)(1-r_1)t^2 R_\infty/(1 - r_1 R_\infty).$$

Similarly, the contributions of the radiation fractions which at the second, third, and following reflections strike upon the surface particles, but are reflected back deeper in the inside, can be summed up with respect to their contribution to R_∞ [183]. The total so obtained is given by the series:

$$R_\infty = 2xr_1 + x(1-2xr_1 t)$$
$$+ x(1-x)(1-2xr_1)t[(1-r_1)t R_\infty/(1-r_1 R_\infty)]$$
$$+ x(1-x)^2(1-2xr_1)t[(1-r_1)t R_\infty/(1-r_1 R_\infty)]^2$$
$$+ x(1-x)^3(1-2xr_1)t[(1-r_1)t R_\infty/(1-r_1 R_\infty)]^3$$
$$+ \cdots$$
$$= 2xr_1 + \frac{x(1-2xr_1)t(1-r_1 R_\infty)}{(1-r_1 R_\infty) - (1-x)(1-r_1)t R_\infty}. \qquad (186)$$

This is the general expression for the absolute reflectance of an infinitely thick layer. R_∞ thus depends via r_1 on the refractive index of the particles, and via t on their absorption coefficient k, and contains no empirical

---

[183] In principle this procedure is therefore not essentially different from that used in summing the contributions of individual layers, each of the thickness of the particle.

168 IV. Theories of Absorption and Scattering of Tightly Packed Particles

parameter like S (scattering coefficient). To calculate the quantities t and r_1 in (186), the following postulates are made:

The transmittance of an individual spherical particle of diameter d is, from (II, 81), given by:

$$T = \frac{2}{(kc_0 d)^2} [1 - (1 + kc_0 d) e^{-kc_0 d}]. \tag{187}$$

In order to take into account the additional reflection at the phase boundary between the particle and air, an internal reflection coefficient r_2 is introduced (see p. 131). Assuming that the internal surface has the same reflecting properties throughout, at the first reflection the fraction $(1 - r_2) T$ emerges from the particle, while the fraction $r_2 T$ is reflected back within the particle. At the second reflection the fraction $(1 - r_2) r_2 T^2$ emerges, at the third fraction $(1 - r_2) r_2^2 T^3$, etc., so that we obtain for the transmittance of the particles, taking into account the infinitely large number of internal reflections, the series

$$t = (1 - r_2) T + (1 - r_2) r_2 T^2 + (1 - r_2) r_2^2 T^3 + \cdots$$
$$= \frac{(1 - r_2) T}{1 - r_2 T}. \tag{188}$$

The remainder, $1 - t$, is absorbed:

$$1 - t = \frac{1 - T}{1 - r_2 T}. \tag{189}$$

For small absorption, that is, small values of $kc_0 d$, the e-function of (187) may be expanded, and broken off at the third term, to give

$$T \cong 1 - \tfrac{2}{3} kc_0 d. \tag{190}$$

If the expression for t in (188) is inserted in (186) we obtain

$$R_\infty = \frac{2xr_1(1 - r_2 T) + x(1 - 2xr_1)(1 - r_1 R_\infty)(1 - r_2) T}{(1 - r_1 R_\infty)(1 - r_2 T) - (1 - x)(1 - r_1)(1 - r_2) T R_\infty}, \tag{191}$$

which is a quadratic in R_∞. Since always $0 < R_\infty < 1$, we only use the negative root. In measuring R_∞, for known values of c_0 and d, the absorption coefficient of the material in question can be calculated if the quantities r_1 and r_2 are also known.

The mean regular external reflectance r_1 for diffuse irradiation, using unpolarized light, is given by (II, 24), and depends on the value of the refractive index (see Table 1). To determine r_2, the related postulate is made:

$$r_{2\,(\text{diff. irrad.})} = 2 \int_0^{\alpha_g} \sin\alpha \, \cos\alpha \, f(\alpha, n) \, d\alpha + 2 \int_{\alpha_g}^{\pi/2} \sin\alpha \, \cos\alpha \, d\alpha. \tag{192}$$

Here α_g is the limiting angle for total reflection, which can be determined for powdered materials to a good approximation if the powder is inserted in media of various refractice indices and at various temperatures. It follows from (192) that

$$r_2 = (1 - \sin^2\alpha_g) + 2 \int_0^{\alpha_g} \sin\alpha \cos\alpha \, f(\alpha, n) \, d\alpha. \tag{193}$$

The dependence of r_1 and r_2 on n has been graphically illustrated by Melamed.

The equations are valid, to emphasize the point again, only for particles of diameter much greater than the wavelength, so that the laws for regular reflection can be used. But in the case of these particles it is doubtful whether the radiation reflected at the internal surface can still be regarded as diffuse. The evaluation of r_2 will therefore be the most important limiting factor in the general applicability of the statistical theory (see also p. 132).

Chapter V. Experimental Testing of the "Kubelka-Munk" Theory

a) Optical Geometry of the Measuring Arrangement

In order to test the applicability of the Kubelka-Munk theory as far as practical conditions are concerned, it is first necessary to investigate how far the assumptions involved in this theory can be fulfilled, and also whether or not the experimental measuring conditions limit or even exclude the applicability of the theory. For reasons discussed on p. 106, the equations of the Kubelka-Munk theory have been derived for diffuse incident irradiation. Integration over the hemisphere is also required for the detection of the transmitted and/or reflected radiation. In principle, suitable measuring devices using two integrating spheres are possible for measuring the transmission, but the energy conditions are very unfavorable, and such devices cannot be designed detecting the reflection. Therefore, we are practically limited to diffusely incident irradiation and directed detection or vice versa, or even to directed irradiation *and* detection. The devices available for these purposes (see also p. 222 ff.) consequently have different optical geometries.

In order to investigate the influence of this geometry on the measured results, the reflectance R'_∞ or R'_0 of different powders relativ to that of freshly prepared MgO as standard was measured by means of three devices [184]. These are the following ones: 1. Zeiss Spectrophotometer Modell PMQ II, with reflection attachment RA 2, with incident irradiation under 45° and with vertical detection ($_{45}R'_0$); 2. the same spectrometer as under 1. with integrating sphere, with diffuse incident irradiation, and with vertical detection ($_d R'_0$); 3. the Beckman spectral photometer DK 2 with integrating sphere, with vertical incident irradiation and diffuse detection ($_0 R'_d$). The measurements are recorded in Table 9. The various reflectance values tabulated for each group were obtained at different wave lengths.

The agreement of the results lies within the error limit of the methods. Only in case of the last three measurements, where noticeable absorption already exists, small systematic deviations arise between the $_{45}R'_0$ and $_0R'_d$ values. Since in these measurements the same angular distribution of the reflection is being assumed for the sample and the standard, it can be concluded that, in this reflection range at any rate, the measured

---

[184] *Kortüm, G.*, and *D. Oelkrug*: Z. Naturforsch. **19a**, 28 (1964).

Table 9. *Reflection measurements as a function of the optical geometry (MgO as standard)*

Substance	$_{45}R'_0$	$_dR'_0$	$_0R'_d$
1. CaF_2; R'_∞	0.970	0.978	0.980
	0.941	0.947	0.950
	0.904	0.906	0.912
2. CaF_2; R'_0	0.885		0.880
	0.856		0.853
	0.827		0.828
3. CaF_2; R'_0	0.897	0.899	
	0.846	0.845	
4. Coloured glass R'_∞	0.860		0.862
	0.821		0.825
	0.755		0.773
	0.691		0.708
	0.632		

results are practically independent of the optical geometry. Therefore, the measurements are suitable for the application of the Kubelka-Munk theory. This is probably the case because the "forward scattering", assumed to be in accordance with *Mie's* theory of single scattering, becomes lost in the transition to multiple scattering, since the latter no longer has a preferred direction of scattering (see also p. 99). Therefore, if we use directed incident irradiation, we can expect that, already after few single scattering processes, the scattering distribution is again isotropic. This observation should nevertheless not be generalized. In many cases, the reflectance with directed incident irradiation is markedly different from the reflectance with diffuse incident irradiation[185]. The reason of this is frequently that the surface is not sufficiently smooth so that shade effects occur in case of directed incident irradiation; other causes can be that the samples possess structure (e.g. textile fabrics), or that the powder is too coarse.

In case of reflectance measurements on dyed papers, *Stenius*[186] experienced something similar. When $R'_\infty > 0.6$, the results within the accuracy of measurement were nearly independent of the optical geometry of the devices used; below $R' \cong 0.6$ only the geometry $_dR'_0$ gave reflectance values which could be interpreted by the Kubelka-Munk theory. $_0R'_d$, $_0R'_{45}$, and $_{60}R'_d$ values showed deviations from the Kubelka-

---

[185] See for instance, *Helwig:* Licht **8**, 242 (1938). — *McNicholas, H. J.:* J. Res. Bur. Std. **1**, 29 (1928).

[186] *S: son Stenius, Å.:* J. Opt. Soc. Am. **45**, 727 (1955).

Munk theory which increased in the order given. This seems to point to the fact that, in case of low reflectance values, i.e., in cases of high absorption, significant doubts concerning the applicability of the Kubelka-Munk theory arise. We shall return to this point later.

The result obtained by *Stenius* is especially conspicuous, since the $_0R'_d$ reflectance values and the $_dR'_0$ reflectance values which should coincide in accordance with Helmholtz's law of reciprocity, do not do so. Various authors have disputed the validity of the reciprocity law for the case of diffuse reflection. Special doubt was cast on the law's validity in cases of partial reflections under certain angles (i.e. $_{\vartheta,\varphi}R'_d = {_dR'_{\vartheta,\varphi}}$). However, according to *Fragstein* [187] the validity of the relation $_dR'_0 = {_0R'_d}$ is beyond doubt. When, nevertheless, noticeable differences among these values are found, this can only be attributed to the fact that completely *diffuse* reflection never exists in practice, but that *regular portions* of surface reflection are always present (see p. 58 ff.). The regular reflection amounts to approx. 4% of the total for glass ($n = 1.5$) according to *Fresnel's* equations with vertical incident irradiation, but to approx. 9.2% when the incident irradiation is diffuse (see p. 12). For these reasons, it is obvious that the same scattering intensity cannot be expected where apparatus geometry is held constant and radiation geometry is reversed. A portion of regular reflection of the sample surface must, therefore, cause also a deviation from the Kubelka-Munk theory, and this does not depend upon whether or not directionally incident or diffusely incident irradiation is being employed. Further, differences in the reflectance values measured by means of different devices can be expected even if these devices operate on basis of an analogous radiation geometry since differences in distance and size of the photoelectric cell, i.e. differences in the solid angle of detection, can cause deviations.

This was established by an investigation in which 4 different gray samples (almost white, light gray, dark gray, and almost black) were tested for their spectral reflectance in the visible region at 30 different laboratories using 14 different apparatuses of various optical geometries [188]. The samples had flat surfaces and gloss, although steps were taken to eliminate regular surface reflection in so far as possible. For most of the experiments MgO served as the comparison standard although in some instances, other standards calibrated against MgO were used. The results of the measurements were divided into two groups depending upon whether or not they were directionally irradiated and observed (0/45 or 45/0) or diffusely irradiated or diffusely observed (0/d or d/0). In both groups deviations of the data to about 10% were observed. This deviation was partially due to the fact that the various MgO standards used had various

---

[187] Fragstein, C. v.: Optik **12**, 60 (1955).
[188] Robertson, A. R., and W. D. Wright: J. Opt. Soc. Am. **55**, 694 (1965).

absolute reflectances. When the R_∞-values were calculated relative to that of the lightest sample, the influence of the differences in absolute reflectance of the MgO standards dropped out, and the scatter in the data dropped by about 4%. In addition, the average values R'_∞ determined from 45/0 and 0/45 apparatuses turned out to be about 1% higher than those obtained from apparatuses with integrating spheres. This can be traced back to the observation that the surfaces of the samples (showing gloss) and of the MgO must show different strong deviations from the Lambert Cosine Law.

When a sample surface is irradiated with parallel light at the angle α one can frequently observe in the direction of the macroscopic reflection angle ϑ a "gloss" which is caused by partly regular reflection. As already discussed in Chap. II, from the purely physical point of view [189] this "gloss" can be defined as the regular component of the remitted light by assuming that a surface which shows "gloss" is composed of innumerable, microscopically small, regularly reflecting components which are inclined to the macroscopic surface in all possible angles (see p. 29). The division of the remitted radiation into a diffuse and a regular part can, according to *Stenius* [190], be described in a simple manner as follows:

A radiation beam I_0 falls on a surface which in part reflects diffusely and in part regularly. The fraction αI_0 may be reflected regularly, and the fraction $(1-\alpha)I_0$ may penetrate into the layer, and of this the fraction I may be diffusely reflected. Then we can write for the diffuse fraction of the reflectance

$$R_{\text{diff}} = \frac{I}{(1-\alpha)I_0}, \quad (1)$$

and for the measured total remission

$$R_{\text{exp}} = \frac{I + \alpha I_0}{I_0}. \quad (2)$$

When I is eliminated from both the equations, we obtain

$$R_{\text{exp}} = (1-\alpha) R_{\text{diff}} + \alpha. \quad (3)$$

In case of disappearing surface reflection, we get $R_{\text{exp}} = R_{\text{diff}}$. When a crystalline powder where the crystalline particles are still large compared

---

[189] Gloss is considered as both a physical phenomenon and a phenomenon which is partly physiological and partly psychological (see *Harrison, V. G. W.:* Definition of gloss. London: 1945). The best agreement with the subjective sensation of "gloss" is obtained when the ratio of directionally reflected light to diffusely reflected light is used for the objective measuring of "gloss" (see *Schlötterer, H.:* Metalloberfläche (Metallic surface) **16**, 49, 81, 141 (1962).

[190] *S: son Stenius, Å.:* Svensk Pappersstidning **54**, 701 (1951); **56**, 607 (1953).

with λ is concerned, the regularly reflected portion can be calculated in accordance with *Fresnel's* equations (see p. 12), and α lies between 4 and 9% according to the refractive index and the type of irradiation (directed of diffuse). If the crystalline particles are distributed statistically, α is constant over the whole of the hemisphere and is measured equivalently by every measuring arrangement. In Fig. 57, the Kubelka-Munk function

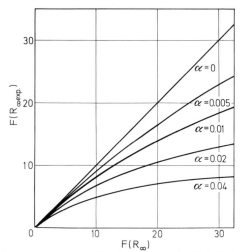

Fig. 57. Influence of regular reflection parts on $F(R_\infty)$ of absorbing powders

(IV, 32) for purely diffuse reflection is plotted as abscissa and $F(R_{\infty exp})$ as ordinate, for various values of α as parameter [191]. When $\alpha = 0$, we obtain a straight line at an angle of 45° of course. When α is increasing, we observe curves tending towards a limiting value, the magnitude of which is determined by α. For instance, we find from the diagram that the heights of two bands where $F(R_\infty)$ values have the ratio 10:30, give, in case of superimposed regular reflection of 2%, measured values which have the ration $0.65:1.3 = 1:2$ (see also p. 65). Only in the left lower part of the diagram the curves can still, as a first approximation, be considered as straight lines (with smaller slope), i.e. the Kubelka-Munk function can still be considered "valid".

For the increase of the measured remission due to the regular fraction, we obtain from equation (3) the following equation

$$R_{exp} - R_{diff} = \alpha(1 - R_{diff}). \qquad (4)$$

As can be seen, the influence of the regular reflection decreases with increasing reflectance, i.e., with decreasing absorption. As *Stenius* points

[191] Kortüm, G., and D. Oelkrug: Naturwiss. **53**, 600 (1966). The influence of $\varkappa$ on α has here been neglected; α increases rapidly in case of large absorption.

out, the applicability of the Kubelka-Munk theory is possible in case of hand-made paper which is submitted to directionally incident radiation, because it has a weak gloss and high remittance. However, when paper is colored and directionally incident radiation is used, deviations from the theory occur so that we are forced to resort to the more general Ryde theory for vertically incident irradiation (see p. 127).

Also in the case of diffuse irradiation, however, systematic deviations from the Kubelka-Munk theory can frequently be observed. To these belongs, for example, the observation that the scattering coefficient S of dyed papers decreases as absorbance increases. *Van Akker*[192] has attempted to explain the observed discrepancies by means of the scattering and absorbing characteristics of various hypothetical models in which the scattering and/or absorbing centers are unevenly distributed. The deviations from the Kubelka-Munk theory are therefore traced back to deficient isotropic scattering, to which irregularities in packing density and radiation loss through total internal reflection can also be added.

b) The Dilution Method

As already shown in the preceding section and in more detail in Chap. II, regular reflection parts can cause considerable deviations from the Lambert Cosine Law and thus also from the Kubelka-Munk theory which presupposes the Lambert Law. On p. 66 it has been shown in detail how the regular parts of the remitted radiation can be eliminated by using linearly polarized radiation and crossed polarization prisms, and how the diffuse part of the reflection can be measured separately. But this method has two considerable drawbacks as follows: a) the polarizer halves the irradiation strengths of the samples, and this is disadvantageous in case of low reflection (in absorption regions); b) the aperture angle even of Glan-Thompson prisms is relatively small so that it is necessary to use approximately parallel radiation.

Furthermore, the transmittance of both calcite prisms and so-called polarization foils (see also p. 231) is limited in the ultra-violet. But apart from the method just mentioned another method exists, namely the so-called "dilution method" which eliminates the regular part of the remitted radiation. In using the "dilution method", the powder under investigation is diluted (by grinding or mixing) with an inactive, non-absorbing standard (MgO, NaCl, $BaSO_4$, SiO_2, etc.) in such excess, that the regular part of the remission is eliminated within the measuring accuracy of the method when measured relatively against the same pure

---

[192] Compare *van Akker, J. A.*: Modern aspects of reflectance spectroscopy, p. 27ff. New York: Plenum Press 1968, and the references cited therein.

standard[193]. This method has stood the test very well and generally makes possible the experimental testing of the Kubelka-Munk theory under conditions adapted to the theoretical requirements.

Dilution can be carried out in various ways as follows: The sample is ground with an excess of the standard in a mortar or ball mill to such an extent that a homogeneous mixture is obtained. Two limiting cases exist here. Either we get a simple homogeneous mixture of the small crystals (e.g. Cr_2O_3 and MgO), or the sample is molecularly adsorbed on the surface of the standard, as is generally the case when organic solic materials are ground with inorganic standards. Therefore in such cases, one measures the reflectance spectrum of the *adsorbed* material, and this reflectance spectrum can be quite markedly different from the reflectance spectrum of the non-adsorbed, pure sample (the following contains examples of this; see p. 256 ff.). The same condition of the sample can be obtained when it is dissolved in an inactive solvent and is adsorbed from there on the standard. Finally, the sample can be adsorbed on the solid standard from the gaseous state. In all three cases, the same reflectance spectrum of the molecularly adsorbed material is then obtainable, if the standard surface is sufficiently large so that the sample is adsorbed molecularly. That means, the spectrum can depend on the degree of dilution. An example for this is shown in Fig. 58 which shows the spectrum of anthraquinone in diluted alcoholic solution as well as the spectra of anthraquinone at various dilutions adsorbed on NaCl[194]. The logarithm of the Kubelka-Munk function $F(R'_\infty)$ (IV, 32) divided by the mole fraction x is plotted as function of the wave number with various dilutions of the antraquinone on the NaCl as parameter. It is evident that, with increasing dilution, the reflectance spectrum of the antraquinone with the two main bands at $29,600 \text{ cm}^{-1}$ and $39,000 \text{ cm}^{-1}$ becomes sharper. At mole fractions below $3 \cdot 10^{-4}$, the spectrum becomes independent of the degree of dilution and resembles the spectrum of the alcoholic solution to a great extent. This is the spectrum of the single molecule adsorbed on the surface and is also obtainable when anthraquinone is adsorbed on NaCl from the diluted solution. Furthermore, it can be seen that the reflection spectrum of undiluted, pure anthraquinone does not resemble the "genuine" spectrum. Especially the short-wave bands are very greatly reduced, and this is (partly) a consequence of the regular reflection part[195].

---

[193] See *Kortüm, G.*, and *G. Schreyer*: Angew. Chem. **67**, 694 (1955).

[194] *Kortüm, G. W. Braun*, and *H. Herzog*: Angew. Chem. **75**, 653 (1963); Intern. Edition **2**, 333 (1963).

[195] Furthermore, a coupling of the π electron system of several molecules occurs here if the molecules lie in positions parallel to the axis of greatest polarizability [see also *Perkampus, H. H.*: Z. Physik. Chem. N.F. **13**, 278 (1957); **19**, 206 (1959)].

Generally, the dilution method offers several advantages in the investigation of reflectance spectra. These advantages are briefly as follows:

1. When the material to be tested is highly absorbing, the measuring range can be shifted to the region $0.1 < R < 0.7$ where the error of the measurements is at a minimum (as can be learned from an error consideration (see p. 251).

2. The scattering coefficient of the mixture is practically given exclusively by the scattering coefficient of the diluting agent, and the latter scattering coefficient of such non-absorbing or slightly-absorbing materials can always be measured [196].

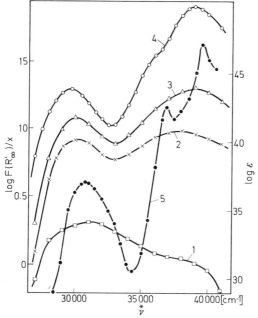

Fig. 58. Reflection spectra of anthraquinone (1) undiluted; (2) adsorbed on NaCl ($x = 1.26 \cdot 10^{-2}$); (3) $x = 5.0 \cdot 10^{-3}$; (4) $x = 1.9 \cdot 10^{-4}$; (5) transmission spectrum in diluted ethanolic solution

3. When measuring relatively against the pure diluting agent possible deviations from the isotropic scattering distribution can be eliminated so that the measurements become independent of the type of measuring arrangements.

4. Regular reflection parts can be neglected when measuring the diluted sample against the pure dilution agent.

---

[196] Kortüm, G., and D. Oelkrug: Z. Naturforsch. **19a**, 28 (1964).

5. When the material to be tested is adsorbed on excess standard in molecular dispersion, as it is normally done in case of organic materials and inorganic adsorbents, the particle-size dependence of the absorption coefficient can also be eliminated (p. 61). This phenomenon will be discussed again later.

c) Concentration Dependence of the "Kubelka-Munk" Function $F(R_\infty)$

For simple experimental testing of the Kubelka-Munk theory, the easily measurable "reflectance" R_∞ of a layer of infinite thickness and the Kubelka-Munk function $F(R_\infty)$ derived from it via Eq. (IV, 32) are especially suitable.

According to (IV, 31), R_∞ is a function of the ratio K/S and not of the individual values of K and S themselves. This is also valid, by analogy, for the rigorous theories of *Chandrasekhar* and *Giovanelli* (see p. 161). In this connection, it is assumed that scattering and absorption are continuous (the dimension of both K and S is $[\text{cm}^{-1}]$), while in reality, absorption and scattering occur only at particles embedded in a matrix. The absorption and scattering should thus be proportional to the concentration of particles at least when they are separated from one another by a sufficient distance. In this case we can write Eq. (IV, 31) $R_\infty = R_\infty \left(\dfrac{K_0 c}{S_0 c} \right) = R_\infty \left(\dfrac{K_0}{S_0} \right)$ which indicates that R_∞ should not depend on concentration. This has been established by measurements of *Blevin* and *Brown*[197] for white and colored pigment suspensions in water and terpentine oil. The relationship is also valid for powdered white standards in air under conditions of various packing densities prepared by subjecting the samples to compressions from 13 to 334 atmospheres. R_∞, especially with colorless substances, has proven to be constant within a measuring accuracy of 1% over an amazingly large concentration range (e.g., with TiO_2 in water suspension in the range from 0.01 to 30 volume per cent at 400 mµ). Similarly large regions of concentration independence of R_∞ in the air matrix were also observed and only at very high concentrations did R_∞ noticeably decrease. The authors assumed that at these concentrations the condition $S = S_0 c$ is not fulfilled any more.

Schatz[198] has reported, to the contrary, that R_∞ of powders in an air matrix is, to a considerable extent, pressure sensitive in the region from 20 to 4800 atm. Indeed, the diffuse reflectance of oxides (e.g., Al_2O_3, MgO, $BaSO_4$, CeO_2, etc.) noticeably decreases with increasing pressure

---

[197] Blevin, W. R., and W. J. Brown: J. Opt. Soc. Am. **51**, 129 (1961).
[198] Schatz, E. A.: J. Opt. Soc. Am. **56**, 389, 465 (1966); **57**, 941 (1967).

especially in the near infrared region. On the contrary, the regular part of the reflection increases slightly with increasing pressure. The observations can be explained by the assumption that by increasing the packing density, the portion of that part of the radiation totally reflected within the particle decreases. This again amounts to the scattering coefficient and concentration no longer being proportional.

Because K and S depend on wavelength in different ways, the expected and/or observed concentration independence of R_∞ is naturally valid only for monochromatic irradiation. The Kubelka-Munk function

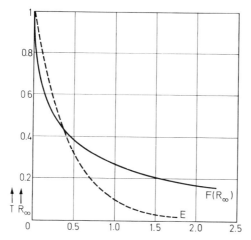

Fig. 59. $F(R_\infty)$ and absorbance E as function of R_∞ and transmittance T respectively

(IV, 32) must therefore likewise be a function of wavelength; it corresponds to the absorbance E in transmittance measurements. When $F(R_\infty)$ is plotted against R_∞ and parallel to it the absorbance E against the transmittance T of transmission measurements, we obtain Fig. 59. It is evident that with high reflectance, $F(R_\infty)$ rises much less steeply than E with high transmittances. Then the two curves intersect, and $F(R_\infty)$ approaches the value ∞ much slower than E does.

When we examine $F(R_\infty)$ as a function of the wave number in a dilution series in a concentration range where no mutual influences among the adsorbed molecules are possible (large surface area of the standard), we get a series of parallel logarithmic curves reproduced in Fig. 60 [199]. These curves refer to flavazin (pyrazolone dye) which was adsorbed on starch from solutions of various concentrations. The (apparent) surface concentrations c in mole/gramm adsorbent can be

---

[199] Schwuttke, G.: Z. Angew. Physik **5**, 303 (1953).

derived by calculation from the concentrations of the initial solutions and final solutions after centrifugation. The fact that these curves coincide when shifted parallel in the direction of the ordinate, constitutes an experimental confirmation of the validity of the Kubelka-Munk function

$$F(R'_\infty) \equiv \frac{(1-R'_\infty)^2}{2R'_\infty} = \frac{K}{S}, \tag{5}$$

since the scattering coefficient S is solely determined by the standard (starch) which is the same for all dilution series. In view of the stipulated preconditions of low concentration and large surface of the standard, the absorption coefficient K should be proportional to the concentration of the adsorbed material in the same way as the Lambert-Beer Law is also valid in solutions. When $F(R'_\infty)$ for a given wavelength λ is plotted against the surface concentration c, we obtain the straight lines of Fig. 61, which show that in fact $F(R'_\infty)$ is proportional to the surface concentration of the adsorbed material. In this case, the Kubelka-Munk function can be written as follows:

$$F(R'_\infty)_\lambda \sim \frac{\varepsilon c}{S} \quad \text{or} \quad \log F(R'_\infty) = \log c + C. \tag{6}$$

The concentration should really be given in mole/l, but with sufficiently high dilution mole/gramm adsorbent or mole fraction can also be used, since under these conditions the various concentration data are proportional to one another.

Various authors have carried out analogous measurements. A further example for the validity of Eq. (6) and also for the obtainable accuracy of photometric concentration determination, possible by means of such measurements (see also p. 250), is to be found in Fig. 62. Here the concentration dependence of $F(R'_\infty)$ according to more recent measurements[200] for pyrene adsorbed on dry NaCl at wave number $\tilde{v} = 29{,}500$ cm^{-1} (the maximum of the longwave absorption band) is shown. The mole fraction x of the pyrene is used as abscissa. Up to $x = 2 \cdot 10^{-4}$ the curve is linear, and than bends. Where this takes place, depends essentially on the grain size and with it on the surface of the adsorbent. In case of very fine grain, the linear range can extend up to $x \simeq 10^{-3}$. Bending of the curve indicates that the surface of the adsorbent approaches saturation by the first mono-molecular covering layer (mono-layer coverage). On the other hand, the saturation concentration allows determination of the specific surface area of the adsorbent (see also p. 272). The mean error in the measured data around the Kubelka-Munk straight line here amounts to approx. 3% and depends on the humid-

---

[200] *Kortüm, G.*, and *W. Braun*: Z. Physik. Chem. N.F. **28**, 362 (1961).

ity content of the samples. Under favorable conditions, the error can be reduced to approx. 2%.

Eq. (6) corresponds to the Lambert-Beer Law $E = \varepsilon c d$ for transmission measurements. Both here and there the law concerned is a limiting law for high dilution. The validity range of this law differs from case to case.

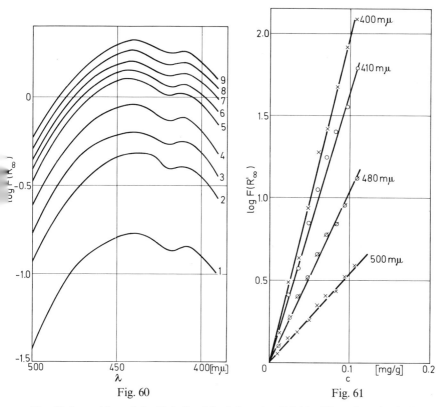

Fig. 60. Logarithm of the Kubelka-Munk function $F(R'_\infty)$ of flavazin, adsorbed on starch, at different surface concentrations from about 0.01 to 0.1 mg/g adsorbent

Fig. 61. Concentration dependence of the function $F(R'_\infty)$ in Fig. 60 at various wavelengths

Monochromatic irradiation is important, since both ε and S are wavelength-dependent. It is further assumed that the "dilution" practically eliminates the regular part of the reflection that may possibly exist.

Strictly speaking, the *absolute* reflectance R_∞ instead of the measured relative reflectance R'_∞ has to be put into Eq. (6). Only if the absorption of the dilution agent and/or of the adsorbent itself is very low, one can observe the linearity which ought to exist, according to Eq. (6), between $F(R'_\infty)$ and c at a given wavelength (Fig. 62). If either the dilution agent or the adsorbent has a noticeable absorption of its own, this absorption, as it has been shown on p. 150, will not be eliminated when the diluted sample is measured against the pure diluting agent as standard since no

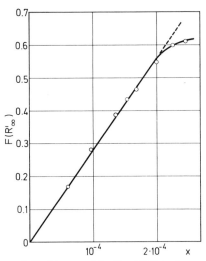

Fig. 62. Kubelka-Munk limiting straight line $F(R'_\infty)$ of pyrene, adsorbed on NaCl, at $\tilde{v} = 29{,}500 \text{ cm}^{-1}$ as function of the mole fraction

logarithmic connection between absorption coefficient and reflectance exists here, as it exists between extinction modulus εc and transmittance. Therefore, when the sample P to be measured is diluted by means of an absorbing diluting agent V, the reflectance of V (the diluting agent) has to be measured separately against a standard (e.g. MgO or another substance) whose absolute reflectance is known. According to Eq. (IV, 144) we can then write

$$F(R_{\infty, P+V}) = \frac{K_P + K_V}{S} = F(R_{\infty, P}) + F(R_{\infty, V}) \tag{7}$$

or

$$F(R_{\infty, P}) = \frac{K_P}{S} = F(R_{\infty, P+V}) - F(R_{\infty, V}).$$

In connection with this, it is assumed that S is practically determined by the diluting agent present in excess. When the directly measured $F(R'_\infty)$ is plotted against c_{sample}, the ratio $F(R'_\infty)/F(R_{\infty, p})$ is no longer a

Concentration Dependence of the Kubelka-Munk Function $F(R_\infty)$ 183

constant but a function of R'_∞ and $R_{\infty,V}$. If we substitute into

$$\frac{F(R'_{\infty,P+V})}{F(R_{\infty,P})} = \frac{F(R'_{\infty,P+V})}{F(R_{\infty,P+V}) - F(R_{\infty,V})} \qquad (8)$$

the explicit expressions with the respective reflectances and rearrange we obtain the equation

$$F(R'_{\infty,P+V}) = F(R_{\infty,P}) \cdot \frac{R_{\infty,V}(1 - R'_{\infty,P+V})}{1 - R'_{\infty,P+V} R^2_{\infty,V}} \, . \qquad (9)$$

When $c_P \to 0$ (or when $R'_{\infty,P+V} \to 1$), $F(R'_{\infty,P+V})$ asymptotically approaches the origin of the coordinates with the slope "zero" and only becomes proportional to $F(R_{\infty,P})$ if $R'_{\infty,P+V} \to 0$:

$$\lim_{R'_{\infty,P+V} \to 0} F(R'_{\infty,P+V}) = R_{\infty,V} \cdot F(R_{\infty,P}) \, . \qquad (10)$$

These connections can be visualized by means of a model test [201]. Cr_2O_3 was diluted with an excess of CaF_2 powder deliberately made impure by adding a little graphite, and was then measured at various concentrations, first against the diluent as standard and second against freshly prepared MgO as standard. Fig. 63 shows the measured results. In the first case a bent curve was obtained, and in the second case a straight line which intersects the ordinate at a point corresponding to $F(R_{\infty,CaF_2})$ and would go through the coordinate origin after deducting this.

As the considerations show, one should exclusively use, when the Kubelka-Munk function is employed quantitatively, the *absolute* reflectance which, in accordance with Eq. (IV, 144), can be derived by calculations from the two measurements: sample + standard against standard and standard against freshly prepared MgO [202, 203]. Even in case of highly reflecting diluents or adsorbents their own absorption cannot always be neglected as the results from the following calculation make evident [204]: According to Eq. (IV, 144) the Kubelka-Munk function of the sample to be measured is given by

$$F(R_{\infty,P}) = F(R_{\infty 2}) - F(R_{\infty 1}) \, .$$

Instead of this, the relative measurement against the pure diluent gives the equation:

$$F(R'_\infty) = \frac{(1 - R'_\infty)^2}{2R'_\infty} \, . \qquad (11)$$

---
[201] Kortüm, G., and D. Oelkrug: Naturwiss. **53**, 600 (1966).
[202] Fujimoto, M., and G. Kortüm: Ber. Bunsenges. **68**, 488 (1964).
[203] Kortüm, G., and V. Schlichenmaier: Z. Physik. Chem. N.F. **48**, 267 (1966).
[204] Kortüm, G., and W. Braun: Z. Physik. Chem. N.F. **48**, 282 (1966).

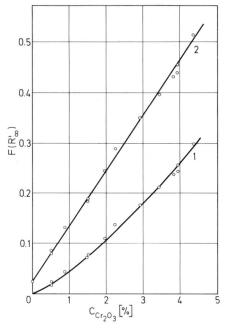

Fig. 63. Relative reflectance of a mixture of Cr_2O_3 with impure CaF_2 in excess, measured against (1) the diluent, (2) freshly prepared MgO in the band maximum at 16,000 cm^{-1}

Consequently the deviation between these two is

$$F(R'_\infty) - F(R_{\infty,p}) \equiv \Delta F(R'_\infty) = F(R'_\infty) - F(R_{\infty 2}) + F(R_{\infty 1}). \qquad (12)$$

The relative error is

$$\frac{\Delta F(R'_\infty)}{F(R'_\infty)} = 1 - \frac{F(R_{\infty 2})}{F(R'_\infty)} + \frac{F(R_{\infty 1})}{F(R'_\infty)}, \qquad (13)$$

where

$$F(R_{\infty 2}) = \frac{(1 - R'_\infty R_{\infty 1})^2}{2 R'_\infty R_{\infty 1}} \qquad (14)$$

and

$$F(R_{\infty 1}) = \frac{(1 - R_{\infty 1})^2}{2 R_{\infty 1}}. \qquad (15)$$

R'_∞ from Eq. (11) amounts to

$$R'_\infty = 1 + F(R'_\infty) - \sqrt{[1 + F(R'_\infty)]^2 - 1}. \qquad (16)$$

The characteristic absorption of the diluent ($R_{\infty 1} \neq 1$) governs the relative error in $F(R'_\infty)$ and by means of the Eqs. (13), (14), (15), and (16) the error can be calculated as a function of $F(R'_\infty)$ and with $R_{\infty 1}$ as parameter. Such a group of curves is reproduced in Fig. 64. It is obvious that the relative errors in case of low diluent absorption (e.g. $R_{\infty \text{NaCl}} = 0.95$ to 0.98) are only of consequence in case of small $F(R'_\infty)$ values, whereas in the case of stronger characteristic absorption (e.g. SiO_2, Al_2O_3, etc.), large $F(R'_\infty)$ values are likeweise considerably biassed.

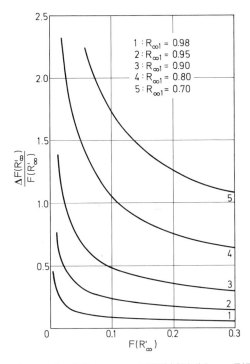

Fig. 64. Dependence of relative error $\Delta F(R'_\infty)/F(R'_\infty)$ on $F(R'_\infty)$ with $R_{\infty \text{ standard}}$ as parameter

The concentration dependence of $F(R'_\infty)$ already depicted in Fig. 63 can also be analytically described by means of the above-mentioned equations as follows [205].

$$F(R'_\infty) = \frac{[R_{\infty 1} - f(x)]^2}{2 R_{\infty 1} f(x)} \qquad (17)$$

with

$$f(x) \equiv 1 + ax + F(R_{\infty 1}) - \sqrt{2[ax + F(R_{\infty 1})] + [ax + F(R_{\infty 1})]^2}. \qquad (18)$$

---
[205] Kortüm, G., and W. Braun: l.c.

V. Experimental Testing of the Kubelka-Munk Theory

In the last equation, x designates the mole fraction or the concentration (e.g. in mole/g diluent or in weight-%) and a stands for the slope of the straight line $F(R_\infty) = ax$ (2 in Fig. 63). Fig. 65 shows this relationship diagrammatically. We observe a change in slope of the curves which is greater the smaller $R_{\infty 1}$ (i.e. the larger the absorption of the diluent). Furthermore, it can bee seen that, as in Fig. 64, the influence of the adsorbent reflectance which deviates from 1 is especially significant when the sample concentration is low, as would be expected.

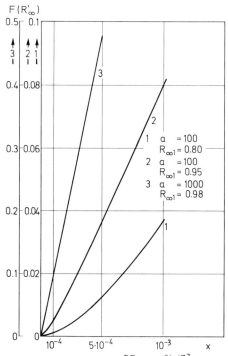

Fig. 65. Calculated function $F(R'_\infty) = \dfrac{[R_{\infty 1} - f(x)]^2}{2 R_{\infty 1} \cdot f(x)}$ with $R_{\infty 1}$ as parameter

d) The Typical Color Curve

A more stringent verification of Eq. (IV, 32) would be possible if the absorption coefficients K were already known from transmission measurements at various wavelengths. As already made obvious from Fig. 58, the K values cannot be derived from the transmittance spectra of the dissolved material of interest since these values are influenced by the respective surroundings of the absorbing substance. Therefore, these transmittance values have to be measured under conditions analogous to those for reflectance measurements. Furthermore, "diluted" substances should be used so that the interference of regular reflection parts can be neglected by relative measurement against the standard. Consequently for such measurements mixed crystals for instance, which have one excess component that does not absorb, should be suitable whose transmittance and reflectance spectra can both be measured. Mixed crystals made up of $KClO_4$ and 0.17 mole-% $KMnO_4$ are suitable for this purpose. Remission curves for various particle sizes have already been shown in

Table 10. *Absorption spectra of* $K(Mn, Cl)O_4$ *mixed crystals at a temperature* $T = 83°\,K$

	1. Transmission						
k	14.5	22.4	20.1	14.1	7.7	4.15	—
$\overset{*}{\nu}$	18,056	18,821	19,594	20,359	21,112	21,890	22,650
$\Delta \overset{*}{\nu}$		765	773	765	753	780	760
	2. Reflection						
$\overset{*}{\nu}$	18,040	18,800	19,570	20,340	21,100	21,880	22,640
$\Delta \overset{*}{\nu}$		760	770	770	760	780	760
$\log \dfrac{1}{R'_\infty}$	1.27	1.45	1.41	1.27	1.05	0.82	—

Average error of $\overset{*}{\nu}$ approx. $20\,\mathrm{cm}^{-1}$.

Fig. 25a, and transmittance spectra have also been measured by various authors [206]. Table 10 shows the position of the main absorption maxima of these mixed crystals and their absorption coefficients k. The main absorption maxima were determined by both methods, and the absorption coefficients k were derived from transmittance measurements.

The Table shows that both measurements agree within the limit of error as far as the position of the bands is concerned. When the $F(R'_\infty)$ values which are derived from the apparent extinctions $\log(1/R'_\infty)$ measured in reflection, are plotted against the k values ascertained by transmission measurements, the straight line reproduced in Fig. 66 is obtained. If accordingly $F(R'_\infty)$ is proportional to the true absorption coefficient k at various wavelengths, this must mean that the scattering coefficient S in Eq. (5) is independent of λ in this wavelength range, i.e.. the reflectance

---

[206] Schnetzler, K.: Z. Physik. Chem. (B) **14**, 241 (1932). — Teltow, J.: Z. Physik. Chem. (B) **40**, 397 (1938); **43**, 198 (1939).

measurement furnishes the so-called *Typical Color Curve* of the substance concerned, and this curve coincides with the true absorption spectrum through parallel shifting in the ordinate when $\log F(R'_\infty)$ is plotted against λ or $\overset{*}{\nu}$. Once more attention should be drawn to the fact that two specifications must be fulfilled in this case as follows:

first, the validity of the Kubelka-Munk function, i.e., diffuse scattering and with this "diluted" systems;

second, within the spectral range investigated, the scattering coefficient S must be independent of the wavelength, i.e., that λ is small in comparison with the dimensions of the scattering particles (see also p. 206).

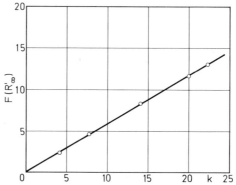

Fig. 66. Testing the Kubelka-Munk function by means of transmittance and reflectance measurements on the same mixed crystals

Both conditions are met by a coarse powder (1 to 2μ) of colored glass [207] in which the rare earth coloring ions are highly diluted in basic glass mass [208]. The transmittance of the glass was measured against a quartz plate having the same thickness. The glass was then pulverized in an agate mortar, and its reflectance measured against ordinary glass used as standard and processed in the same way. Fig. 67 shows that the logarithmically plotted transmittance spectrum and the logarithmically plotted reflectance spectrum cover each other to a large extent, so that the "typical color curve" was also obtained here. Small systematical deviations in the short-wave part of the spectrum can be attributed to a small λ dependence of the scattering coefficient, as will be pointed out later. These measurements show (as already Table 10) that, under suitable conditions, reflectance spectroscopy is able to furnish results which equal those of transmittance spectra.

[207] "Didym" filter glass "BG 36" of Schott & Gen. Mainz, West Germany.
[208] *Kortüm, G., W. Braun,* and *G. Herzog:* Angew. Chem. **75**, 653 (1963). — Intern. Ed. **2**, 333 (1963). — *Oelkrug, D.:* Dissertation, Tübingen 1963.

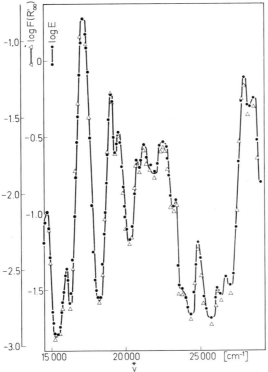

Fig. 67. Transmission spectrum ●—● and reflection spectrum △—△ of a Didym filter glass

e) Influence of Cover Glasses

One of the main applications of reflectance spectroscopy is observing spectra of adsorbed substances (see p. 256 ff.). It was found that substances adsorbed on active surfaces frequently react readily with oxygen or water vapor. In such case the measuring samples have to be covered by quartz or glass plates to exclude external influences, i.e., we have here a system consisting of two homogeneous layers (Fig. 53), whose reflectance R', referred to a standard, was previously calculated (see also p. 135). In order to check the result of this calculation, the reflectance R'_∞ of Cr_2O_3 diluted with CaF_2 or MgO was measured against MgO as standard, at various wavelengths, at various degrees of dilution, and with directionally incident irradiation ($\alpha = 45°$); the same samples were used once without and once with quartz cover plates[209]. The results are shown in Fig. 68 a. The $F(R'_\infty)$ values with quartz cover plates always lie above

---

[209] *Kortüm, G.*, and *D. Oelkrug*: Naturwiss. **53**, 600 (1966).

the $F(R_\infty)$ values without quartz cover plates. The cover plates undoubtedly reduce the reflectance of the sample; however, the position of the maxima is maintained. In accordance with Eq. (IV, 115a) we obtain for the Kubelka-Munk function for the case of infinite layer thickness

$$F(R'_{\infty\,exp}) = \frac{\left(1 - \dfrac{R_{\infty 3}(1 - R_{12})}{1 - R_{12}R_{\infty 3}}\right)^2}{\dfrac{2R_{\infty 3}(1 - R_{12})}{1 - R_{12}R_{\infty 3}}} = \frac{F(R_{\infty 3})}{(1 - R_{12}R_{\infty 3})(1 - R_{12})}. \quad (19)$$

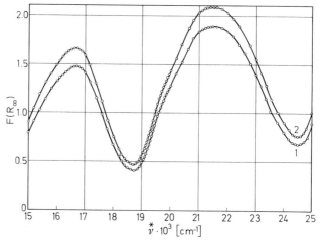

Fig. 68a. Reflection spectrum of Cr_2O_3/MgO, compared with MgO as standard without (1) and with (2) quartz cover plate

Here R_{12} is the reflection of the quartz cover plate with diffusely incident irradiation (see also p. 13), given by Eq. (II, 25). When the two functions $F(R'_\infty)$ with cover plate and $F(R_\infty)$ without cover plate derived directly from measured values are plotted against each other, a very good approximation to a straight line, but with a slope of less then 45° is obtained. When the difference $F(R'_\infty)$ (with cover plate) $- F(R_\infty)$ (without cover plate) is converted to absolute values by calculation and then plotted against $F(R_\infty)$ (without cover plate), we get the points shown in Fig. 68b. The drawn curve was calculated by means of Eq. (19), and the best agreement with the measured points here results, if R_{12} is made 0.10 instead of 0.155 as can be expected from Eq. (II, 25) for $n_{quartz} = 1.5$. This means the measurements confirm the theory qualitatively but not quantitatively. Judd[210] has already drawn attention to this and described various

[210] Judd, D. B., and K. S. Gibson: J. Res. Natl. Bur. Std. **16**, 261 (1936).

possibilities for lowering the theoretically calculated R_{12}. If we consider that at large angles of incidence R_{12} becomes considerable (at grazing incidence $R_{12} = 1$), it is probable that part of the radiation leaves the sample laterally or is absorbed by the sample dish. This reduces the influence of the cover plate.

As can be concluded from Eq. (19) for low absorption $(R_{\infty 3} \to 1)$

$$F(R_{\infty 3})_{\text{with cover plate}} = \frac{F(R_{\infty 3})_{\text{without cover plate}}}{(1 - R_{12})^2} \tag{20}$$

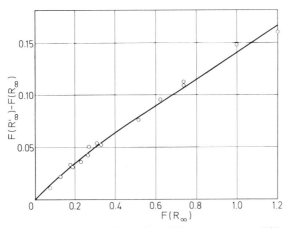

Fig. 68 b. Reflection spectrum of Cr_2O_3: $F(R_\infty)_{\text{with cover glass}} - F(R_\infty)_{\text{without cover glass}}$ as function of $F(R_\infty)$

and for high absorption $(R_{\infty 3} \to 0)$

$$F(R_{\infty 3})_{\text{with cover plate}} = \frac{F(R_{\infty 3})_{\text{without cover plate}}}{1 - R_{12}}. \tag{21}$$

Consequently the Kubelka-Munk function of a covered sample has a larger slope for low values of $F(R_\infty)$ than for high values of $F(R_\infty)$, although Fig. 68 b shows that the deviations from a straight line are small. From these measurements it can be seen, that reflectance spectra should be measured without covering glasses if no special reasons require their use.

f) Scattering Coefficients and Absorption Coefficients

The Kubelka-Munk theory presupposes monochromatic radiation and therefore does not consider that the scattering coefficient depends on the wavelength. Since multiple scattering of densely packed particles of various shapes and sizes is concerned here, even the Mie theory of simple scattering can only serve as guidance for the dependence of S on λ. For this reason, the attempts to establish the relationships between the scattering coefficient and λ must be based on experiments. The Kubelka-Munk function $F(R_\infty)$ already allows us to estimate the scattering coefficients when the absorption coefficients are derived from simultaneous transmittance measurements as exemplified by the case of the filter glass mentioned on p. 189. Since S simultaneously depends on the grain

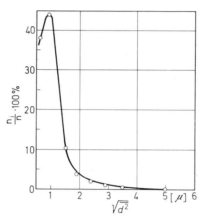

Fig. 69. Grain size distribution of ground Didym filter glass with $\sqrt{\overline{d^2}} = 1.1$

size, it is necessary to produce samples having a grain size range as narrow as possible that can be measured microscopically. Fig. 69 shows such a grain size range of ground "Didym" filter glass; at least 500 particles were measured for this distribution function [211]. The average grain size was derived from it by means of the equation $\sqrt{\overline{d^2}} = \sqrt{\Sigma n_i d_i^2 / \Sigma n_i}$. When the reflectance spectra of a number of such samples having varying values of $\sqrt{\overline{d^2}}$ are measured under conditions as uniform as possible (same spectral band width, same standard, etc.), and when the absorption coefficients K are derived from the simultaneously measured transmittance spectrum [212] of Fig. 67, we obtain the dependence of the scattering coefficient S on the wave number at various grain sizes or, conversely, the dependence of the scattering coefficient on the grain size at a given wavenumber. Fig. 70 and Fig. 71 shows these relationships as follows: the scattering coefficient is inversely proportional to the

---

[211] *Kortüm*, G., W. *Braun*, and G. *Herzog*: Angew. Chem. **75**, 653 (1963); Intern. Ed. **2**, 333 (1963).

[212] Because of Eq. (IV, 16) the K_T values derived from transmittance measurements must be doubled.

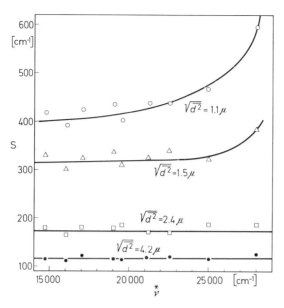

Fig. 70. Dependence of the scattering coefficient S on the wave number at various grain sizes, measured on glass filter powder

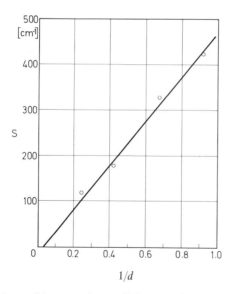

Fig. 71. Dependence of the scattering coefficient S on the average grain size, measured on glass filter powder

average grain size, (a result observed several times[213]); in case of coarser powders the scattering coefficient is, in the investigated spectral range, independent of $\tilde{v}$, but in case of finer powders S distinctly increases with the wavenumber although the measurements here are uncertain. As already pointed out previously, this of course causes a small distortion of the "typical color curve" shown in Fig. 67.

This method for ascertaining scattering coefficients by means of K_T values derived from transmittance spectra is of course applicable only if these spectra are accessible without difficulties. Ordinarily this is not possible, except in case of glasses or mixed crystals. Apart from all this, this method only provides estimated values of the scattering coefficients even if the factor 2, conditioned by Eq. (IV, 16), is taken into consideration. No doubt, proportionality between K_R and K_T can be expected when the absorbing molecules and ions in each case have the same state in reflection measurements and transmission measurements (as it is the case when colored glasses are being used), but the above-mentioned factor of 2 has been derived under the ideal conditions of a homogeneous medium (K and S are everywhere constant) and an isotropic scattering distribution. In fact areas of large K (substance) alternate with areas of small K (air), i.e., the packing density is lower in comparison with the one in the transmittance measurements. Moreover, the path really travelled within the powder is changed in an undefined manner due to total reflection at the small crystals. Consequently it is possible that the theoretically calculated factor 2 is not in keeping with the actual conditions.

An experimental comparison between K_T and K_R can be carried out as follows[214]: Analogous to Fig. 67, the transmittance spectrum and the "typical color curve" in reflection of a colored glass (BG 24 made by *Schott*) were measured, the average grain size being approx. 5μ. Both spectra nearly coincide. As will be described below, the scattering coefficients are derived from R_∞ and from the transmittance of a thin layer of powder. However, S could not be determined in the relevant absorption range since there the transmittances were too low for evaluation. Therefore, measurements were only made in the regions adjacent to the bands, i.e., in the region from 10,000 to 14,500 cm^{-1} and in the region from 22,500 to 28,000 cm^{-1}. S was easily interpolated between them because of its small wavelength dependence[215]. K_R values calculated on the basis of

---

[213] See also G. *Kortüm*, and P. *Haug*: Z. Naturforsch. **8a**, 372 (1953).

[214] *Kortüm, G.*, and D. *Oelkrug*: Naturwiss. **53**, 600 (1966).

[215] Great significance cannot be attached to this interpolation because, as with Mie-scattering, the scattering coefficient will depend on $m = n/n_0$, the ratio of the refractive indices of the particles and the matrix, and n has anomalous dispersion in the absorption region (see p. 307).

Eq. (IV, 32) with these S values can then be compared with K_T values derived from transmittance measurements (Table 11).

The ratio K_R/K_T found here is larger than the theoretical value of 2 and is constant in the range from 15,000 to 22,000 cm^{-1}. How far this phenomenon is based on causes discussed above or how far it has to be attributed to an incomplete isotropic scattering distribution, cannot be decided definitely. A ratio $K_R/K_T < 2$ should be more probable since the sample is less densely packed in reflection than in transmission. On

Table 11. *Scattering coefficients and absorption coefficients of the colored glass BG 24*

$\tilde{v} \cdot 10^{-3}$ [cm^{-1}]	S [cm^{-1}] (interpolated)	K_R [cm^{-1}]	K_T [cm^{-1}]	K_R/K_T
14.5	468	4.8	1.67	2.9
15	471	18.2	6.35	2.9
16	474	65.3	24.73	2.6
17	477	81.9	29.93	2.7
18	479	61.7	22.67	2.7
19	482	55.2	20.12	2.8
20	485	37.2	13.45	2.8
21	488	25.2	9.02	2.8
22	490	11.5	3.93	2.9

the other hand, it is fairly certain that at a grain size of approx. 5 μ, an isotropic scattering distribution no longer exists for the single scattering process. The scattering is partly caused by reflection at the grain boundaries. Furthermore, total reflection partly takes place when the radiation leaves the crystal/air interface so that the path of light is lengthened into what is equivalent to an apparent increase in K. In any case, the fact that the K_T values measured in transmittance are proportional but not identical to K_R values measured in reflection must be taken into account.

All examples cited thus far and concerning the testing of the Kubelka-Munk theory are based on the condition of "infinite layer thickness". But a stricter testing of this theory is only possible if the scattering coefficients and the absorption coefficients are measured separately, e.g. exclusively by reflection measurements. For that purpose, the so-called "scattering power" Sd has to be measured as function of R_0 (reflectance before black background) and of R_∞ (reflectance in case of infinite layer thickness) as can be learnt from the diagrams calculated via the Kubelka-Munk theory (e.g. see Fig. 48). For calculating these diagrams the

Eqs. (IV, 43) and/or (IV, 56) which are equivalent, are used. Previous investigations of this kind were mainly carried out by *Dreosti*[216] and *Judd*[217] who both used enamel, white paint, silicates, opaline glass, and mastic emulsions for their investigations. Subsequently, a number of newer investigations[218] will be discussed which will simultaneously give particulars regarding the dependence of the scattering coefficients on wavenumber and grain size.

Measurements of R_0 and R_∞ were performed on calcium fluoride powders of various layer thicknesses with MgO as comparison standard

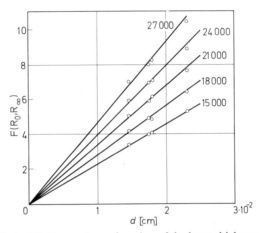

Fig. 72. $F(R_0, R_\infty)$ of CaF$_2$ powder as function of the layer thickness d with various wave numbers as parameter

under directed irradiation. For measuring R_0 the powders were filled into black sample dishes having depths of 0.1 to 0.3 mm, and were rolled as level as possible. The layer thickness was determined by means of a microscope. The reflectance of black background amounted to approx. 0.02 and could be neglected. The measured relative values R' were converted into absolute values R by means of the ϱ-values of MgO (see p. 146). When the function $F(R_0, R_\infty)$ derived by calculation from these measurements is plotted, according to Eq. (IV, 56), against the layer thickness d used, a straight line is obtained for each given wavenumber. This line intersects the origin (zero) and has a slope which gives the scattering coefficient S. This amounts to a verification of Eq. (IV, 56) and confirms the previous experience that the directionally incident irradia-

---

[216] *Dreosti, G. M.:* Dissertation, Utrecht 1930.

[217] *Judd, D. B.:* J. Res. Natl. Bur. Std. **19**, 287, 317 (1937).

[218] *Oelkrug, D.:* Dissertation, Tübingen 1963.

tion is not noticeable since near the surface an isotropic scattering distribution already exists (see also p. 171). These measurements are shown in Fig. 72. S, the slope of the straight line, depends here on the wavenumber to a very large extent.

By means of Eq. (IV, 61), the scattering power Sd can also be derived from R_∞ and from the transmittance T. In order to test the validity of Eq. (IV, 61), filter papers having a constant layer thickness $(150 \pm 2 \mu)$[219] were used for model tests. Fifteen of these papers placed upon one another served for measuring R_∞, whereas T and R_0 were each measured by 1 to 5 papers placed one behind the other[220]. Black velvet was used as background for the R_0 measurements, and its reflectance was neglected since it only amounted to 0.005. The directed irradiation was at an angle of 0°. The functions $F(R_0, R_\infty)$ and $F(T, R_\infty)$ derived by calculation from the measurements are, in Fig. 73, plotted against the number of papers placed one after another. Likewise here we obtain a straight line intersecting the coordinate origin, i.e., both measuring methods lead to the same scattering coefficient S given by the slope. From these observations we conclude that directionally incident irradiation again did not bias the measured results in case of these somewhat rough and lustreless papers. Such biases would be noticeable by deviations of the data from the straight line.

Stenius [221] obtained somewhat different results when testing handmade papers. He used the Eq. (IV, 37) and (IV, 40) for deriving S by

---

[219] *Kortüm, G.*, and *D. Oelkrug: Z. Naturforsch.* **19a**, 28 (1964).

[220] In accordance with Eqs. (IV, 75a) and (IV, 77a), we can write for two *equal* homogeneous layers placed one on top of the other the following equations:

$$T_{1,2} = \frac{T_1^2}{1 - R_1^2} \quad \text{and} \quad R_{1,2} = R_1 + \frac{T_1^2 R_1}{1 - R_1^2}. \tag{22}$$

In accordance with Eq. (IV, 66) the relationship between T, R_0 and R_∞ can be expressed as follows:

$$a \equiv \frac{1}{2}\left(\frac{1}{R_\infty} + R_\infty\right) = \frac{1 + R_0^2 - T^2}{2R_0},$$

where, for a given sample, a is constant for any layer thickness. Therefore, for a layer 1 and a double layer 1, 2 it must hold that

$$\frac{1 + R_{01}^2 - T_1^2}{2R_{01}} = \frac{1 + R_{01,2}^2 - T_{1,2}^2}{2R_{01,2}}. \tag{23}$$

When the above mentioned expressions for $T_{1,2}$ and $R_{01,2}$ are inserted here, it can be seen that the equation is satisfied. This means that one homogeneous layer could, for the sake of theoretical reasoning, be replaced by two (or even more) sublayers without the theoretical equation being altered.

[221] *S: son Stenius, Å.: Svensk Papperstidning* **54**, 663, 701 (1951).

calculation from measurements of R_0 and R_∞. Furthermore, he substituted the so-called area weight W of the paper in grams per m² for the layer thickness. The corresponding s and k are then called the *specific* scattering and absorption coefficients respectively. When s and k are plotted as function of W for various papers made from the same pulp, the curves of Fig. 74, measured at 456 mµ, are obtained. For a given wavelength, s and k are not constant, but increase or decrease with

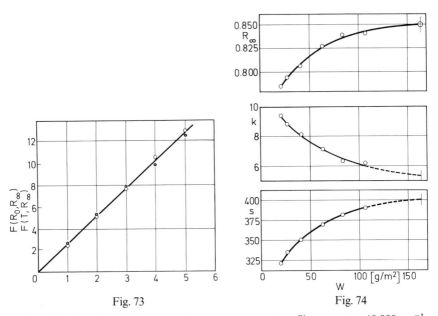

Fig. 73

Fig. 74

Fig. 73. Reflectance and transmittance measurements on filter papers at 19,000 cm⁻¹.
● Reflection; ○ Transmission

Fig. 74. The "specific" scattering coefficient s of hand-made paper as function of the so-called area weight at 456 mµ

increasing W. In accordance with Eq. (IV, 32) this means that R_∞ too (after conversion to absolute values) is not constant, but depends on W. Since s and k cannot be measured for large values of W, they have to be extrapolated to $W \to \infty$ whereas R_∞ can be measured directly. For these values of s and k extrapolated to $W \to \infty$, the Kubelka-Munk function (IV, 32) is then valid again.

When interpreting these results, we must assume that these papers do not represent homogeneous layers, but that the inner parts of the layers are more densely packed and, therefore, scatter more strongly then the outer parts. For this reason, we can imagine that such paper is composed of numerous very thin layers parallel to the surface, and that these layers for their part are homogeneous layers as far as s and k are concerned. In connection with this, the Kubelka-Munk theory for the multi-layer model (see p. 123) has to be applied [222]. It can be shown that in the case of a symmetrical sequence of such thin layers with reference to the central layer, the Kubelka-Munk theory remains valid. Since papers do not possess this symmetry (different properties of the gauze side and the felt side) they have to be folded to obtain the symmetrical sequence mentioned above. Such folded papers then obey again the Kubelka-Munk theory (see p. 126). Due to their inhomogeneity, however, papers are generally not suitable for testing the theory. Nevertheless, they are frequently used for this purpose since paper forms freely suspended layers whose reflectance values R_0 and R_∞ and also transmittance T can easily be measured. Powders are more suitable for testing the theory due to their homogeneity and do not make the measurement of R_0 and R_∞ (see Fig. 72) difficult. For measurement of T, however, a firm transparent base or even a small vessel with two windows has to be used. This in turn requires the use of the complicated multi-layer theory in order to eliminate the influence of the windows on T (see p. 191).

For observing the "typical color curve" (see p. 188), the substance to be tested is diluted by or adsorbed on non-absorbing standards. Since in the concentration range $10^{-4} < x < 10^{-2}$ the absorbing substance does not influence the scattering conditions (see p. 177) and it is not expected that the absorbing substance changes the scattering characteristics of the diluent, the scattering coefficients can be determined using the pure diluent only, and the results thus obtained can be adopted for the absorbing sample. This is favorable in as much as the S values of absorbing samples are generally not accessible since the measured effects are too small. For this reason it pays to measure scattering coefficients of such white standards as functions of wavenumber, grain size, and refractive index, and use these dependencies to decide whether or not the multiple scattering of the powders obeys laws which are similar to the laws calculated and confirmed by *Rayleigh* and/or *Mie* for single scattering. The measurements given subsequently [223] are listed in accordance with increasing grain size in Table 12. The grain sizes were determined and averaged partly by means of the light microscope, partly by means of the electron microscope, and also were to some extend derived by calculations from the specific surfaces as determined by the "BET" method (adsorption of nitrogen).

Particles having a very small diameter $(d \ll \lambda)$ are difficult to make. Aerosils (Degussa) are best suited for this purpose. By varying the

---
[222] *S: son Stenius, Å.*: l.c.
[223] *Kortüm, G.*, and *D. Oelkrug*: Z. Naturforsch. **19a**, 28 (1964).

Table 12. Scattering coefficients of the Kubelka-Munk theory

Substance tested	Specific surface according to BET [m²/g]	Average grain size d [μ]	Making and preparing	Quantities measured	Power α of the wave-number dependence of $S(S = \text{const. } \tilde{\nu}^\alpha)$	Grain size
Aerosils	376	0.01	heated at 600° C for 1 hr.	$T_{1,2,3}, R_\infty$	3.6	Grain size $< \lambda$
	294	0.015			3.5	
	196	0.02			3.2	
	106	0.04			3.0	
	38	0.08			2.6	
Calciumfluoride		≅ 0.2	precipitated	$T_{1,2}, R_\infty$	≅ 1	Grain size ≅ λ
SiO₂–Al₂O₃ cracking catalyst	520 (inner surface)	0.2–0.4	heated at 600° C for 2 hrs. ground for 20 hrs.	$T'_{1,2}; R_\infty$	≅ 1	
Sodium chloride	8	≅ 0.4	heated at 600° C for 2 hrs.	R_0, R_∞	≅ 1	
Magnesiumoxide		0.1–0.2	freshly prepared on black background	R_0	≅ 1	
Quartz powder		5, 10	ground quartz glass	R_0, R_∞	< 1	Grain size $> \lambda$
Glass powder		2.5, 3, 7, 15	ground window glass	R_0, R_∞	≅ 0	
Sodium chloride pulverized subt.		15–25	heated at 400° C for 2 hrs.	R_0, R_∞	≅ 0	
Color glass BG 23 (Schott & Gen.)		≅ 5	ground	$T_{1,2}, R_\infty$	< 1	

manufacturing procedures, they can be produced in form of powders which have a very narrow range of particle sizes (according to Fig. 69) and an average grain size between 0.1 and 0.01 μ. The scattering coefficients derived for $K \to 0$ or, respectively, $R \ge 0.95$, from measurements of T and R_∞ by calculation based on Eq. (IV, 56) are double-logarithmically plotted against the wavenumber $\overset{*}{\nu}$ in Fig. 75. Since the Rayleigh equation (III, 19), according to which $S' \sim \lambda^{-4} \sim \overset{*}{\nu}^4$, should be valid in the case of such small particles and single scattering, this diagrammatic representation ought to permit, from the slope of the respective straight line, the derivation of the power of the $\overset{*}{\nu}$-dependence. Actually, a linear dependence of S on $\log \overset{*}{\nu}$ is found. With decreasing grain size, the $\overset{*}{\nu}$-dependence increases noticeably, whereas the power 4, valid for single scattering, will not be reached completely [224]. This ultimately means that, even in case of multiple scattering of such small particles, similar conditions exist as in case of single scattering.

Likewise the considerable increase of the scattering coefficients with increasing grain size, as evident in Fig. 75, is in keeping with this conception. According to *Rayleigh*, and in agreement with Eq. (III, 20), S should be proportional to the third power of the radius of the particles, i.e. proportional to their volume. Of course it is necessary here to hold N constant and with it also G/ϱ, the so-called *packing density*. On the other hand, according to Eq. (III, 25) the so-called "Turbidity" S' is proportional to the number N of particles for the case of a constant volume v of particles in condensed systems. The same conditions are found when multiple scattering is concerned. When, in case of a given aerosil, S is plotted against the packing density G/ϱ, i.e., according to Eq. (III, 20) against N [225], we obtain two straight lines as they are shown in Fig. 76 for two samples and two wavelengths. However, the observed dependence of the scattering coefficient on the grain size does not agree with the linear dependence on r^3 or v which, according to *Rayleigh*, should be expected in the case of small particles, just as the observed dependence on the wavenumber is not determined by $\overset{*}{\nu}^4$ but by $\overset{*}{\nu}^\alpha$ where $\alpha < 4$ (see Fig. 75). Therefore, the laws of single scattering at small particles cannot simply be adopted in the case of multiple scattering. All things considered, these measurements allow us to draw the conclusion that, as far as small particles are concerned, the scattering process in the case of dense packing can be described in a manner similar to the process of single scattering. The main reason for this is the fact that in both cases an isotropic scat-

---

[224] In accordance with a refined theory expounded by *Houghton, H. G.* and *J. A. Stratton*: Phys. Rev. **38**, 159 (1931), lower wavelength exponents are indeed to be expected in this grain size range. See also *Luck, W.*: Diplomarbeit Berlin 1945.

[225] Aerosils can easily be compressed into smaller volumes since they form very loose powders.

V. Experimental Testing of the Kubelka-Munk Theory

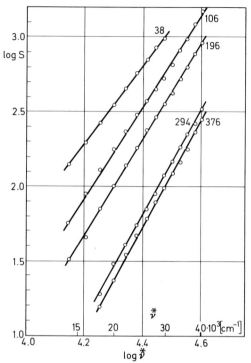

Fig. 75. Scattering coefficients of aerosils as function of the wave number; packing density $G/\varrho = 0.107$; specific surface in m²/g as parameter

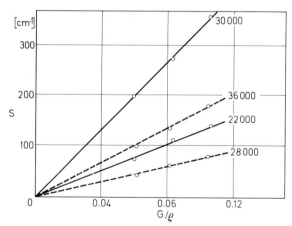

Fig. 76. Scattering coefficients of aerosil, dependent on the packing density G/ϱ with various wave numbers as parameter. ○----○ aerosil 376 m²/g, ○——○ aerosil 196 m²/g

tering distribution exists, so that the scattering coefficients according to *Rayleigh* and those according to *Kubelka-Munk* should, on basis of their definitions, be proportional to one another. Moreover, the energy scattered at single particles is relatively low so that the mutual interaction among the scattered waves cannot be of great importance. Nevertheless, differences occur since with dense packing the particles can often form larger aggregates. Through this, the effective scattering diameter of the particles is increased, and with it the scattering coefficient is also larger. In reality, the scattering coefficient measured in case of small aerosils is 1–2 powers of 10 higher than that calculated by *Rayleigh*[226].

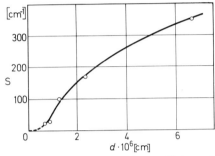

Fig. 77. Grain size dependence of the scattering coefficient of aerosil at 20,000 cm^{-1}

When particles whose grain sizes are similar to the light-wavelength are concerned (second group on Table 12), the measured scattering coefficient S is, in most cases, approx. proportional to the wavelength ($\alpha \cong 1$). Fig. 78 gives an example in which the measured scattering coefficients of MgO (grain size approx. 0.1 μ) and of a cracking catalyst SiO_2–Al_2O_3 (grain size approx. 0.3 μ) are reproduced as function of $\tilde{v}$. Likewise the most finely pulverized CaF_2 (grain size approx. 0.2 μ) shows the same characteristics. In all these cases S was obtained from measurements of T and R_∞. It is noteworthy that the specific surface of the cracking catalyst, which was determined via the "BET" method, amounted to approx. 600 m^2/g, i.e., by far the largest part of the surface consisted of "internal surface". This surface which is formed by pores, crevices, etc. obviously does not influence the scattering coefficient which is exclusively determined by the external grain size as was already mentioned

---

[226] The aggregation of aerosil particles that results from the application of pressure has also been verified by observing glossy peaks in the regular reflection (see p. 44). In addition, according to theory [*Giovanelli, R. G.*: Progr. Optics **2**, 111 (1963)] aggregation of particles will increase the ratio a/σ and thereby diminish the reflectance R_∞.

in case of aerosils. All three substances tested have S values of the same magnitude corresponding to the comparable grain sizes. However, one can recognise a definite dependence of the magnitude of S on the refractive indices of the three substances ($n_{CaF_2} = 1.45$; $n_{SiO_2-Al_2O_3} = 1.55$; $n_{MgO} = 1.75$). This is in keeping with the Mie theory of single scattering [see also Eqs. (III, 49a) and (III, 52)]. The large scattering power of MgO is consequently conditioned by its high refractive index. Likewise the

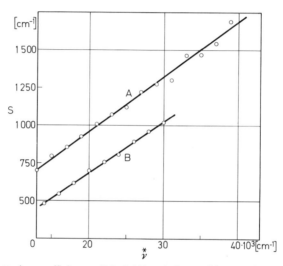

Fig. 78. Scattering coefficients of MgO (A) and of a cracking catalyst SiO_2-Al_2O_3 (B), as functions of wave number

observed linear dependence of the scattering coefficient on λ^{-1} or $\overset{*}{\nu}$ approximately corresponds to the requirements of the Mie theory in this particle size range. But this does not mean that the results of the Mie theory can, without caution, be employed for multiple scattering. This becomes evident when we consider that S and S' are defined in completely different ways (see also p. 109). Above all consideration must be given to the fact that the geometrically undefined particles here are in poly-dispersed form, whereas the Mie theory has been formulated for spherical particles. When changing from the spherical shape to small discs or small rods, an increase of scattering power becomes theoretically necessary. The magnitude of this increase depends on the refractive index [227], and the broader the distribution curve of grain sizes, the more the conformity with the law for uniform grain size becomes blotted out.

---

[227] *Gans, R.:* Ann. Physik **37**, 881 (1912).

Another case which frequently occurs is shown in Fig. 79. The scattering coefficients of dry NaCl (average grain size 0.4 µ) which are derived by calculation from measurements of R_0 and R_∞ also initially increase in linear fashion with $\overset{*}{\nu}$, but from 30,000 cm^{-1} onwards the rise is greater. This is caused by the fact that the refractive index of NaCl strongly increases in this wavenumber range ($n_{30,000} = 1.589$; $n_{30,600} = 1.622$; $n_{40,000} = 1.656$). Already in the case of single scattering, the refractive index in first approximation according to Eq. (III, 49a) takes the form $\left(\dfrac{n^2 - 1}{n^2 + 2}\right)^2$ in the scattering formula. In the case under

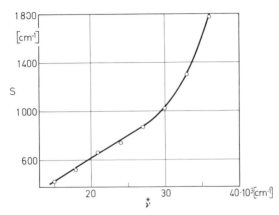

Fig. 79. Scattering coefficients of NaCl as functions of wave number

consideration, this means that in the range from 30,000 to 36,000 cm^{-1}, S increases by 10%, and that this increase is superimposed on the influence of the wavenumber. This influence would still be considerably larger when multiple scattering is concerned.

Finally, particles with an average grain size large in comparison to λ (third group in Table 12) do not show, in accordance with the theory of single scattering, wavelength dependence of the scattering power (see also p. 92), and, at a given wavelength, S is inversely proportional to the grain size in accordance with Eq. (III, 53). As already seen from Fig. 70 and as the systematic measurements of the scattering coefficients of glass powders (calculated from R_0 and R_∞ values) confirm, in the case of multiple scattering S is also inversely proportional to the particle diameter. The wavenumber dependence of S with various average particle diameters as parameter is shown in Fig. 80, which confirms the measurements reproduced in Fig. 40. That means that multiple scattering likewise fulfills the Mie theory. For reasons given above, a numerical

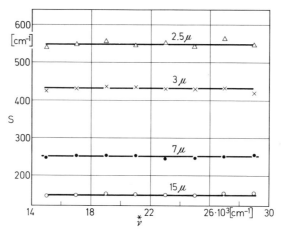

Fig. 80. Scattering coefficients of glass powder in dependence on the wave number; parameter: average particle diameter

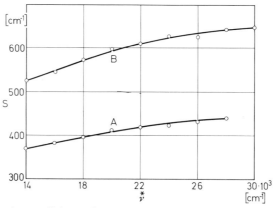

Fig. 81. Scattering coefficients of quartz powder in dependence on the wave number; parameter: average particle diameter

agreement cannot be expected of course, since uniform grain sizes do not exist. For instance, the scattering coefficients of quartz powder increase by 10–20 % in the same grain size range and the same wavenumber range, since grinding of the hard, brittle quartz glass produces a very broad grain size distribution which even includes particles below 1 μ (Fig. 81). The scattering coefficient of NaCl powder which has an average grain size of 20–25 μ, is likewise independent of the wavenumber at 35,000 cm^{-1} and then increases. The letter can also be attributed to the influence of the dispersion of the refractive index.

When S is known, the absorption coefficients K of pulverized substances can be calculated by means of the Eq. (IV, 32). All "white standards" examined so far have finite K values, a fact which can nearly always be traced back to impurities removable only with great difficulty (see also p. 149). In some cases such impurities can be partly eliminated by appropriate pre-treatment. For instance, the reflectance R_∞ of aerosils (at 35,000 cm^{-1}) is increased from 75% to 90% and the

Table 13. *Absorption coefficients of aerosils*

$\tilde{\nu} \cdot 10^{-3}$	specific surface in sq. meter per gram				
	38	106	196	294	376
14	0.2	0.1	0.1		
16	0.2	0.1			
18	0.3	0.1	0.1	0.1	0.1
20	0.2	0.1	0.1		
22	0.3	0.1	0.1		
24	0.3	0.2	0.1	0.1	0.1
26	0.4	0.2	0.1	0.1	0.2
28	0.4	0.1	0.2	0.1	0.2
30	0.3	0.1	0.2	0.1	0.2
32	0.5	0.3	0.2	0.1	0.1
34	0.4	1.0	0.2	0.2	0.1
36	0.8	2.4	0.4	0.5	0.2
38	4.2	5.2	1.6	1.6	1.2
40	7.7	10	4.6	4.3	4.4

reflectance R_∞ of NaCl (at 37,000 cm^{-1}) from 78% to 87% through roasting at 600° C. On the other hand, during the grinding process abrasion can lead to increased impurity of the powders. Table 13 shows the absorption coefficients in cm^{-1} of the aerosils of Fig. 75 [228]. These absorption coefficients have the same order of magnitude and only increase noticeably from 36,000 cm^{-1} onwards, It is possible that this phenomenon is caused by impurities of SiO which has a band spectrum at 41,000 cm^{-1} [229].

As a further example, Fig. 82 shows the absorption coefficients of the four glass samples of Fig. 80; these glass samples have different grain sizes, and the absorption coefficients were, in accordance with Eq. (IV, 32), derived by calculations from the scattering coefficients given in Fig. 80. Within the accuracy of measurement, the measured points coincide.

[228] *Oelkrug, D.:* Dissertation, Tübingen 1963.
[229] *Mohn, H.:* 100 Jahre Heraeus, Hanau. Wiss. techn. Band, p. 331 (1951).

Furthermore, the shape of the curve is comparable with the transmission spectrum.

For experimental reasons, the methods described cannot be used for determining the scattering coefficients of strongly absorbing substances because the transmittances of even very thin layers practically become zero, and because R_0 and R_∞ hardly differ (see also p. 240). However, strongly absorbing substances can be tested by means of the

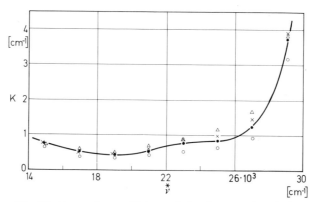

Fig. 82. Absorption coefficients of the glass powder of Fig. 80

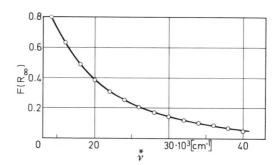

Fig. 83. Reflection spectrum of aerosil + graphite

dilution method described on p. 175. It is still necessary here to demonstrate that the scattering power of such a sample is practically given exclusively by the scattering of the diluent which exists in surplus, and that the addition of the absorbing substance does not essentially alter this. If this is the case, it suffices to measure the scattering coefficient of the pure diluent in order to obtain the absorption coefficient of the sample by means of Eq. (IV, 32).

To make such a sample, aerosil having a specific surface of 196 m²/g and also having known scattering coefficients (Fig. 75), was mixed with fine graphite powder in the ratio $10^3:1$. Over the whole accessible spectral range, graphite has a fairly constant absorbance, whereas the scattering coefficients of the aerosil strongly increase with the wave-

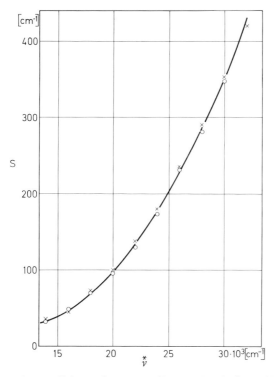

Fig. 84. Scattering coefficients of pure aerosil 196 ($\times$) and of aerosil + graphite ($\bigcirc$) at the same packing density $G/\varrho = 0.107$

number. Consequently, it must be expected that, in accordance with Eq. (IV, 32), $F(R_\infty) = K/S$ of the mixture decreases with increasing wavenumber. This is in fact the case (Fig. 83). Due to the low concentration of the graphite and the relatively low scattering power of the aerosil, the scattering coefficients can be derived from measurements of T of a thin layer and from R_∞. As Fig. 84 shows, these scattering coefficients practically coincide with those of the pure aerosil[230]. With this it has been demonstrated that the scattering power of such a diluted

---

[230] *Kortüm, G.*, and *D. Oelkrug*: Z. Naturforsch. **19a**, 28 (1964).

sample is practically fixed by the scattering of the diluent alone. This was also confirmed by other authors [231]. The K-values (Fig. 85) derived by calculation from S-values show a trend which is opposite to that of the $F(R_\infty)$-values. This is an instructive example for demonstrating how much, in extreme cases, "typical color curves" can be distorted by the wavenumber dependence of the scattering coefficients.

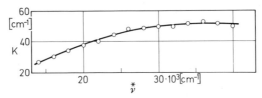

Fig. 85. Absorption coefficients of graphite from reflection measurements on aerosil-graphite mixtures

g) Influence of the Scattering Coefficients on the "Typical Color Curve"

As the example graphite/aerosil shows, the expression $\dfrac{(R_\infty - 1)^2}{2R_\infty} = f(\overset{*}{v})$ characterising a substance and derived from measurements of the "reflectance" deviates more from the real spectrum $K = f(\overset{*}{v})$ of the substance in question the more the scattering coefficients of the substance, or those of its diluent, depend on the wavenumber. Consequently, the typical color curves reflect the spectrum correctly only if the scattering coefficients are independent of $\overset{*}{v}$. In accordance with the conclusions arrived at in the last section, this happens if the grain sizes are a multiple of the wavelength, i.e., the substance to be tested and/or the diluent should always be tested in the coarse-grained state. But then other perturbations of the spectrum can easily occur because of large regular-reflection parts; small surfaces in cases where a substance has to be tested in the adsorbed state; no homogeneous mixture in cases where the dilution has to be brought about by grinding and where the grinding time is too short for thorough mixing; etc. For all these reasons, it is preferable that the samples be tested in a finer state (grain sizes from 0.1 to 1 μ). But it then must be expected that, due to a scattering coefficient dependent on the wavenumber, the typical color curves are increasingly flattened towards the ultra-violet. This means that the relative intensities of the individual bands are not true. When these intensities are required for characterizing the spectrum or when they are used for photo-

---

[231] *Judd, D. B.*, and collaborators: J. Res. Natl. Bur. Std. **19**, 287 (1937).

metric purposes, the dependence of the scattering coefficients as a function of $\overset{*}{v}$ must be measured and with this K has to be derived by calculation from $F(R_\infty)$.

In case of still finer grain, we can expect that, apart from the band flattening, an increasing red-shift of the maxima of the typical color curve exists. Through this, masked maxima which are only marked by a "shoulder" can, for instance, be obliterated altogether. This red-shift can be calculated in the following way:

The position of a maximum results from Eq. (IV, 32)

$$\frac{dF(R_\infty)}{d\overset{*}{v}} = \frac{S\dfrac{dK}{d\overset{*}{v}} - K\dfrac{dS}{d\overset{*}{v}}}{S^2} = 0. \tag{24}$$

When the shape and position of $\overset{*}{v}_0$ of an absorption band is described by the Lorentz relation [232]

$$K = \frac{K_{max} b^2}{4(\overset{*}{v}_0 - \overset{*}{v})^2 + b^2}, \tag{25}$$

where b stands for the width at half of maximum band intensity, and the scattering coefficient depends on $\overset{*}{v}$ as

$$S = \overset{*}{v}{}^\alpha, \tag{26}$$

inserting Eq. (25) and Eq. (26) together with their derivations into Eq. (24), gives the relative band shift

$$\frac{\overset{*}{v}_0 - \overset{*}{v}}{\overset{*}{v}_0} = \frac{1 - \sqrt{1 - c^2 \alpha(\alpha + 2)}}{\alpha + 2}, \tag{27}$$

where $c \equiv b/2\overset{*}{v}_0$. When this equation is plotted as function of $b/\overset{*}{v}_0$ for various powers α of the wavenumber dependence of S as parameter, the curves of Fig. 86 are obtained. When, for instance, a band at $\overset{*}{v}_0 = 30{,}000$ cm^{-1} has a value of $b = 6{,}000$ cm^{-1}, which can frequently happen in a reflection measurement, Fig. 86 gives, for $b/\overset{*}{v}_0 = 0.2$ and for $\alpha = 4$, a relative red-shift of the band amounting to somewhat more than 0.02, i.e., approx. 650 cm^{-1} (for $\alpha = 1$ which ordinarily occurs, approx. 150 cm^{-1}). In case of narrower bands, the red-shifts decrease quickly so that the wavenumber dependence of S here does not bring about considerable errors.

When the grain is very fine, apart from band flattening and red-shift due to the λ-dependence of the scattering coefficients, account must

---

[232] See, for instance, *Kortüm, G.*: Kolorimetrie, Photometrie und Spektrometrie, 4. Aufl. (4th edition), p. 48 ff. Berlin-Göttingen-Heidelberg: Springer-Verlag 1962.

be taken of a further source or error, which is based on the fact that the condition of "infinite layer thickness" might not be satisfied in certain circumstances [233]. Simple relations formulated on basis of the Kubelka-Munk theory can tell whether or not this is the case. When the reflectance R_0 of a thin, partly still transparent, non-absorbing layer $(K=0)$ is concerned, we can write in accordance with Eq. (IV, 57a) and dependent

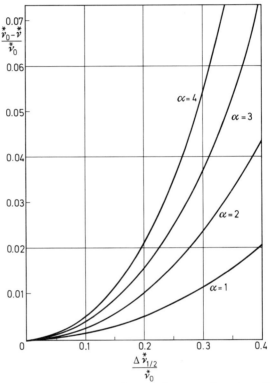

Fig. 86. Relative shift of an absorption maximum as function of the half width and of the wave number dependence of S

on the scattering coefficient S and the layer thickness d, the following expressions:

$$R_0 = \frac{Sd}{Sd+1} \quad \text{and} \quad R_0 + T = 1. \tag{28}$$

The scattering coefficients of substances used normally as standards or as diluents lie in the range from 300 to 1,000 cm^{-1} (see also Fig. 78), so that in case of the usual layer thickness of 3 mm, R_0 has values between

---

[233] Kortüm, G., and D. Oelkrug: Naturwiss. **53**, 600 (1966).

0.989 and 0.997. In such cases, we can assume with good approximation that a 3 mm layer already constitutes an "infinitely thick" layer.

In case of very fine-grained substances (e.g., aerosils discussed on p. 203 and having scattering coefficients lying in the visible between 10 and 50 cm^{-1}) layer thicknesses between 20 and 100 mm would consequently be required in order to obtain a R_0 of 0.99. A greatly reduced scattering ($S \sim d^3$) which exists in case of very small particles favors a deep penetration of the radiation into the sample and causes a sample which is irridiated on a small area to show reflection from a much larger area. Consequently, sample plates with a diameter of 4 to 20 cm have to be used in order to realize a condition where the radiation is completely rescattered into the backward hemisphere again. Even photometric spheres of corresponding aperture do not make possible quantitative measuring here, i.e., and "infinite" layer thickness is unobtainable in such cases. However, adding an absorbing substance to such a weakly scattering standard soon creates a "finite" layer thickness since the penetration depth of the radiation greatly decreases with increasing absorption. Therefore the aerosils for instance, can serve an diluent and/or adsorbent, but cannot be used as a comparison standard.

h) Particle-Size Dependence of the Kubelka-Munk Function

It has already been pointed out in Chapter II g that the absorption coefficient in heterogeneous systems is a function of the particle size, and that this effect can be observed in transmission when scattering is suppressed, as far as possible, by selecting suitable refractive indices of particles and surrounding medium. Since, as discussed on p. 195, K_T and K_R are proportional to each other, K_R must also be dependent on the particle size in case of reflectance measurements. As Fig. 23 shows K_T decreases with increasing particle diameter d, and this decrease becomes greater with increasing extinction modulus kc_0 of the particles. This is valid for all sorts of packing densities and means that the spectrum gets flattened by the particle size dependence of K_T. The reverse dependence exists in case of reflectance measurements on weakly absorbing substances: With increasing d absorption also increases (Figs. 25a and 28), even if the influence of the regular reflection part is eliminated by measuring with linearly polarized radiation. As already mentioned on p. 61, the explanation of this phenomenon is given by the fact that the particle size dependence of K_R and S overlap here so that the dependence of K_R/S and with this of the Kubelka-Munk function on d is actually observed.

When, analogously to the Mie theory of single scattering, S is assumed proportional to d^{-1} for $d \geq 1\,\mu$ (this is also experimentally confirmed

– see Fig. 70), $F(R_\infty) = K/S$ as function of d can be derived by calculation from Eq. (II, 80) and $S \sim 1/d$. For instance, $kc_0 = 10^4$ for the absorption maximum of $KMnO_4$ at 19,000 cm^{-1}. For this value of kc_0, a quantity proportional to $F(R_\infty)$ is plotted against log$d$ in Fig. 87 [234]. $F(R_\infty)$ first increases with increasing d and practically becomes independent of d

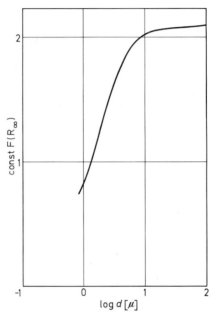

Fig. 87. Grain size dependence of $F(R_\infty)$

for $d > 10\mu$. This was actually confirmed by measuring with linearly polarized radiation (see also Fig. 30) [235], and this confirms the fundamental correctness of these deliberations.

When a strongly absorbing sample is diluted with an excess of a non-absorbing diluting agent, K_R becomes smaller and the scattering of the mixture is practically determined solely by the diluent. Increasing the concentration of the absorbing substance in the mixture merely increases the packing density which, in accordance with Fig. 23 for $P \ll 1$, hardly influences the ratio $K/K_{(d \to 0)}$ so that, as far as this case is concerned, the particle size dependence of absorption holds good.

When the substance to be tested is, in the molecularly dispersed state, adsorbed on a standard, the adsorbate forms a unit with the particles

---

[234] Kortüm, G., and D. Oelkrug: Naturwiss. **53**, 600 (1966).
[235] Kortüm, G., and J. Vogel: Z. Physik. Chem. N. F. **18**, 230 (1958).

of the adsorbent so that the particle size of this unit is decisive for the absorption coefficient. On the other hand, the concentration c_0 of the absorbing substance can likewise be related to the unit adsorbent/adsorbate. Consequently, in case of the normal coverings which are at best mono-molecular (10^{-5} to 10^{-3} g/g), this concentration is very low so that the grain-size dependence of the absorption coefficient K_R becomes negligible.

When the results of the critical weighing of the Kubelka-Munk theory are summed up, the following conclusion is reached: Provided that the measuring conditions are suitable, the "typical color curve" of a sample which is, at the same time, diffusely scattering and absorbing can be established by means of the theory (Kubelka-Munk theory). When this "typical color curve" is plotted logarithmically, it coincides with the true absorption spectrum through parallel shifting in the $\log F(R_\infty)$-coordinate. Moreover, it is assumed that the regular reflection parts are eliminated through sufficiently high dilution of the substance in question with a suitable "white" standard (relative measurement), and that the scattering coefficients of the standard are independent of the wavenumber in the spectral range used. When the latter is not assured, the scattering coefficient of the standard has to be separately measured at the proper grain size. If possible, reflectance values $R'_\infty < 0.6$ should not be measured since, in case of higher absorption, deviations from the Kubelka-Munk theory apparently occur, and the causes of these deviations have not yet been clarified. Likewise for this reason, dilution of the measuring sample with a suitable standard is recommended. Apart form all this, the dilution method offers a number of advantages, and these advantages shall once more be listed here:

1. The scattering coefficients of the mixture are exclusively given by those of the diluent in practical cases. Since the latter scattering coefficients can always be measured, the absorption coefficients K_R can be derived from the Kubelka-Munk function by calculation.

2. When measuring relatively to the pure diluent, possible deviations from the isotropic scattering distribution can be eliminated so that the measuring becomes independent of the geometry of the measuring arrangement.

3. The reflectance of the diluent can easily be measured against freshly prepared MgO so that conversion of the relative measured values into absolute ones is always possible. This is especially important for all quantitative evaluations of the Kubelka-Munk function.

4. Regular reflection parts are practically unimportant when measuring the diluted sample against the pure diluent under identical conditions.

5. When the sample to be tested is adsorbed monomolecularly on a standard which is in excess (as it is nearly always done in the case of

organic substances), the particle size dependence of the absorption coefficient is also eliminated.

In order to experimentally test the *Melamed*[236] statistical theory of diffuse reflectance a method similar to that based on Fig. 67 was employed, i.e., the transmission spectrum of a color glass was compared with the reflectance spectrum of powders of varying average grain size made from this glass. The absorption coefficients of the powders were, in accordance with Eq. (IV, 191), derived from the measured values of R_∞ by means of this statistical theory and the agreement was generally fairly good. Likewise other authors[237] tested the validity of the Melamed theory and found that it is satisfactory.

Companion[238], to the contrary, found considerable deviations between theory and experiment with V_2O_5 powders. For x_b, he had to use the factor 0.1 instead of 0.284, the factor for close packing of spheres. This infers that the small crystalls, because of their flat faces, can lie together more closely than the equal sized spheres which were used as a model for the derivation of the theory.

The theory cannot be applied to adsorbed substances and the precondition that the particle dimensions d must have the magnitude $d \gg \lambda$ leads to results biased by regular reflection of the sample surface (the validity of *Lambert's* Cosine Law is assumed here).

---

[236] *Melamed, N.:* J. Appl. Phys. **34**, 560 (1963).

[237] *Poole, C. P.,* and *J. F. Itzel, Jr.:* J. chem. Phys. **39**, 3445 (1963).

[238] *Companion, A. L.:* Theory and applications of diffuse reflectance spectroscopy. Developments in applied spectroscopy, Vol. 4. New York: Plenum Press 1965.

Chapter VI. Experimental Techniques

a) Testing of *Lambert's* Cosine Law

To determine the "indicatrix" of a remitting surface (see p. 29) a goniophotometer can be used to measure the radiation flux at various angles of reflection and azimuths, for given (and usually variable) angles of incidence, α. Numerous such photometers have been described [239]. The older apparatus, naturally, works by the visual photometric principle of measurement with a reference beam. The accuracy of such measurements amounts, therefore – after removal of numerous possible systematic errors – to, at most, 1–2%. Frequently used, for example, was the well-known *Pulfrich* photometer of Zeiss, with suitable additional equipment [240]. The position of the sample relative to the optical axis of the photometer could be adjusted with the help of a vertical and a horizontal axis and could be read on the divided circles with a vernier to 1'. In this way a quarter of the indicatrix sphere could be measured. Also, the angle of incidence of the radiation could be changed by shifting the light source on a third axis. The insertion of a nicol prism allows measurements to be made with linearly polarized light. The sample is arranged to give an image on the measuring diaphragm of the Pulfrich photometer so that there is a structureless field of vision. This image then falls on a second nicol prism. The measurements are made by adjusting the diaphragm in the reference beam to obtain the same brightness. For example, MgO or $BaSO_4$ serves as a reference standard at 45° angle of incidence and 0° angle of measurement. The use of a second lamp for the reference beam represents here, for instance, a systematic source of error, because the energy distribution (color temperature) of both lamps will generally be different.

Modern photoelectric apparatus are based on the same principle, but the reference beam path is not required, and the radiation density of the sample is measured by using the photocurrent from a barrier layer cell. Fig. 88 shows a diagram of the Zeiss goniophotometer. The

---

[239] See, e.g., *Messerschmidt, J. B.*: Ann. Physik **34**, 867 (1888). — *McNicholas, H. J.*: J. Res. Natl. Bur. Std. **1**, 29 (1928); **13**, 211 (1934). — *Weigel, R. G.*, and *G. Ott*: Z. Instrumentenk. **51**, 1, 61 (1931). — *Slater, J. M.*: J. Opt. Soc. Am. **25**, 218 (1935). — *Moon, P.*, and *J. Laurence*: J. Opt. Soc. Am. **31**, 130 (1941). — *Harrison, V. G. W.*: J. Sci. Instr. **24**, 21 (1942) and numerous other papers.

[240] *Falta, W.*: Jenaer Jahrb. 1954.

image of the filament of the lamp G is formed by the lens K on the (exchangeable) aperture, B_1. This aperture, with the objective O_1, forms the illumination collimator that can rotate on the axis about the point M. A similar collimator leads the remitted light to the barrier cell, Ph. The collimators fix the aperture angle of both light beams at a definite value (0.24–1°, according to the size of the diaphragm), because the ideal case of parallel beams cannot be achieved, since the diaphragms cannot be made extremely small because of the needed intensity. The image of a fixed diaphragm connected with lens K is projected by a field lens

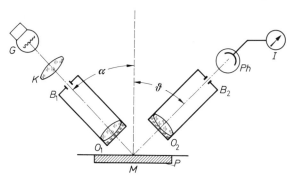

Fig. 88. Schematic drawing of the goniophotometer of Carl Zeiss

onto the sample, so that the illuminating field in perpendicular illumination ($\alpha = 0°$) is round and structureless. For angles $\alpha > 0°$, the field of illumination is elliptical. A similar diaphragm is placed in the measuring collimator in front of the photocell, onto which the image of the sample is projected. Because it is somewhat smaller than the field of illumination, only the illuminated part of the sample can be measured. A colored filter can be inserted between the diaphragm and the lamp and/or the photocell. $BaSO_4$ or a black glass plate with a reflection of 4.2% for small $\alpha = \vartheta$ can be used as a standard for diffuse or regular reflection. Systematic errors occur here e.g. from non-proportionality between illumination intensity and photoelectric current, that always exists in barrier cells[241]. Thus, the internal resistance of the galvanometer used to measure the photo-current must be as small as possible. By applying a multiplier and an optical compensator, the method can easily be modified to give a null-method of much better accuracy[241].

Instead of directed illumination, the sample can be illuminated with diffuse light by using an integrating sphere or a "milk glass" hemisphere,

---

[241] Cf. *Kortüm, G.*: Kolorimetrie, Photometrie und Spektrometrie, 4th Ed. Berlin-Göttingen-Heidelberg: Springer 1962; also, *Kortüm, G.*, and *R. Hamm*: Ber. Bunsenges., **72**, 1182 (1968).

uniformly illuminated from outside. The remitted radiation at various angles ϑ and φ is measured and the indicatrix is thus determined. Such measurements have been carried out, for example, by *McNicholas*[242]. A recording goniophotometer has been described by *R. S. Hunter*[243].

b) The Integrating Sphere

The Kubelka-Munk theory assumes diffuse irradiation of the measured sample. As was shown on p. 172, in many instances, the reflectance does not depend on whether the illumination is directed or diffuse. But when the sample has a structure or gloss, or roughness, which cause shadows when illuminated by directed irradiation, diffuse irradiation must be used. The so-called integrating sphere[244], is therefore used, mentioned above, where absolute reflectance values are dealt with (see p. 138).

First, it can be shown that a sample illuminated by diffuse light with an integrating sphere remits considerably more radiation than by a single reflection. This is a result of the multiple reflection by the walls of the sphere. If the sample is irradiated at a given angle with irradiation intensity S, and if the reflectance of the sample is denoted as R_p, the radiation density received by the detector through the aperture $\Delta\omega$ in the direction ϑ (cf. Fig. 11), is given, from Eq. (II, 55) by

$$B_{1p} = R_p S \Delta\omega. \qquad (1)$$

If the sample, however, is placed into the opening of the photometer sphere, so that it forms part of the inner surface of the sphere (Fig. 52), which may have a mean reflectance R_K, then the radiation density, according to (IV, 127), falling under the same conditions on the detector due to the multiple scattering by the sphere walls, is defined by[245].

$$B_p = R_p R_w S \cdot \frac{1}{1 - R_K} \Delta\omega, \qquad (2)$$

---

[242] *McNicholas, H. J.*: J. Res. Natl. Bur. Std. **1**, 29 (1928).

[243] *Hunter, R. S.*: Modern aspects of reflectance spectroscopy, p. 226 ff. New York: Plenum Press 1968.

[244] *Sumpner, W. E.*: Proc. Res. Phys. Soc. London **12**, 10 (1892). — *Ulbricht, T.*: Elektrotech. Z. **21**, 595 (1900). For an outstanding summary review of theory, error possibilities, and methods of preparation of a reflecting layer see *Wendlandt, W. W.*, and *H. G. Hecht*: Reflectance spectroscopy. New York: Interscience Publishers, John Wiley & Sons 1966. Instead of the integrating sphere, a hemisphere can also be used to obtain even more favorable radiation output. In this regard, see *Derksen, W. L.*, et al.: J. Opt. Soc. Am. **47**, 995 (1957).

[245] If there is no screen in the sphere to prevent directly reflected light from P falling on the measuring surface, the same sum as in Eq. (IV, 126) is calculated, that is Eq. (IV, 127) has an extra term.

in which $\dfrac{1}{1-R_K}$ is the so-called "sphere factor". R_w, the reflectance of the sphere surface, usually coated with MgO [246], is of the order of magnitude of 1, and R_K is about 0.95, so that 20-times the radiation density is received as compared to that recorded by the detector in the absence of the sphere. This is especially important for strongly absorbing samples. The mean reflectance R_K of the sphere, according to Eq. (IV, 129), is composed of that of the sphere surface R_w and that of the sample R_p, each multiplied by the corresponding surface area. If f_p is the area of the sample, f_E is that of the entrance window and f_m that of the exit window for the radiation, and F is the total surface area of the sphere, then

$$R_K = \frac{(F - \Sigma f) R_w + f_p R_p}{F}. \tag{3}$$

If a standard is substituted for the sample (substitution method), we have analogously to Eq. (2):

$$B_{st} = R_{st} R_w S \frac{1}{1 - R_{K'}} \Delta\omega. \tag{4}$$

$R_{K'}$ is different from R_K, so that Eq. (5) is obtained, analogous to Eq. (3):

$$R_{K'} = \frac{(F - \Sigma f) R_w + f_p R_{st}}{F}. \tag{5}$$

The mean reflectance of the surface depends on how large the difference is between R_p and R_{st}. Only for an infinitesimally small area, f_p, $R_K \cong R_{K'}$. Thus, from Eqs. (2) and (4),

$$\frac{B_p}{B_{st}} = \frac{R_p}{R_{st}} \cdot \frac{1 - R_{K'}}{1 - R_K}. \tag{6}$$

The measured quantity B_p/B_{st} is not equal to the "relative reflectance" R' when measured by this substitution method in the integrating sphere, because the conditions of irradiation for both measurements are different. The factor

$$f \equiv \frac{1 - R_{K'}}{1 - R_K} \tag{7}$$

is termed the "sphere error". The reflectance of the sample, relative to the standard, is given, from Eqs. (6) and (7) as

$$R' \equiv \frac{R_p}{R_{st}} = \frac{B_p}{B_{st}} \cdot \frac{1}{f}. \tag{8}$$

---

[246] In the vacuum-UV (500 to 2000 Å), sodium salicylate is used instead, with a background of a MgO pigment which has a good reflectance in the λ range where the salicylate is strongly fluorescent. See *Heaney, J. B.*: J. Opt. Soc. Am. **56**, 1423 (1966).

The sphere error can be considerable. When MgO is used as a lining and a standard, so that $R_{st} = 0.98$, and calculations are made using the following sphere dimensions, which correspond to a practically used instrument: $r = 6.3$ cm, $f_p = f_E = f_M = 7$ cm^2, and $R_{K'} = 0.9526$, sphere errors for various values of R_p can be obtained. These are given in Table 14. The smaller R_p, the larger, naturally, is the sphere error. It depends, also, on the magnitude of the sample surface relative to that of the sphere. The smaller f_p and the larger F, the smaller is the error. But, as R_K

Table 14. *Sphere error in reflection measurements using the substitution method with an integrating sphere*

R_p	0.900	0.800	0.700	0.600	0.500	0.400	0.300	0.200
R_K	0.9515	0.9501	0.9487	0.9473	0.9459	0.9445	0.9431	0.9417
f	1.025	1.055	1.085	1.114	1.144	1.173	1.203	1.233
$1/f$	0.9753	0.9479	0.9220	0.8975	0.8743	0.8522	0.8313	0.8113

decreases with increasing F, as a consequence of Eq. (3), a compromise must be reached, and a medium sized sphere must be used. For special measurements, integrating spheres as large as 1 m in diameter have been used [247]. The sphere error is usually eliminated in the usual measuring apparatus by proper guidance of the beams.

Integrating spheres are almost always lined with freshly prepared MgO [248]. The base must also have a high reflectance (e.g. roughened Al or MgO mixed with water glass), so that radiation passing through the MgO layer, will not be absorbed.

c) Measuring Apparatus

Reflectometers, or remission photometers, have frequently been described in the literature. Photoelectric methods are used exclusively today. The usefulness of the various apparatus depends on the method of measurement used. Here, distinction must be made between direct reading methods, compensation methods, substitution methods and scintillation methods [249a]. The two latter methods are independent of

---

[247] *Blevin, W. R.*, and *J. Brown:* J. Opt. Soc. Am. **51**, 129 (1961).

[248] For a description of the technique of preparing MgO, see *Dimitroff, J. M.*, and *D. W. Swanson:* J. Opt. Soc. Am. **46**, 555 (1956).

[249a] *Kortüm, G.:* Kolorimetrie, Photometrie und Spektrometrie, 4th. ed. Berlin-Göttingen-Heidelberg: Springer 1962.

the properties of the detectors and amplifiers, so they produce the most reliable results. However, they are, as a rule, commercially not available. An example of a modern apparatus is Zeiss' electrical remission photometer (Elrepho). Its principal features are described in two perpendicular projections in Fig. 89.

The sample A is placed into the opening of the integrating sphere, and is indirectly irradiated diffusely through the sphere by two lamps L. Directed radiation is cut off by the screen S. The reflected radiation

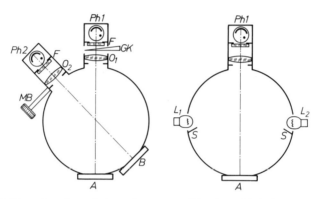

Fig. 89. Photoelectric remission photometer (Elrepho) of Zeiss in two perpendicular cross sections

emerges perpendicularly ($_dR_0$), i.e., the objective O_1 forms the image of the sample on the cathode of the photocell, Ph_1. In the same way, an image of the standard B is formed by the objective O_2 on the photocell Ph_2.

The adjustable measuring diaphragm MB in the reference beam is used for measurement. Its position can be read accurately. The wedge filter GK can reduce the irradiation intensity at the photocell to a tenth. F are color filters (value of half width, about 30 mμ). First, a second standard of known reflectance is substituted for the sample, the measuring diaphragm is adjusted to this value, and, using the filter wedge, both photocurrents are balanced. Then the standard is replaced by the sample, and compensated with the measuring diaphragm, so that its reflectance relative to the standard can be read from the measuring diaphragm. The reference beam should compensate for the lamp intensity fluctuations. It is hence a two-cell method with optical compensation, i.e., a kind of null method. The reproducibility of the measurement is $\pm 0.2\%$. A "milk-glass" disc that has been standardized against MgO is used as a standard, since it is much more constant than MgO itself. The "sphere error" described on p. 221 is eliminated in this apparatus because the irradiation conditions for A and B are always the same. If the irradiation intensity is changed when exchanging the sample and standard, this also happens for the reference beam B, so that the difference in photocurrent for both detectors does not change. It is necessary to assume that the irradiation intensities and the photocurrents are proportional to each other, and that the proportionality factor for both detectors is the same, i.e., this is not an actuall null method.

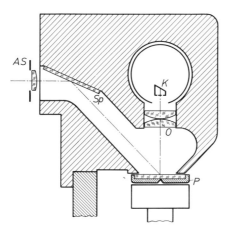

Fig. 90. Remission attachment RA 2 for Zeiss spectrophotometer PMQ II

Disregarding special reflection measuring instruments, each photoelectric spectrophotometer can, in principle, be used for spectral reflection measurements. Indeed, most commercially available instruments already have a suitable attachment available. As in transmission spectrometry, we distinguish between single- and double-beam instruments. Fig. 90 schematically describes, as an example, the remission attachment RA 2 for the Zeiss spectrophotometer. The apparatus is a single beam instrument, and works on the deflection principle. The primary radiation from the monochromator falls at 45° on the exchangeable sample P. A lens forms an image of the monochromator prism at the sample surface, so that a change in the slit width only changes the irradiation intensity. The objective O forms an image of the sample at the cathode of the multiplier ($_{45}R_0$). By exchanging a standard for the sample, a deflection is obtained that gives a measure of the relative reflectance R'_∞. As a rule, a larger effective slit width is used than in transmission measurements, so that the errors owing to the relatively large band width $\Delta\lambda$ and to insufficient monochromaticity of the radiation are generally larger than those for absorption measurements in transmission [249 b].

The optical geometry $_{45}R_0$ can lead to systematic errors when measuring samples with structures (textiles, paper, powders with non flat surfaces) (cf. p. 171). In such instances, we must irradiate diffusely and observe perpendicularly, i.e., we use the optical geometry $_dR_0$. This is provided for in the remission attachment RA 3 for the Zeiss PMQ II which can be used in two positions.

In the position A (see Fig. 91 a), with optical geometry $_dR_0$, the sample is diffusely irradiated by the screened lamp through the integrating sphere, and the image is formed by the optics 4, 5, 6, and 7 on the entrance slit of the monochromator. The spectrally resolved remitted radiation falls, by choice, on one of the various detectors that are sensitive to a given spectral range. The sphere wall serves as a reference standard, with the help of a tilted mirror 4 in place of the sample. The sphere error is thus eliminated, but no other reference standards can be used, unless they are first calibrated against the sphere wall (MgO).

In the position B (cf. Fig. 91 b), the optical geometry is $_0R_d$, i.e., reciprocal. The lamp is replaced by multiplier 11, the detector box is exchanged for the radiation source unit 12 and the radiation passes through the apparatus in the reverse direction. This arrangement is to be preferred when the sample is photochemically sensitive, because the sample is irradiated only with a narrow wavelength range as in the case of RA 2. Both arrangements should be equivalent, because of the reciprocity law $_dR_0 = {_0R_d}$ (see p. 172). The attachment RA 3 (unlike the RA 2) cannot be used in the ultra violet region (limit 380 mµ) because no quartz optics are used and because the coating of the sphere absorbs radiation. Recently, however, a version of the RA 3 has been produced, with quartz optics and a roughened integrating sphere of aluminium, on which freshly prepared MgO can be deposited.

Another remission attachment for the PMQ II spectrophotometer has been developed for the qualitative and quantitative evaluation of thin layer chromatograms in the visible and ultraviolet spectral regions [249c].

[249 b] For the use of the Beckman spectrometer for measuring reflectances, these errors, and the possibility of reducing them have been studied in detail [cf. *Hammond, H. K.*, and *I. Nimeroff*: J. Opt. Soc. Am. **42**, 367 (1952)]. A detailed discussion of sources of error in reflectance measurements is given by *Derksen, W. L.*, et al.: J. Opt. Soc. Am. **47**, 995 (1967). Apart from the correctness of the reference standards, the sources of errors are the same as those which play a role in transmission measurements, in particular the limited monochromaticity of the radiation and the effective spectral width (cf. *Kortüm, G.*: Kolorimetrie, Photometrie und Spektrometrie, 4th Ed. Berlin-Göttingen-Heidelberg: Springer 1962).

[249c] *Jork, H.*: Cosmo Pharma **3**, 33 (1967). — *Stahl, E.*, and *H. Jork*: Zeiss Information Bulletin **16**, 52 (1968) and the literature given therein.

It consists of a mechanical stage on which the thin layer plate is horizontally placed and monochromatically irradiated perpendicularly from above. The remitted radiation is observed at an angle of 45° and transmitted to the detector (geometry is $_0R_{45}$). The direction of the radia-

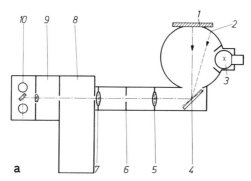

Fig. 91a. Remission attachment RA 3 for Zeiss spectrophotometer PMQ II; Arrangement A, optical geometry $_dR_0$. *1* Sample; *2* Spherical wall; *3* Electric bulb; *4* Tilting mirror; *5, 7* Lenses; *6* Diaphragm; *8* Monochromator; *9* Changer for cuvettes; *10* Detector housing

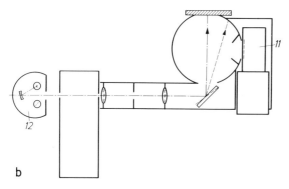

Fig. 91b. Remission attachment RA 3 for Zeiss spectrophotometer PMQ II. Arrangement B, optical geometry $_0R_d$. *11* Detector; *12* Radiation source

tion path can also be reversed. In this case the chromatogram is irradiated with the continuum from an incandescent or hydrogen lamp ($_{45}R_0$). This arrangement is conveniently used to investigate fluorescing samples (see p. 233). A potentiometric recorder and also, if need be, an integrator can be coupled to the movement of the stage to record the spectra. This chromatogram-spectrophotometer is designed for easy localization of the substance zones in order to directly observe the absorption and/or fluorescence spectra of the individual spots and thereby analyze them.

Also, for double beam spectrophotometers, attachments for reflection measurements have been developed. These apparatus usually measure the ratio of the radiation fluxes remitted by th sample on the one hand and the reference standard on the other; the value of the standard is set at 100%. The logarithm of this ratio (i.e. the apparent absorbance of the sample) or also the Kubelka-Munk function $F(R'_\infty)$ directly, is recorded as a function of λ or $\tilde{\nu}$. These instruments also usually are of the modified deflection type with double beams and only one detector[250]. Here, too we must assume a linear characteristic of the amplifier and detector.

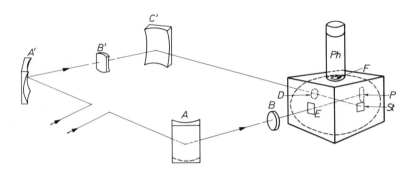

Fig. 92a. Remission attachment for Model 14 Cary spectrophotometer with integrating sphere and optical geometry $_0R_d$. ABP sample beam; A'B'C'St reference beam; P sample; St standard; D sample position for measuring the transmittance in case of diffuse radiation; F exit of radiation to photoelectric cell Ph

Examples of this type of measuring method are, e.g., the attachments to the Beckman DK, to the Bausch & Lomb Spectronic 505 and the Cary 14 spectrophotometers. The beam path of the latter is described schematically in perspective in Fig. 92a.

Monochromatic radiation, modulated by a 30 Hz commutator, falls, alternately, perpendicularly on the sample and the standard. It is scattered diffusely in an integrating sphere and then passes through the opening F in the sphere onto the detector. The ratio of the two photocurrents is registered. The optical geometry is thus $_0R_d$. The sample and standard are simultaneously within the integrating sphere, so there is no "sphere error" (cf. p. 220). Screens inside the sphere prevent directly reflected radiation from the sample or standard from falling on the detector.

---

[250] See *Savitzky, A.*, and *R. S. Halford:* Rev. Sci. Inst. **21**, 203 (1950), and also *Kortüm, G.:* Kolorimetrie, Photometrie und Spektrometrie, 4th ed., p. 309. Berlin-Göttingen-Heidelberg: Springer 1962.

With the same arrangement, the total transmission of diffusely scattering layers can be measured for perpendicular irradiation. The sample and standard are replaced by compressed MgO discs, and the thin layer being studied is placed in the entrance for the reference beam so that all the radiation passing through reaches the walls of the integrating sphere. If the MgO disc in the reference beam is replaced by a "radiation trap", the unscattered portion of the transmitted radiation is absorbed. Thus, only the diffuse transmission of the thin layer is measured, which is important for *Duntley's* theory (cf. p. 127ff.).

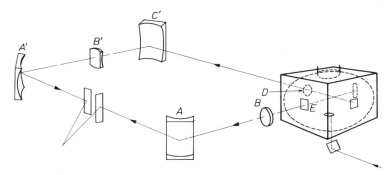

Fig. 92b. Remission attachment for Model 14 Cary spectrophotometer with integrating sphere and optical geometry $_dR_0$. Reverse path of beams

Furthermore, the integrating sphere can be replaced by a so-called "ring collector" in which the optical geometry $_0R_d$ is replaced by $_0R_{45}$. Likewise, the sample is perpendicularly irradiated, but only the radiation reflected at $45 \pm 7°$ is directed by the ring mirror to the detector and compared with the radiation of the reference beam. For strong absorption by the sample, the reference beam can be attenuated. Then the sample is replaced by a standard and the measurement is repeated, i.e., the relative reflectance is measured.

Also in the double-beam apparatus, the radiation path can be reversed (as in the RA 3 attachment to the Zeiss single beam instrument). This converts the optical geometry from $_0R_d$ to $_dR_0$. The relevant radiation path, e.g., for the Cary spectrophotometer, is shown in Fig. 92b. The wall of the integrating sphere is irradiated from below by a mirror and a H_2- or tungsten lamp. The sample and standard thus are diffusely irradiated and the reflected radiation is collected perpendicularly. The image of each is, using a shutter and the optical system, formed alternately on the slit of the monochromator. This method is preferable when the sample is photochemically stable and at the same time fluorescent, because in this instance, the reflectance spectra are less affected by fluorescence.

Recently an improved attachment with an integrating sphere of 25 cm diameter for use in the spectral range from 250 mµ to 2.5 µ has been developed for the Cary 14 spectrophotometer [251]. This attachment does not need to be removed when going from reflectance to transmittance measurements and also has micrometer alignment capabilities to provide for the measurement of the specular reflection of small crystals.

The Hardy spectrophotometer [252] in conjunction with an integrating sphere (Fig. 93) operates by a scintillation method, which is a genuine null method of reflectance measurements. The remitting surfaces B and C serve as sample and reference standard. If, for example, the sample B has a flat, polished surface, then the regularly reflected portion of the radiation arrives at B'. If B' consists, e.g., of MgO, this portion is diffusely scattered and the total remission is obtained. If B' is a "radiation trap", the regularly reflected portion is eliminated, so that only the diffusely scattered portion of the radiation from B is measured, relative both to the scattering of C and C', which consist of MgO [253]. The diffuse and regular components of the remission from powders can be separately obtained in this way. The powder is compressed so that the surface exhibits gloss at the regular reflection angle. However, the regular portion cannot completely be determined in this way because the crystallites in the surface of the compressed powder do not lie completely parallel (cf. also Fig. 14 and 15c) [254].

The so-called scintillation method is based on the following principle: Radiation from the slit S of the double monochromator is passed through a Rochon-prism N_1 and becomes linearly polarized, and then by a Wollaston prism W the radiation is resolved into two equal,

---

[251] *Hedelman*, and *W. N. Mitchell*: Modern aspects of reflectance spectroscopy, p. 158ff. New York: Plenum Press 1968.

[252] *Hardy, A. C.*: J. Opt. Soc. Am. **25**, 305 (1935). — *Gibson, K. S.*, and *H. S. Keegan*: J. Opt. Soc. Am. **28**, 372 (1938). — *van den Akker, I. A.*: J. Opt. Soc. Am. **33**, 257 (1943).

[253] The diffusely and regularly reflected portions can be separated from each other in a similar way in the reflection attachment to the Perkin Elmer 4000 A and 350. [*Anacreon, R. E.*, and *R. H. Noble*: Appl. Spectroscopy **14**, 29 (1960).]

[254] Numerous apparatus that almost always operate by the deflection method have been described for the relative or absolute measurement of the *regular* reflectance of plane surfaces. In this regard, see, for example, *Bennett, H. E.*, and *W. F. Koehler*: J. Opt. Soc. Am. **50**, 1 (1960). — *Reid, C. D.*, and *E. D. MacAlister*: J. Opt. Soc. Am. **49**, 78 (1959). — *Shaw, J. E.*, and *W. R. Blevin*: J. Opt. Soc. Am. **54**, 334 (1964). Preferable methods are those in which the irradiation is perpendicularly incident so that the results of the measurements are independent of the polarization state of the incident radiation.

mutually perpendicularly plane polarized beams. This radiation is then analysed by a second, rotating, Rochon-prism N_2. Thus each of the beams is alternately transmitted and quenched. The phase shift is 180° and the modulation frequency is 60 Hz. The radiation scattered diffusely by the integrating sphere is recorded by the detector situated behind U. By twisting the first Rochon prism, the intensity ratio of both beams can be changed as desired, so that the absorption of the scattering sample

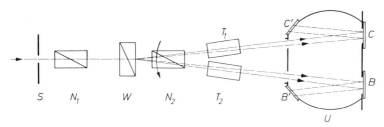

Fig. 93. Hardy spectrophotometer for remission measurements

can be compensated and calculated from the angle of rotation. As long as the intensities of both beams are different, a pulsed direct current is produced in which the alternating current component is amplified and applied to the outer coil of a motor which turns the prism N_1 until equal flux for both beams is achieved. Then the alternating component of the photocurrent disappears. This procedure of optical balancing of the absorption does not involve any assumptions. In particular, it is independent of the properties of the detector and amplifier, so that very reliable values are produced [255].

For measuring reflectance spectra at *extremely high or low temperatures* to the temperature of boiling nitrogen, special cuvette containers or sample holders must be constructed [256] in which the sample plates can be fixed. It is essential, that the entire sample inclusive of the cover glass must be kept at constant temperature so that, e.g., no easily volatile component is preferentially concentrated at the cover glass by sublimation. Such a housing, using the same optics as in the RA 2 for example (cf. p. 223), is attached to the spectrophotometer. Constructions of such cuvette containers have been described, e.g., for fluorescence measurements [257], and these can be modified for reflectance measurements by

---

[255] See *Kortüm, G.:* Kolorimetrie, Photometrie und Spektrometrie, 4th ed. Berlin-Göttingen-Heidelberg: Springer 1962.

[256] See, e.g. *Wendlandt, W., P. H. Franke,* and *J. P. Smith:* Anal. Chem. **35**, 105 (1963). — *Frei, R. W.,* and *M. M. Frodyma:* Anal. Chim. Acta **32**, 501 (1965).

[257] *Kortüm, G.,* and *H. Bach:* Spectrochim. Acta **21**, 1117 (1965).

making trivial alterations. Also, the lower part of the RA 2 attachment with the sample holder can be separated and thermally isolated from the remaining apparatus. These lower parts are best heated to given temperature with a stream of nitrogen. This flows also over the sample surface so that no temperature gradient arises in the sample. The sample is covered at a certain distance with a quartz disc.

Measurements at low temperatures can be simply carried out. A metal block with a sample holder cut in it is immersed in a Dewar vessel

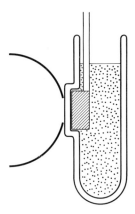

Fig. 94. Reflection cuvette for measurements at low temperatures

containing a cooling bath. Radiation with the geometry $_{45}R_0$ is applied and analysed using two inverting prisms [258]. This has the advantage that the sample does not have to be covered. It is possible also to use a sealed quartz cuvette containing the sample. It is immersed in a quartz Dewar vessel with plane-parallel windows and filled with liquid nitrogen (Fig. 94). The cell is fixed between the windows. The cuvette is diffusely irradiated from an integrating sphere and its image formed on the monochromator slit or the detector [259] ($_dR_0$). A low temperature cell for the reflectance attachment for the Unicam SP 540 spectrophotometer has been described by *Symons* and *Trevalion* [260]. In addition, the McPherson Company reports a double beam apparatus especially designed for low temperature reflectance measurements in the spectral region from 1050 Å to the visible. The radiation reflected from the standard or the sample respectively is alternately conducted via a fiber optic to the multiplier.

[258] *Kortüm, G.,* and *H. Schöttler:* Z. Elektrochem. **57**, 353 (1953).
[259] *Oelkrug, D.:* Ber. Bunsenges. Physik. Chem. **70**, 736 (1966); **71**, 697 (1967).
[260] *Symons, M. C. R.,* and *P. A. Trevalion:* Unicam Spectrovision **10**, 8 (1961).

In more recent times, recording spectrophotometers for remittance measurements were applied also for color measurements and calculations of color recipes and color adjustments. This is especially important for the textile, plastics and paint industries (cf. p. 301 ff.). The first such instrument was the Hardy apparatus described above, but modified by inclusion of a curved plate. Thus, it is not the R'_∞ nor the apparent extinction $\log(1/R'_\infty)$ that is recorded, but the Kubelka-Munk function which is measured directly as a function of λ [261]. Since then, a complete series of recording (and also non-recording) color measuring instruments has appeared on the market [262], some of which have an analog computer directly attached for the integration of the normal color values. A new instrument has been announced by Zeiss (the DMC 25) [263], which differs from the other apparatus in that the optical geometry for irradiation and reflection is variable. For excluding the mirror reflection (gloss), the remission attachment allows irradiation at 45° and measurement at 0° ($_{45}R_0$). The sample is irradiated by two radiation sources at various azimuths, in order to avoid shadows cast by rougher surface structures. With this attachment, exchangeable integrating spheres of various sizes are provided, to ensure the diffuse irradiation of the measured sample and the directed observation at 8° or (when the radiation paths are reversed) a directed irradiation at 8° and an integrating observation by the integrating sphere. The range of measurement is from the near IR to the mid-UV region. The attached analog computer is used to integrate the normal color values. The complete measurement and computation process is fully automatic.

d) Measurements with Linearly Polarized Radiation

As was shown in Section II g, the regular portion of the remission can be eliminated by placing the sample between two crossed polarization foils or prisms, and by ensuring that the angle between the plane of incidence and the plane of oscillation of the radiation (the azimuth) is 0° or 90°. This means, of course, that the incident beam is directed, and the analysis is carried out perpendicularly so that the optical geometry is e.g. $_{45}R_0$. Fig. 95 shows the remission attachment (RA 2) for this type of measurement which can be used in conjunction with the Zeiss spectrophotometer (Fig. 90) in which two polarization foils P and A have been incorporated. These foils consist of stretched, iodine

---

[261] *Pritchard, B. S.*, and *E. I. Stearns:* J. Opt. Soc. Am. **42**, 752 (1952). See also *Derby, R. E.:* Am. Dyestuff Rep. **41**, 550 (1952); also the Cary spectrometer can be supplied with a $F(R_\infty)$ scale.
[262] See e.g. *van den Akker, J. A.*, et al.: Tappi **35**, 141 A (1952).
[263] *Loof, H.:* Zeiss Information Bulletin (in press).

colored polyethylene sealed between 1 mm-thick glass plates. The transmittance of the crossed foils, which lies between 0.01 and 0.001%, is thus sufficiently small. Owing to their characteristic absorption they can only be used in the visible spectral range [264].

For measurements in the UV up to about 250 mµ, a remission attachment was developed for the Zeiss PMQ II spectrophotometer in which polarization prisms made from calcite are used in place of polarization

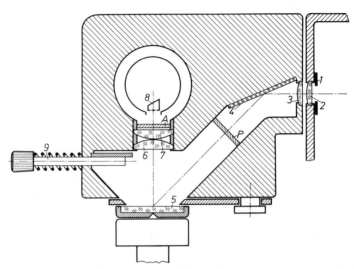

Fig. 95. Zeiss remission attachment RA 2 with built-in polarization foils. *1* exit aperture of the monochromator; *2, 3, 6, 7* optical system; *4* deflection mirror; *5* sample; *8* multiplier cathode; *P* polarizer; *A* analyzer

foils (Fig. 96). The polarizer consists of a Glan prism (in which two half prisms are separated by a thin layer of air) with an angular aperture of 7° and the analyzer is a Glan-Thompson prism with a 30° aperture angle [265].

e) The Measurement of Fluorescent Samples

If the measuring radiation causes fluorescence in an absorption band of the sample, all the undispersed fluorescence radiation is included in the measurement as an apparent remitted radiation when the irradiation is monochromatic. Thus the measurement can be greatly in error.

---

[264] Manufacturer: E. Käsemann, Optische Werkstätten, Oberaudorf/Inn. Type used: Ks-MIK. Recently, UV transmitting polarization filters have been described: *Makas, A. S.*: J. Opt. Soc. Am. **52**, 43 (1962).

[265] Supplied by the firm, B. Halle, Berlin-Steglitz.

In such instances, the sample must be placed in front of the monochromator, so that the fluorescence radiation, which is always at a longer wavelength than the exitation radiation (mirror symmetry!) can be resolved from the remission and thus be eliminated. This is only possible, however, in regions where the absorption and fluorescence bands do not overlap. Monochromatic irradiation can also be used, of course (for sensitive

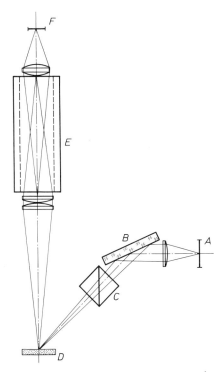

Fig. 96. Remission attachment for measurements between crossed polarization prisms. *A* exit aperture of the monochromator; *B* deflection mirror; *C* polarizer; *D* sample; *E* analyzer; *F* multiplier

samples), and the remitted and fluorescent radiation can be resolved by a second monochromator [266], but this requires the use of more sophisticated measuring techniques.

In many instances it is desirable to be able to jointly measure the remission *and* the fluorescence of a sample. This is the case for example when dealing with a fluorescent pigment or a so-called "optical clearing agent" which causes weakly colored media to appear "white" as a result

---

[266] *Donaldson, R.*: Brit. J. Appl. Physics **5**, 210 (1950).

of additional short wavelength fluorescence under irradiation from day light, for example. In this case also, the sample must be arranged before the monochromator and irradiated with a continuum. The radiation flux $I(\lambda)$ measured behind the monochromator relative to that of a standard (MgO) is composed of the reflected part and a fluorescence part excited by the short wavelength region of the incident irradiation. It is directly obvious that $I(\lambda)$ in this case, in contrast to pure remission measurements, depends on the radiation flux and the spectral energy distribution of the light source in the excitation range of the fluorescence and additionally on the quantum efficiency of the fluorescence. $I(\lambda)$ can thus be strongly influenced by the choice of light source.

For such special measurements, non-recording [267] as well as recording [268] apparatus have been developed. More usual apparatus with reversible radiation paths can, however, also be used for these purposes [269]. In these cases the fluorecence is, as a rule, excited with a xenon lamp of high irradiation intensity fitted with suitable filters to absorb the short wave UV. The spectral energy distribution of the combination corresponds somewhat to ordinary daylight. Additional UV-suppression filters serve to separate the fluorescence and remission from one another insofar as the sample is not excited to fluorescence by radiation in the visible region (see also p. 304).

f) Influence of Moisture on Reflectance Spectra

As has already been shown in Chap. III, the scattering power of small particles depends on the ratio of particle size to wavelength and on the ratio of the refractive indices of the particles, n, and the surrounding medium, n_0. This is true, also, for multiple scattering. If, e.g., in the pores of a dried powder the air is replaced first by water vapor, and then by capillary condensed water, the ratio n/n_0 and the scattering coefficient decrease, and thus, from Eq. (IV, 32) $F(R_\infty)$ increases. It is generally observed that moist samples are visually darker than dried samples. It is therefore expected, as is confirmed by experience, that reflectance spectra obtained from dilution methods (p. 177) can be affected by moisture if either the substance of study or the diluent or both are hydrophilic and easily take up water. In such instances, it is frequently necessary to dry very carefully the samples and the diluent (standard) and to protect them against penetration of moisture by use of cover glasses while spectra

---

[267] *Schultze, W.:* Farbe **2**, 13 (1953).

[268] *Koch, O.,* and *K. Bunge:* Chem. Ing. Techn. **32**, 810 (1960).

[269] For example, the "Elrepho" described on p. 222. See also *Berger, A.:* Zeiss Information Bulletin (in press).

are being measured (cf. p. 190). An experiment using K_2CrO_4 and NaCl as easily water soluble substances and Cr_2O_3 and CaF_2 as slightly water soluble substances has confirmed this[270]. For example, the reflectance spectrum of Cr_2O_3 diluted with CaF_2 is measured against pure CaF_2 as a standard. This gives the curves in Fig. 97, when the highly dried sample is placed in water-vapor-saturated air for long periods, and then placed in vacuum over P_2O_5. Under the influence of this "condensed moisture", the bands are considerably increased in intensity, but without changing

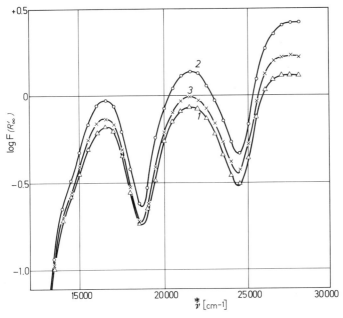

Fig. 97. Reflection spectra of Cr_2O_3 + excess of CaF_2 against CaF_2 as standard. *1* without water (dry); *2* after 76-hour standing in air which is saturated with water vapor; *3* after 7-days over P_2O_5 in vacuum

their position. This effect is almost completely reversible as the moisture is removed. The slightly incomplete reversibility can be assumed to arise because of the great difficulty in removing condensed water from the smallest pores completely. If the same measurements are made with K_2CrO_4 and NaCl, the analogous spectral changes are no longer reversible. This may be due partly to recrystallization effects, resulting in a grain

---

[270] *Kortüm*, G., W. *Braun*, and G. *Herzog:* Angew. Chem. **75**, 653 (1963); Internat. Ed. **2**, 333 (1963).

enlargement. This has been known to cause an additional increase in the absorbance because of the larger penetration depth of the incident radiation (cf. p. 61). Such effects cannot be made reversible by subsequent drying. Some times a powder that has been pressed loosely into the measuring cell forms a solid disc under the influence of moist air. This is an indication that such a recrystallization has occured.

As these experiments show, it is advisable to choose water-insoluble diluents or standards, provided that their application is not excluded on other grounds. For quantitative, e. g., photometric measurements, or for reflectance measurements on adsorbed substances, it is nevertheless necessary, however, to exclude moisture as far as possible, making sophisticated experimentation necessary[271]. The diluent, dependent on its chemical constitution, should generally be ignited for a long time (400–1000° C) or even until constant weight is achieved, and is then allowed to cool in a vacuum dessicator. This is transferred to a so-called "glove box", which already contains the weighed sample and the grinding apparatus which have been dried previously. The sample and the diluent, or only the latter when preparing the standard, are poured into the grinder, and the box is made air-tight with a suitable seal (e.g. a Teflon ring). After grinding, the grinder is returned to the glove box, and the diluted sample and the standard are placed in the measurement plates. For this purpose, the glove box contains the quartz cover plate, the packing medium, equipment for transfering the sample (forceps, spatula, etc.) and a hot plate, so that the sample holder can be prepared for measurement in the complete absence of moisture.

Glove boxes can easily be constructed in the laboratory. They consist of an air-tight container, on which is sealed a well-fitting, overlapping glass or Plexiglass window for observation. The manipulations that are necessary are carried out using two rubber gloves leading from fitted rings, with which they are attached, into the box. Highly-dried CO_2 is introduced slowly into the bottom of the box, and it escapes from the roof of the box. The box contains, also, a dish of P_2O_5 as the drying agent. About an hour before the sample preparation, the hot plate is switched on, and the sample holder, quartz cover plate and the instruments are again dried by heating, with constant flushing of the box with dry CO_2. When the heating current has been switched off and the plates have cooled, the preparation of the samples for measurement can be started.

Recently, various types of glove boxes (evacuable or non-evacuable, with or without an entrance chamber) of various shapes and sizes have become commercially available[272].

[271] *Kortüm, G., J. Vogel*, and *W. Braun:* Angew. Chem. **70**, 651 (1958).

[272] See, e.g., Mecaplex AG, Grenchen, Switzerland, sold through Dr. H. Rumm & Co., Augsburg, West Germany.

g) Preparation of Samples for Measurement

As in single scattering, the scattering coefficient for multiple scattering depends not only on the ratio n/n_0, but also very strongly on the particle size. And, of course, so does the associated spectrum $F(R_\infty)$, as has already been pointed out (Chap. II g). The particle size used in the dilution method (p. 175) is, in turn, dependent on the duration of grinding, and the hardness of the material being ground. The *grinding process* for the material used to dilute the sample to be measured with the standard, carried out in a mortar or a ball mill, must be standardized if it is used for comparative or quantitative photometric measurements. A porcelain grinding device, of 50 to 250 cm^3 capacity with two large or six smaller smooth porcelain balls is found to be suitable[273]. For harder materials, agate grinders with grinding balls of agate or hard metal (tungsten carbide and cobalt) are preferable. For a series of measurements at various dilutions, the same type of grinder with similar balls should always be used if the measurements are to be quantitatively reproducible. The particle size of the ground material depends essentially on the type and number of balls, on the speed of revolution of the mill, and on the duration of grinding. The standard must be treated exactly the same as the sample, and for this purpose it is best to use a twin mill. To achieve a homogeneous mixing of sample and standard or to obtain a uniform adsorption of the substance to be studied on the surface of the standard, the grinding process is carried out for at least six hours, or preferably overnight. The effect of duration of grinding on the apparent extinction $\log \frac{1}{R_\infty}$ of a mixture of K_2CrO_4 and $BaSO_4$ is shown in Fig. 98. The extinction tends to a limiting value so that it does not increase indefinitely with continued grinding[274]. This indicates that the finest particles agglomerate so that a stationary state of particle sizes and of particle size distribution is reached in the region of 2 down to about 0.2 µ.

A long grinding time causes abrasion of the grinder and its balls. However, systematic experiments have shown that this abrasion does not affect the measurements within the accuracy of the method if polished porcelain balls are used that have been in use for a long time. Agate balls, and especially Widia steel balls always show abrasion, particularly when they are new. This is to be borne in mind and the ground standards are to be compared with freshly prepared MgO.

---

[273] Suitable ball mills are produced by, e.g., the firms of Ludwig Hormuth, Wiesloch (Baden) and Alfred Fritsch, Idar-Oberstein, West Germany.

[274] *Kortüm, G.*, and *G. Schreyer:* Angew. Chem. **67**, 694 (1955).

To obtain even finer particles, the "vibrator" devised by v. Ardenne [275] can be used. In this device, a steel spring vibrates, in a suitable vessel with two or three balls, in resonance in front of the core of an electromagnet carrying a 50 Hz alternating current. A reasonably homogeneous particle size of less than 1 μ is obtained in a short time. To obtain very small particles, the mechanical pulverising method must be avoided and the "lyophilisation" method can be used. This can be applied only to soluble substances and is carried out as follows: The substance concerned is dissolved in water or another suitable, crystallisable solvent. The

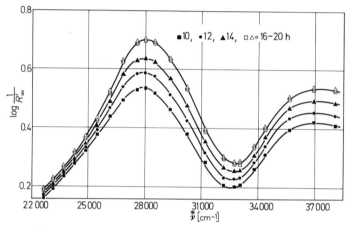

Fig. 98. Influence of grinding time on the reflection spectrum $K_2CrO_4/BaSO_4$; ■ 10 hours, ● 12 hours, ▲ 14 hours, □ ○ △ 16–20 hours

solution is cooled rapidly by immersion in a CO_2-acetone or liquid N_2 bath, so that it solidifies, and the solvent is sublimed in vacuum. According to the rate of freezing out and the concentration of the solution, uniform particle sizes of as little as 0.1 μ or less can be obtained. This method can also be applied to the formation of a sample-diluent (e.g. K_2CrO_4 and KBr) mixture in any desired ratio. The dry powder can be measured immediately (e.g. against pure KBr as standard). The smallest particles that are as yet accessible are the aerosils (already mentioned on p. 43), produced by the hydrolysis of $SiCl_4$ vapor (see Table 12). These have been useful as a diluent for the investigation of certain problems [276].

---

[275] v. Ardenne, M.: Angew. Chem. **54**, 144 (1941); Kolloid-Z. **93**, 158 (1940).

[276] See also Wagner, E., and H. Brunner: Angew. Chem. **72**, 744 (1960). Very fine particle sizes can also be similarly obtained with TiO_2.

As well as the particle size, the nature of the sample surface also has an effect on the measured reflectance spectrum. The surface should, if possible, be smooth and even, but should have no gloss. Roughness and cavities cause shadowing effects in inclined directionally irradiation (RA 2) and thus give a falsely large absorption. The sample should be compressed into the measurement plates either with a suitable glass or metal stamp (slightly etched) or a quartz plate subjected to slightest twisting motion, or the surface should be made even by placing a glazed paper over the sample and rolling it flat with a thick glass rod. For very loose samples, the dependence of the spectrum on the packing density must be considered (cf. p. 61), and the same amount must be weighed into the sample holder for all photometric measurements. Sometimes, mixtures are found that exhibit gloss even after minimal pressure. This is so for Cr_2O_3–MgO but not for Cr_2O_3–CaF_2. In such instances, the sample must first be rolled smooth, and with a sieve a uniformly-thick layer applied to eliminate the gloss. Samples with *surface structure* (paper, textiles, etc.) should always be measured using diffuse irradiation. For very inclined directionally irradiation, differences of many per cent are obtained between the reflectances measured in two mutually perpendicular positions of the sample.

The sample holder should always be filled to the brim. *Different distances of the sample surface from the sphere wall*, or particularly from the image-forming lens (Fig. 95) cause differences in reflectance of up to 5%, probably because the effective solid angle of radiation remitted by the sample changes with the distance of the sample from the lens. In principle, it should make no difference whether the *dilution medium* or another non-absorbing standard is used as a reference standard, as long as an isotropic scattering distribution can be assumed. It is known, however, that this is not completely attained in the uppermost powder layer, so that the effective angular distribution of the remitted radiation can differ for the various standards (because of their various particle sizes and crystal forms). It is recommended, therefore, that the diluent is always used as the reference standard.

Layer thicknesses of 2–5 mm generally suffice for the measurement of R'_∞, depending always on packing density, only the very fine particles of the aerosils (Table 12) require layer thicknesses of up to 10 mm [277]. In this instance, errors can arise because laterally scattered light does not leave the sample surface quantitatively.

To determine the *scattering coefficient*, we must measure R_0 or T in addition to R_∞ (cf. p. 195). The choice of layer thickness for the measurement of R_0 or T depends on the ratio of scattered and absorbed radiation

---

[277] In the red spectral region, even this layer thickness is insufficient for the finest aerosil particles.

by the sample [278]. If we intend to measure R_0, the accuracy with which S can be determined depends in the first place on the difference between R_0 and R_∞ which should not be smaller than 0.05. If T is to be measured, it should lie in the region 0.5–0.05. For larger transmittances in directed irradiation, part of the light passes through the sample without being scattered so that the Kubelka-Munk theory can no longer be used. For non-absorbing substances (standards), the layer thicknesses are best restricted to between 0.1 and 0.3 mm. As the absorption increases, the difference between R_0 and R_∞ decreases rapidly, as does the transmittance. In this case the reflectance method rapidly becomes less useful than the transmittance method, as can be seen from the following example. If $S = 500$ cm^{-1} (cf. e.g., Fig. 80), $d = 0.01$ cm and $R_\infty = 0.600$, then Eq. (IV, 57) gives $R_0 = 0.598$, i.e., $R_\infty - R_0$ is smaller than the accuracy of measurement of the method. For transmittance measurements under the same conditions, Eq. (IV, 60) gives $T = 0.08$, which is still possible to measure with sufficient accuracy.

To be able to measure the scattering coefficients of more strongly absorbing substances, much thinner layers must be prepared. This is limited by two difficulties: the granular nature of the material results in non-planarity of the layer, so that great errors are introduced. Furthermore because the particle size of the powder cannot be made arbitrarily small, the layer thickness can reach only a certain limiting value to be consistent with the theory which assumes that the particle size must be small in comparison with the layer thickness. On this basis the scattering coefficients of feebly absorbing substances only can be measured [279]. For the dilution method, described on p. 175, only the scattering coefficients of the diluents or standards are of interest, so this limitation is only of slight importance.

To measure R_0, the powder is placed in a sample holder 0.1–0.3 mm deep. With a piece of glazed paper and a round glass rod, the surface can be made even. The layer thickness can be determined by using a micrometer scale microscope which is focussed accurately at the position of the sample surface and at the brim of the holder, and, analogously, the depth of the empty holder is measured. For accuracy of reading of about 1μ, the sample thickness can be determined with an accuracy of only 5μ. However, this error can be lessened by taking measurements at many positions and using the average value.

---

[278] See *Oelkrug, D.:* Thesis, Tübingen 1963.

[279] Thus a method, recently given by *Caldwell, B. P.:* J. Opt. Soc. Am. **58**, 755 (1968), for calculating the Kubelka-Munk coefficients (S, K, R_∞, R_0) from transmission measurements on 2 layers with thickness ratio of 1 : 2, is of limited applicability.

A matte black lacquer layer can be used as a background in the bottom of the holder. This only reflects about 2% of the incident light so that this small reflectance by the base can be neglected in practice.

As was shown on p. 136, transmittance measurements on absorbing substances should be carried out, if possible, using only one quartz plate, because exact values cannot be obtained using two covering plates. Furthermore, it should be considered that the Kubelka-Munk theory cannot be applied to transmittance measurements on very thin layers and under directed irradiation, as is generally used [280]. The theory discussed in Chap. IV d, should be applied.

When *selecting a standard* for diluting the sample to be measured, we must know if the standard will react with the sample. Acids adsorbed on MgO or $CaCO_3$ show the spectra of the salts, whereas salts adsorbed on acidic Al_2O_3 reflect the superimposed spectra of the acid and salt. Many standards that can be obtained in a highly dried condition cannot be used for the adsorption or dilution of organic substances. For example, silica gel is already so active after an hour's drying at 600° C that aromatic compounds are decomposed when ground with the silica gel. Fig. 99 gives an example [281a, 281b]. Apparently, the heating of the silica gel

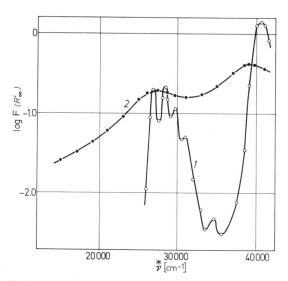

Fig. 99. Reflection spectra of anthracene on silica gel. *1* SiO_2 not dried beforehand; *2* SiO_2 dried for one hour at 600° C

---

[280] Otherwise, two photometer spheres must be used, which is experimentally rather inconvenient.

[281a] *Kortüm, G.,* and *W. Braun:* Z. Physik. Chem. N. F. **28**, 362 (1961).

[281b] *Briegleb, G.,* and *H. Delle:* Z. Physik. Chem. N. F. **24**, 359 (1960).

produces Si–O–Si bridges. These are mechanically broken during the grinding process (or possibly because of local overheating) to form active centers $-O_3Si^+$ or O_3Si-O^- or, also, radicals, which react with the organic compounds [281c].

h) Adsorption from the Gas Phase and from Solution

Although the grinding method is very useful for a simple mixing or dilution process, it is often not so suitable for the study of the spectra of adsorbed molecules (cf. p. 256 ff.) as the example in Fig. 99 shows. Moreover, the grinding also frequently changes the particle size of the

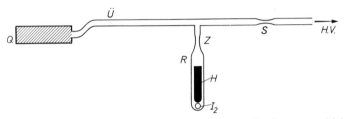

Fig. 100. Device for examining the reflection spectra of substances which have been adsorbed from the gaseous phase (side view)

adsorbent, which is troublesome if the study of the effect of surface coverage at constant surface area is desired. In such instances, it is preferable that the substances to be studied be adsorbed onto the standard from the gas phase or from solution.

Naturally, *adsorption from the gas phase* is possible only when the substance concerned is already a gas, or has sufficient vapor pressure at an easily accessible temperature. The great advantage of this method is that the adsorbent is heated in a sealed cuvette in high vacuum. Thus, it is freed of all other adsorbed substances, and is then coated with the substance under investigation in the same apparatus. Fig. 100 shows a diagram of such an apparatus that was used, e.g., for the study of the spectrum of adsorbed iodine [282].

The quartz cuvette Q, filled with the adsorbent, is connected, *via* a connecting piece U (Quartz/Jena glass), to a temperature controlled tube R which contains a sealed ampoule of iodine. The whole apparatus is heated to 580–600° for several hours in a high vacuum and is sealed at S with the pump still running. A similar cuvette (without R) serves as a

---

[281c] *Weyl, W. A.:* Research **3**, 230 (1950). — *Benson, R. E.,* and *J. E. Castle:* J. Physic. Chem. **1958**, 840.

[282] *Kortüm, G.,* and *H. Koffer:* Ber. Bunsenges. **67**, 67 (1963).

reference standard. First, the difference in the reflectance of the two cuvettes if determined. Then the iodine ampoule is broken with the hammer H. The temperature of the cuvette, t_1, and that of the tube, t_2 are kept constant with appropriate thermostats, so that always $t_1 > t_2$. For the standardization of the adsorption equilibrium, this can be continued for several hours depending on the particle size and the packing density of the standard, and then the reflectance spectrum is measured. There is always excess iodine in the temperature controlled tube so that the vapor pressure is fixed by t_2. When this tube is sealed at Z, measurements can be made at constant iodine coverage and at various temperatures. Thus the association equilibrium $2I_2 \rightleftharpoons I_4$ at the surface can be determined. As this is the behaviour of a "sample with a cover glass", the measurements can be evaluated using Eq. (IV, 114–115a).

Adsorption from the gas phase is experimentally relatively simple. However, apart from the long time required for stabilising the equilibrium, another limitation is that it is not applicable to substances with low vapor pressures, because either the covering thickness on the surface is very small or – at higher temperatures – decomposition or other reactions with the adsorbent take place. *Adsorption from solution* therefore is more versatile, but it is experimentally more intricate. However, this procedure also prevents the undesirable complications resulting from the grinding process. A suitable apparatus[283] is described schematically in Fig. 101. L is the solvent reservoir, which is connected via the vacuum stop cock H_1 and a cold trap K_1 with the pump, and also via H_2 with cold trap K_2 that serves as a solution container. A contains the adsorbent, which is attached to H_3 and cold trap K_3 in the high vacuum line. It is shown below on a larger scale. The lateral connecting tube goes to the container in which the adsorbent is heated. By turning the joint through 90°, the adsorbent falls into the container A, and is there brought into contact with the solution by tipping the whole apparatus. The solution was previously prepared in K_2. The adsorbent and the solution are mixed with a magnetic stirrer. Then, after the adsorption, the powder forms a sediment on the quartz plate fixed underneath. This sample is then irradiated from below and the reflectance is measured with an adjustable remission attachment (optical geometry $_{45}R_0$). The solvent is purified by repeated freezing with liquid nitrogen and melting in L, so that it is thoroughly degassed, and then distilled over into K_2, which contains the sample, having been previously weighed out and also highly vacuum dried and degassed. In this way, the solution can be prepared in the sealed apparatus in the absence of air or moisture. The amounts of sample, solvent (e.g. hexane) and adsorbent can be chosen so that the

---

[283] *Kortüm, G.*, and *V. Schlichenmaier:* Z. Physik. Chem. N.F. **48**, 267 (1966). — *Kortüm, G.*, and *M. Friz:* in press.

244 VI. Experimental Techniques

equilibrium lies almost completely on the side of the adsorbent, so that, for known specific surfaces of the adsorbent, the coverage can be calculated.

The measurement can be made in two different ways. Either the solvent is removed again completely in high vacuum after the adsorption. This is again an example of "sample with cover glass", and the

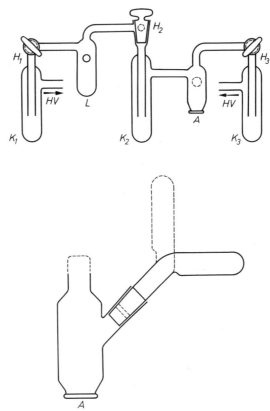

Fig. 101. Device for examining the reflection spectra of substances which have been adsorbed from solution

measurements can be evaluated using Eqs. (IV, 114–115a). Thus it must be measured against a standard that also has a cover glass. Alternately, the solvent can be left with the sample in the cuvette. Thus the measured effects are larger because the relative refractive index and also the scattering coefficient are smaller. Of course, the Kubelka-Munk theory is no longer rigorously applicable (cf. p. 131), and the method should be limited to such instances where only qualitative results are of importance;

the position of the absorption bands is also in this instance correct. However, the measurements can be evaluated according to (IV, 113).

i) Measurements in the Infrared

Although reflectance measurements can be extended without difficulty to the near IR (ca. $2\,\mu$)[284], with the usual instruments and a suited multiplier or PbS cell, experiments have shown that considerable experimental difficulties are met with in the longer wavelength region of the IR. There are two reasons for this. First, the IR sensitive detectors are also sensitive to the thermal radiation of the surroundings (ca. 300° K),

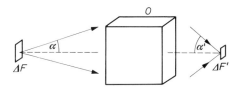

Fig. 102. Optical projection of a surface ΔF onto a smaller surface $\Delta F'$

and this causes high background noise which increases with the square root of the detector surface[285]. Thus the smallest possible surfaces should be used. Whereas the surface area of a multiplier is of the order of 1 cm², a thermoelement or a bolometer surface area is of the order of 0.5 mm². Second, there is the difficulty that the radiation reflected diffusely by the sample must be collected onto a small detector surface as completely as possible. To obtain an image of a plane ΔF, using an optical system, on a smaller plane $\Delta F'$, the generally valid relationship

$$\sin^2 \alpha' \geq \frac{\Delta F}{\Delta F'} \sin^2 \alpha \qquad (9)$$

holds, where α and α' signify the half aperture angles (Fig. 102). For diffuse irradiation ($\alpha = 90°$), it follows from the inequality of Eq. (9), that no diminished image can be obtained, i.e., the sample surface too must be made as small as possible.

Experiments using the integrating sphere (e.g. with a NaCl lining) for reflectance measurements in the IR, failed[286]. It may be assumed that

---

[284] Cf. e.g. *Hoffmann, B. K.*: Chem. Ing. Techn. **35**, 55 (1963).

[285] Cf. e.g. *Kortüm, G.*: Kolorimetrie, Photometrie und Spektrometrie, 4th ed. Berlin-Göttingen-Heidelberg: Springer 1962.

[286] *v. Hirschhausen, H.*: Diplomarbeit, Tübingen 1961.

the wall of the sphere is ideally reflecting ($R_\infty = 1$) and that the irradiation entrance and the detector surface are small compared with the sphere surface. When the absorbing sample is replaced by a standard, the fraction of the total radiation that falls on the detector is given by the ratio (detector surface)/(detector surface + irradiation opening) and is of the order of 1/2. However, in reality, the reflectance of the sphere is not 1 but is markedly less, more so in the IR than in the visible region. If it is assumed that the integrating sphere has a diameter of only 5 cm, and that $R_\infty = 0.95$, the absorption of the wall (5%) corresponds to an additional entrance surface of $0.05\,\pi d^2 \cong 4$ cm^2. If the detector surface $\cong$ irradiation entrance = 0.01 cm^2, the above ratio becomes 1/400, i.e., even without absorption by the sample only a vanishingly small fraction of the radiation falls on the detector. Thus integrating spheres cannot be used for measurements in the IR.

These optical difficulties can be overcome rather easily, if the sample to be measured is placed inside a hohlraum reflectance spectrophotometer[287]. Such an apparatus was developed by *Perkin-Elmer*[288] (Model 13/205). Its radiation paths are schematically described in Fig. 103. The principle of measurement is as follows: A part of the inner surface of the hohlraum radiator which can be heated to 400–1100° C, is replaced by the water cooled sample. Because the sample surface and the entrance are small compared to the inner surface of the radiator, the sample obtains radiation which has the density and the spectral distribution of a black body. It reflects a part of this radiation through a (small) exit in the base. A reference beam from the wall of the cavity leaves by the same opening. Both beams are passed, via a system of mirrors, to a double beam spectrophotometer where their intensity ratio is measured in the usual way. Although this method seems so simple, it has one considerable limitation: Because the sample is not in radiation equilibrium with its surroundings, it is heated and, depending on its absorption and its thermal conductivity it may or may not be possible to cool it so well that its surface temperature does not exceed the mentioned limiting value of 50° C. Self emission of the sample as a result of high temperatures and temperature fluctuations of the cavity walls can lead to considerable systematic errors.

The only, as yet satisfactory method which collects the radiation diffusely reflected by the sample on a small area detector, uses the optical

---

[287] *Gier, J. T.*, et al.: J. Opt. Soc. Am. **44**, 558 (1954). — *Starr, W. L.*, and *E. Streed*: J. Opt. Soc. Am. **45**, 584 (1955). Details of the hohlraum type infrared reflectance spectrophotometer are given by *Keith, R. H.*: in "Modern aspects of reflectance spectroscopy", p. 70ff. New York: Plenum Press 1968.

[288] Perkin-Elmer Instrument News **10** (4), 1 (1959). — *Reid, C. D.*, and *E. D. McAlister*: J. Opt. Soc. Am. **49**, 78 (1959).

characteristics of the rotation ellipsoid [289]. The sample P is placed at one focus of the half ellipsoid with the Al mirror surface, and the detector R is at the other focus. The sample is irradiated monochromatically through an opening in the ellipsoid (Fig. 104) so that warming up is not to be suspected. In the arrangements reported earlier, the ellipsoid was replaced by a spherical mirror, which is more readily accessible, and the sample and detector were placed at the same, and as small as possible, distance from the mid-point of the sphere [290]. With this arrangement, the

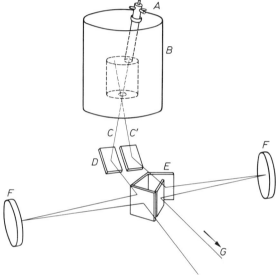

Fig. 103. Path of radiation in the reflection spectrometer, Model 13/205, from Perkin-Elmer, Norwalk, Conn., U.S.A. *A* sample plate; *B* Hohlraum; *C* sample beam; *C'* reference beam; *D* mirror; *E* deflecting mirror; *F* spherical mirrors; *G* to spectrometer

detector surface must again be made rather large because of spherical aberration. If quantitative measurements are desired, the apparatus must here again be calibrated with a comparison standard of known reflectance. Apparently quite significant systematic errors occur during the measurements so that absolute measurements are not possible without calibration [291].

[289] *Paschen, F.:* Ber. Berl. Akad. Wiss. **27** (1899). — *Coblentz, W.:* Bull. Natl. Bur. Std. **9**, 283 (1913). — *Sanderson, J. A.:* J. Opt. Soc. Am. **37**, 771 (1947). — *Derksen, W. L.*, and *T. I. Monahan:* J. Opt. Soc. Am. **42**, 263 (1952).

[290] Such a device, incorporated in a double beam spectrophotometer, has been produced by Beckman, Inc.

[291] *Kronstein, M.*, et al.: J. Opt. Soc. Am. **53**, 458 (1963).

An IR reflectance spectrophotometer that actually incorporates a rotation ellipsoid has been developed by reconstructing the Perkin-Elmer 12 C spectrophotometer with CaF_2 prisms[292]. The thermo-element, a reversing mirror and a concave mirror were removed from the production model. The path of the rays is shown in Fig. 105. The radiation is projected immediately behind the exit slit of the monochromator via the plane mirror M_6 onto the spherical, concave mirror M_7. The latter mirror forms the image of the exit slit, reduced in the ratio 2 : 1,

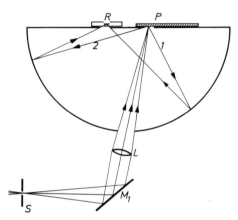

Fig. 104. Optical projection of the diffusely reflecting sample onto the detector by means of a rotation ellipsoid

on the half-elliptical mirror M_8 at the focus of the ellipsoid where the sample is placed. The radiation reflected diffusely by the sample is reflected by the ellipsoid onto the detector at the other focus. The spherical mirror M_7 is placed in a window cut out of the half ellipsoid, just opposite the sample. This makes it possible to cover the sample with a window. The radiation reflected regularly from this window is projected back to mirror M_7 and does not reach the detector. The latter is of gold-doped germanium[293] with a detector surface 1×5 mm. The spectral sensitivity is sufficient for measurements up to 7.6 µ. Thus it covers the region of the ground state vibrations of most organic molecules. Furthermore, it is possible that the half-ellipsoid can be combined with a double beam spectrophotometer. The measurement can be arranged in such

---

[292] *Kortüm, G.*, and *H. Delfs*: Spectrochim. Acta **20**, 405 (1964). A similar arrangement was given by *Brandenberg, W. M.*: J. Opt. Soc. Am. **54**, 1235 (1964). In addition, the focusing characteristics of elliptical and spherical mirrors were investigated.

[293] Philco (Landsdale, Penn., USA).

a way that it is not made against a standard but that the radiation flux itself is used as a reference, and absolute reflectance values are thus obtained. Because, however, all detectors respond worse the greater the angle of incidence of the radiation (according to *U. White*[294], at an aperture angle of 2π, only about 60% of the diffuse incident radiation is registered), this arrangement also requires calibration with a sample of known reflectance (MgO).

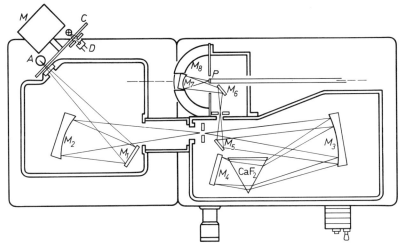

Fig. 105. Path of radiation of the Perkin-Elmer 12C spectrometer, converted to a reflection spectrometer with rotation ellipsoid. *A* Globar; M_1—M_6 optical system; M_7 concave mirror for irradiating the sample *P*; M_8 elliptical mirror; *M* 50 cycle motor; *C* 800 cycle chopper; *D* scanner

An IR reflectance spectrophotometer with a rotation ellipsoid has also been described by *Blevin* and *Brown*[295] that uses for reflection only that part of the ellipse separated by a plane through the secondary axis. This has the advantage that the aperture angle that focuses the incident radiation at the detector amounts to less than $\pi/2$ instead of 2π and thus eliminates large angles of incidence at the detector. On the other hand, the sample, which is brought into the other focus point of the mirror with the help of a bracket, partially screens the radiation between the elliptical mirror and the detector. This must be corrected for. Calibration is made by moving an aluminum mirror in place of the sample by means of the bracket. The reflectance of this standard must be known beforehand. A bolometer with NaCl windows and a relatively large detection

---

[294] *White, U.*: J. Opt. Soc. Am. **54**, 1332 (1964); see in addition *Keegan, H. J.*, and *V. R. Weidner*: J. Opt. Soc. Am. **55**, 1567 (1965); **56**, 540 A (1966).

[295] *Blevin, W. R.*, and *W. J. Brown*: J. Sci. Instr. **42**, 385 (1965).

surface of 12.6 × 6.5 mm, which implies a considerable radiation background noise, serves as the detector. The measuring region lies between 0.7 and 14 μ. Because of the multiple reflection between the mirror and the detector, this case also requires further calibration by means of a sample of known reflectance for quantitative measurements.

A method given by White[294] for reflectance measurements in the IR attempts to combine the advantages of optical characteristics of rotation ellipsoids and of diffuse irradiation. It is true, here also the ellipsoid is replaced by a less favorable spherical mirror. A radiation source and the sample are placed at the focus points so that the latter is diffusely irradiated by the radiation from the mirror. The reflected radiation is conveyed to the detector through a hole in the hemisphere via a mirror system. Since the sample can be heated in the event studies at higher temperatures are desired, the radiation is modulated by a chopper between the source and the sample so that the emission and reflection of the sample can be distinguished. The reference radiation is taken directly from the source itself so that a double-beam spectrometer can be used. The sample holder has temperature control capabilities so that samples can be investigated in a large temperature range (-175 to $1000°$ C). The emitted radiation serves for temperature determination.

k) Discussion of Errors

It is assumed that the accuracy of the measurement is not affected by the scatter of the instrument readings (e.g., the galvanometer deflection), but depends only on the intensity difference dR_∞ to which the detector just responds. Hence, the relative error in the Kubelka-Munk function can be obtained by differentiation of Eq. (IV, 32) with respect to R_∞:

$$\frac{dF(R_\infty)}{F(R_\infty)} = -\frac{1+R_\infty}{1-R_\infty} \cdot \frac{dR_\infty}{R_\infty}. \tag{10}$$

Even when the absolute error dR_∞ is only 0.003, a limit, below which it is only seldom possible to proceed, this makes the relative error of K/S at $R_\infty = 0.9$ about 6%, and at $R_\infty = 0.1$ about 4%. It is occasionally stated that a relative accuracy of 0.1–0.2% can be attained in remission measurements, but this statement has no foundation. The relative error reaches a minimum at $R_\infty = 0.414$; its dependence on R_∞ is given in Fig. 106, where the minimal error is set equal to 1. It is apparent that the most favourable range for measurement lies[296] between $0.2 < R < 0.6$, and the error at larger or smaller reflectances increases very rapidly. In reality, this error is still larger, because the absolute reflectance of MgO, relative to which all values are expressed, is only inaccurately known (cf. p. 146).

---

[296] See also Nickols, D. G., and S. E. Orchard: J. Opt. Soc. Am. **55**, 162 (1965).

For the accuracy with which the scattering coefficient S can be determined from the measurements of R_∞ and R_0, the difference between R_∞ and R_0 is decisive, as has already been pointed out on p.240. Stenius[297] has shown that the ratio of the relative error in S to the relative error in R_∞ at constant R_0 is

$$-\frac{dS/S}{dR_\infty/R_\infty} = -\frac{R_\infty}{S}\left(\frac{\partial S}{\partial R_\infty}\right)_{R_0}. \qquad (11)$$

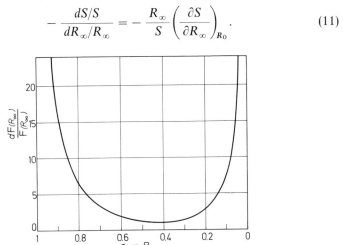

Fig. 106. Relative error of the Kubelka-Munk function in dependence on R_∞ (minimum = 1)

This ratio is plotted in Fig. 107 as a function of R_0 and R_∞ in the form of a group of curves. It can be seen that this ratio becomes smaller as the difference between R_0 and R_∞ increases. At $R_0 = R_\infty$, it becomes infinitely large. Moreover, the error likewise increases rapidly with increasing absorption as has already been indicated.

The expression analogous to (11) is

$$-\frac{dK/K}{dR_\infty/R_\infty} = -\frac{R_\infty}{K}\left(\frac{\partial K}{\partial R_\infty}\right)_{R_0}. \qquad (12)$$

The ratio of the relative error in the absorption coefficient to the relative error in R_∞ at constant R_0 is likewise shown in Fig. 108 as a function of R_0 and R_∞. Comparison with Fig. 107 shows that the ratio $R_\infty dK/K dR_\infty$ is larger than the ratio $R_\infty dS/S dR_\infty$ by a factor of ten under otherwise identical conditions. Thus, e.g., the point $R_\infty = 0.8$ and $R_0 = 0.6$ has a value of about 0.8 from the curve in Fig. 108, but has a value of ten from the curve in Fig. 107. An error of measurement of 1% in R_∞ causes, in this instance, an error of only 0.8% in the scattering coefficient, but an error

---

[297] S: son Stenius, Å.: Svensk Papperstidning **54**, 663, 700 (1951).

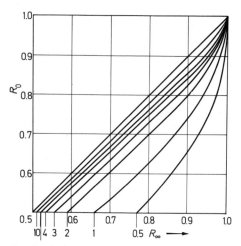

Fig. 107. The ratio of relative error of the scattering coefficient S to the relative error of the reflectance R_∞ as function of R_0 and R_∞

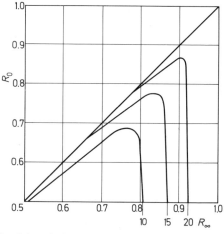

Fig. 108. The ratio of the relative error of the absorption coefficient K to the relative error of the reflectance R_∞ as function of R_0 and R_∞

of 10% in the absorption coefficient. This is typical of the extreme sensitivity of the absorption coefficient to errors in measuring R_∞. Moreover Fig. 108 shows that the error ratio $\dfrac{dK/K}{dR_\infty/R_\infty}$ is not reduced with increasing reflectance, but increased, as indeed would be expected. It is practically impossible to obtain very small K values from the measurement of R_0 and R_∞.

Chapter VII. Applications

The possibilities for applying reflectance spectroscopy in industry and research are practically unlimited, and a great number of examples can be found in the literature in which the method has been applied successfully. Even so, however, all the advantages that the method offers have not been fully realized. It is not the goal of this book to enumerate or even give examples of all the possible applications. We would like to limit ourselves to such problems that cannot be solved by other methods, or are difficult to solve. In this way, not only can the applications and importance of reflectance spectroscopy be shown, but it can be compared with spectroscopic transmission methods. Moreover, these examples can provide an incentive for applying the method to other problems which are difficult to solve.

a) The Spectra of Slightly Soluble Substances, or Substances that are Altered by Dissolution

Typical examples of such substances are inorganic pigments. With a few exceptions, these are heavy metal oxides, sulphides, hydroxides, sulphates, silicates, chromates and carbonates that are prepared by precipitation or hydrolysis in aqueous solutions of the corresponding salts. The typical color curves of these compounds are often rather different from the transmission spectra of the corresponding salts. This is because the "color" of such pigments depends not only on the absorption coefficient K, but also on the scattering coefficient S. It has been shown that S depends on the refractive index (crystalline form), the size and shape of the particles, and the wavelength, so that the mutual effect of all these factors can give the same pigment rather different colors. A well-known example are the iron oxide pigments. Their color scale stretches from yellow to red and brown, and even to black. In addition to other substances that might be present, these pigments contain $\alpha\text{-}Fe_2O_3$, $\gamma\text{-}Fe_2O_3$, $\alpha\text{-}FeOOH$, $\gamma\text{-}FeOOH$ and Fe_3O_4 in various amounts, and different shades of color are obtained depending on the composition and the particle sizes. To characterise such pigments, it is preferable to plot the Kubelka-Munk function $F(R'_\infty) = K/S$ rather than the remission intensity R'_∞, measured against a highly reflecting standard (MgO, CaF_2, NaCl), as a function of λ or $\tilde{\nu}$. Usually it is not necessary to convert

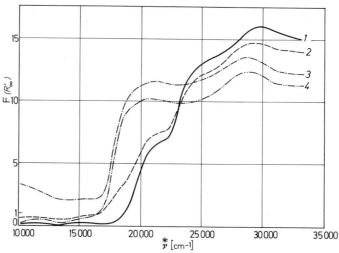

Fig. 109. $F(R'_\infty)$ of four different undiluted ferric oxide pigments recorded against CaF_2 as standard at high dispersion (Cary Model 14). *1* yellow; *2* brown; *3* brown; *4* red

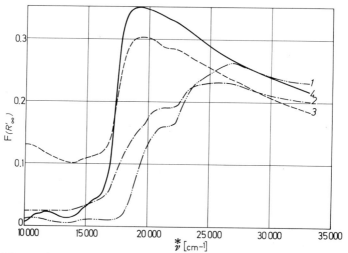

Fig. 110. "Typical color curves" of the same pigments as in Fig. 109 approx. 1:1,000 diluted with CaF_2, recorded against CaF_2 as standard

the measurements to absolute values R_∞. Fig. 109 shows such a plot for four, similarly-colored iron oxide pigments[298]. Even though pigments 2 and 3 make almost the same impression on the eye, their spectra

---

[298] According to the measurements of *M. Schranner*.

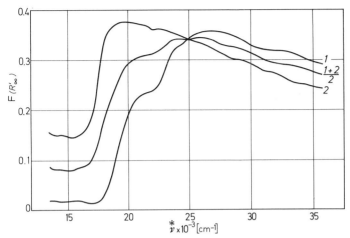

Fig. 111. Reflection spectra of two ferric oxide pigments diluted approx. 1:1000 with CaF_2 and reflection spectrum of the 1:1 mixture of these two diluted ferric oxides measured against CaF_2 as standard. Additivity of the spectra

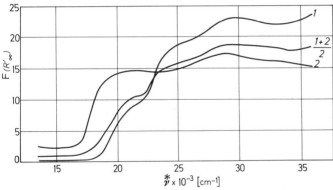

Fig. 112. Reflection spectra of the undiluted pigments and of their 1:1 mixture measured against CaF_2 as standard. No additivity of the spectra

are rather different. These curves are not "typical color curves" in the sense discussed on p. 188, because the Kubelka-Munk function is valid only for diluted substances, i.e., not at higher absorption. If these pigments are diluted with the standard in the ratio of ca. 1:1,000, or more, it is possible to obtain "typical color curves". If they are plotted on a logarithmic scale, they can be shifted along the ordinate until they coincide. They are shown for the same four pigments in Fig. 110. The amazingly large differences in the ratios of intensities of the individual bands can be distinguished when compared with Fig. 109.

The Kubelka-Munk function is valid only when the pigment is sufficiently diluted with a non-absorbing standard. This is proved by the fact that only when dilution is sufficient, the values of $F(R'_\infty)$ are additive (to a sufficient approximation) when a mixture of two pigments is prepared. This is shown in Figs. 111 and 112 [299]. The typical color curve of a mixture of diluted components in the ratio 1 : 1 shows, to a good approximation, the additivity [300] as compared to the typical color curves of both components. This is demonstrated by the isosbestic point and by the equal distances between the three curves (Fig. 111). On the other hand, mixtures of pure pigments do not show additivity (Fig. 112), i.e. the Kubelka-Munk equation is no longer valid under these conditions.

b) Spectra of Adsorbed Substances

Until recently, there was only one method available that could distinguish the changes in the spectrum of a molecule caused by adsorption on a solid surface. This was the "scattered transmission" method, which is based on the transmittance of very thin layers of pulverized material onto which the substances to be studied are adsorbed in a previous operation. This method proved to be particularly useful in the mid IR region [301] for the following reasons: As was shown on p. 201, it can be assumed that, for sufficiently small particles, also for multiple scattering the efficiency factor Q_{st} decreases to the first approximation with increasing wavelength as λ^{-4}, as in *Rayleigh* scattering. Comparison of Eqs. (III, 52) and (III, 56) shows that the absorption cross section Q_{abs} decreases, to the same approximation, only as λ^{-1}. This means that, with increasing λ, the scattering media which are composed of very small particles are getting more and more transparent (photography in the IR!), yet the absorption cross section is less affected by this change in λ. Thus, it is possible to record good absorption spectra with scattered transmission, especially in the mid IR region. In the near IR, and even more so in the visible and UV regions, it is difficult to apply this technique. In addition to the impossibility of evaluating quantitatively the measurements in scattered transmission because the contri-

---

[299] *Kortüm, G.*, and *D. Oelkrug*: Naturwiss. **53**, 600 (1966).

[300] That the additivity is not exactly within the limits of error can be explained by the differences in the scattering coefficients of both components and of the mixture. This occurs because the mixture was additionally ground whereas the pure components were mixed only by lengthy shaking in a test-tube.

[301] See, e.g. *Eischens, R. P.*: Z. Electrochem. **60**, 782 (1956). — *Terenin, A.*, and *L. Roev*: Spectrochim. Acta, **15**, 274 (1959). — *Sheppard, W.*: Spectrochim. Acta **14**, 249 (1959). — *Succhesi, P. J., J. L. Charter*, and *J. C. Yates*: J. Phys. Chem. **66**, 1451 (1962). — *Basila, M. R.*: J. Phys. Chem. **66**, 2233 (1962) and numerous other publications.

bution from scattered light can not be ascertained, there is no information available about the thickness of the layer. However, such a quantitative evaluation is possible, even in the visible and UV regions, by using the Kubelka-Munk theory. Thus, it can be expected that reflectance spectroscopy will be particularly useful for measuring the spectra of adsorbed molecules in the UV, visible and near IR regions of the spectrum. This has been verified in practice, by recording in this way the spectra of molecules absorbed on colorless adsorbents. The results were so good that the method can be considered as a standard method for evaluating the effects of adsorption forces.

Adsorption under the influence of the so-called Van der Waals forces (dispersion, dipole and mutual induction effects) generally only slightly changes the electronic spectrum of a substance. This was proved for *substantive dyes* on various fibers. Compared with transmittance spectra obtained from aqueous solutions, reflectance spectra show only small shifts – the so-called medium effects. Because of the sensitivity of the eye, however, different shades of colors on fibers can be distinguished when using different fibers. It is only when a strong interaction between adsorbent and adsorbed substance due to chemisorption exists that much larger spectral changes are observed. This type of interaction depends on the system concerned, and it can often be explained after the reflectance spectrum has been recorded. Shifts of charges (acid-base interactions, redox reactions, reversible cleavage reactions, tautomerisation, etc.) can change the spectra of adsorbed molecules as compared with free molecules in the gas phase or in solution. Several examples of such interactions will be given in the following discussion.

Acid-Base Reactions between the Adsorbed Substance and the Adsorbent

Weitz et al.[302] have shown for the first time that numerous organic compounds change color when they are adsorbed on active surfaces. In many cases, it could be proved by reflectance measurements that the acidic adsorbent forms a colored salt with the basic adsorbed substance. The spectrum of the salt is different from that of the free base and is shifted to longer wavelengths[303].

p-Dimethylaminoazobenzene (DMAB) is an example of such a compound that has been studied in detail. It changes from yellow to red when it is adsorbed onto silica gel, dried α-Al$_2$O$_3$ or γ-Al$_2$O$_3$,

---

[302] *Weitz, E., F. Schmidt,* and *F. Singer:* Z. Elektrochem. **46**, 222 (1940); **47**, 65 (1941).

[303] *Schwab, G.-M.,* and *E. Schneck:* Z. Physik. Chem. N.F. **18**, 206 (1958). — *Schwab, G.-M., B. C. Dadlhuber,* and *E. Wall:* Z. Physik. Chem. N.F. **37**, 99 (1963).

Bentonite, etc. The same color change occurs when an ethanolic solution of DMAB is acidified. Thus the adsorbent shows acidic properties. This is because of its ability to release protons (e.g. from an SiOH group) and its function as a Lewis acid (because of its partially filled outer electron shell). The latter is particularly true for the Al_2O_3-SiO_2 catalyst (Bentonite). When DMAB is chemisorbed, the formation of the salt of the dye at the adsorbent surface can be described by the following proton-exchange reactions:

$$\begin{array}{c}
\text{Ph}-N=N-\text{Ph}-NR_2 \longleftrightarrow \text{Ph}=N-N=\text{Ph}=NR \\
+H^+ \downarrow \\
\text{Ph}-\overset{+}{N}(H)=N-\text{Ph}-NR_2 \longleftrightarrow \text{Ph}=N(H)-N=\text{Ph}=\overset{+}{N}R
\end{array}$$

The more symmetrical degree of contribution of both limiting resonance hybrids to the electronic state of the cation in which one electron is mobile, decreases the energy difference between the ground and first excited state. This results in a red shift in the first absorption band. As the extended acid-base theory of *Ebert* and *Konopik*[304] allows cations to be considered as acids, the adsorption of DMAB on salts like $BaSO_4$ and $CaSO_4$ can be expected to produce a color shift in the same direction. This has been observed[305] after the salts had been heated to remove completely any adsorbed water.

Chemisorption is restricted, naturally, to a monolayer of the adsorbed molecules. This can be discerned from the concentration dependence of the reflectance spectrum (Fig. 113). When the surface coverage is small, only the bands of the red form are seen, but as the coverage increases, the bands of the yellow form also appear. The red band does not increase further after a certain concentration is reached, but the yellow band increases continually. This corresponds to adsorption resulting in the formation of second and subsequent layers and means that both chemisorption and physical adsorption take place simultaneously. However, the chemisorption only occurs as is permitted by the acid-base reaction directly in the first layer. If the values of $F(R'_\infty)$ for the maximum of the band of longest wavelength (which arises from the chemisorbed species), as measured against a pure $BaSO_4$ standard, are plotted against the molar ratio of adsorbed DMAB, the curve shown in Fig. 114 is obtained. The curve has the shape of a Langmuir adsorption isotherm. At low concen-

---

[304] *Ebert, L.,* and *N. Konopik:* Öst. Chem. Z. **50**, 184 (1949).
[305] *Kortüm, G., J. Vogel,* and *W. Braun:* Angew. Chem. **70**, 651 (1958).

Spectra of Adsorbed Substances 259

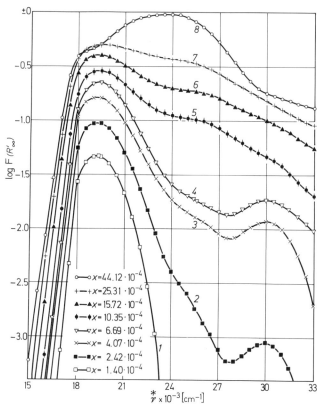

Fig. 113. Reflection spectra of *p*-dimethylamino-azobenzene adsorbed on dry BaSO$_4$ at different concentrations (mole fractions)

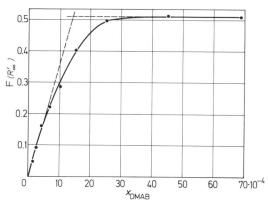

Fig. 114. Adsorption isotherm of the *p*-dimethylamino-azobenzene on BaSO$_4$ as adsorbent; isotherm derived from reflection measurements

17*

trations, $F(R'_\infty)$ increases almost linearly with x, as has already been shown in Fig. 62. At higher concentrations, a limiting value is reached. Extrapolation of the linear parts gives a value of $x_{sat.} = 14 \cdot 10^{-4}$ for the mole ratio at complete chemisorption. Similar spectra for DMAB on $CaSO_4$ and $MgSO_4$ have the red bands at the same wavelengths, but their intensity increases in the sequence Ba^{2+}, Ca^{2+}, Mg^{2+} for a given mole ratio of DMAB.

When humid air comes into contact with the sample, the red color changes instantenously to yellow, and the reflectance spectrum is then

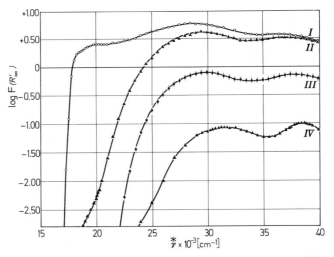

Fig. 115. Reflection spectra of HgI_2, adsorbed on dry MgO. (I) $x = 1 \cdot 10^{-2}$, (II) $x = 5 \cdot 10^3$, (III) $x = 1 \cdot 10^{-3}$, (IV) $x = 1 \cdot 10^{-4}$

identical with that of the yellow form. When placed in a vacuum dessicator with $CaCl_2$ the solid gradually becomes red again. This means that the acid-base surface reaction is reversible. Seemingly, water is a stronger base than DMAB, and displaces the dye from the surface. This reaction is so sensitive that it can be used for the detection of traces of water.

Other acid-base reactions, in which, by way of contrast, the adsorbent is a base and the adsorbed substance is an acid, are the adsorption of several heavy metal iodides on alkali or alkaline earth halides or oxides. Tetrahedral or planar complexes are formed at the surface, which can be recognized by the change in color. HgI_2 is an example that has been studied in detail[306, 307]. A heterogeneous mixture of red HgI_2 with a

---

[306] *Kortüm, G.*: Trans. Faraday Soc. **58**, 1624 (1962).
[307] *Griffiths, T. R.*: Anal. Chem. **35**, 1077 (1963).

large excess of, for example, dried MgO, changes, over several days in contact with the air, into a homogeneous yellow powder. At the same time, the Debye-Scherrer lines of HgI_2 vanish. This can be explained by assuming that the HgI_2 spreads on the surface of the adsorbent as a monolayer. The degree of coverage depends on the mole ratio. The reflectance spectrum (Fig. 115) differs completely from that of HgI_2, and the band of the red, crystalline HgI_2 at $18{,}000\ cm^{-1}$ can only be observed at high concentrations. This band can then overlap those of the adsorbed form, which is the spectrum of tetrahedral $[HgI_4]^{2-}$.

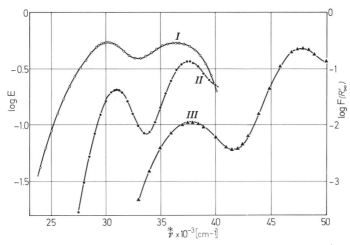

Fig. 116. (I) Reflection spectrum of HgI_2 adsorbed on KI; $x = 1.5 \cdot 10^{-4}$; (II) Absorption spectrum of HgI_2 in 10 n KI-solution; (III) Absorption spectrum of HgI_2 in water

The latter is very similar to that of HgI_2 observed in concentrated KI solutions (Fig. 116). Thus, a similar complex is formed at the surface by an acid-base interaction. Both free Hg^{2+} and the HgI_2 molecule absorb in the shorter wave region in aqueous solution. It is interesting that adsorption occurs on alkali fluorides without the formation of complexes. At all concentrations, reflectance spectra can be obtained that are identical with that of free HgI_2; the Debye-Scherrer diagram of crystalline HgI_2 also remains unchanged. It seems that fluoride ions are not sufficiently strong bases to cause complex formation.

BiI_3 behaves in the same way as HgI_2. For instance, Fig. 117 shows the adsorption of BiI_3 on dried KI, and the variation of the reflectance spectrum with concentration. The spectrum is rather similar to that obtained in aqueous solution; both are different from the spectrum of solid BiI_3.

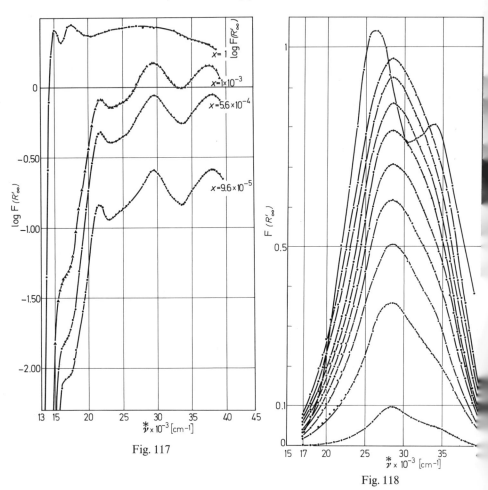

Fig. 117. Reflection spectrum of the pure BiI_3 and of BiI_3 adsorbed on dry KI at various mole fractions

Fig. 118. Reflection spectra of iodine adsorbed on dry NaI at different saturation vapor pressures of the iodine. Curve with two maxima after the admission of humid air

Charge Transfer Complexes

A transfer of electrons between the adsorbent and the adsorbed substance different to that observed in the acid-base interactions takes place during the formation of "charge-transfer" bonds[308]. In these interactions, the adsorbent can act either as an electron donor and the adsorbed substance as an acceptor, or *vice versa*. To this class belong, for instance, the $n\sigma$ molecular complexes which are formed by adsorption of I_2 from the gas phase onto alkali halides:

$$Hal^- + I_2 \rightleftharpoons (I_2Hal)^-.$$

These complexes have two typical spectral bands that are specific for this type of compound. In this type of interaction, the negative ions of the adsorbent act as electron donors, and the adsorbed iodine is the electron acceptor. Fig. 118 shows the reflectance spectra of I_2 adsorbed on dried NaI under various saturation pressures of the solid iodine[309]. Here the two bands have coalesced into one band, but this band shows a pronounced asymmetry. The addition of a little water vapor causes the band to split. If the value of $F(R'_\infty)$ corresponding to the maximum of the band is plotted against the vapor pressure of I_2, no Langmuir-adsorption isotherm is obtained as it was in Fig. 114. Instead, the dependence has the form of a Freundlich isotherm, $F(R'_\infty) = 1.36 p_{I_2}^{0.36}$, as can be shown by plotting $\log F(R'_\infty)$ against $\log p_{I_2}$. This gives a linear plot with slope 0.36. Thus there are active centres on the surface which have various interaction energies with I_2. In an analogous manner, the adsorption of I_2 or Br_2 vapor on dried alkali halides leads in general to trihalogen complexes of the types X^-I_2 or X^-Br_2 on the surfaces ($X^- = I^-, Br^-, Cl^-$). The spectra of these complexes agreed very well with those observed in solution[310].

Adsorption of I_2 or Br_2 on highly dried oxides such as Al_2O_3, $Al(OH)_3$, MgO, and CaO allows the observation of the analogous complexes OH^-I_2 and OH^-Br_2, respectively. The absorption maxima of these complexes lie between 43,000 and 46,000 cm^{-1} [311]. However, the adsorption of I_2 on dried aerosil gives a linear relationship between $F(R'_\infty)$ of the maximum of the I_2-band (19,300 cm^{-1}), even at much higher partial pressures[312]. Therefore, in this instance, the I_2 is only physically absorbed, and the surface of the adsorbent behaves homogeneously. A further maximum at about 32,000 cm^{-1}, observed in the vapor and in solution also, can be ascribed to I_4.

When the alkali halide or oxide adsorbents are not completely dried and still hold water molecules adsorbed on their surfaces, additional long wavelength bands are observed besides the charge-transfer bands. In these cases, the shapes of the entire spectra are time dependent. An example of this is given in Fig. 119 which shows the spectrum of I_2

---

[308] *Mulliken, R. S.*: J. Physic. Chem. **56**, 801 (1952). — *Briegleb, G.*: Elektronen-Donator-Acceptor-Komplexe. Berlin-Göttingen-Heidelberg: Springer 1961.

[309] *Mackensen, M. v.*: Diplomarbeit, Tübingen 1963.

[310] *Kortüm, G.*, and *H. Vögele*: Ber. Bunsenges. **72**, 401 (1968).

[311] *Kortüm, G.*, and *M. Grathwohl*: Ber. Bunsenges. **72**, 500 (1968).

[312] *Kortüm, G.*, and *H. Koffer*: Ber. Bunsenges. **67**, 67 (1963).

vapor adsorbed on undried MgO as a function of time. We observe a weak, only slightly time dependent, band at 43,800 cm^{-1} which can be ascribed to the OH$^-$I$_2$ complex. The low intensity of this band is a reflection of the fact that the MgO surface is largely saturated with ad-

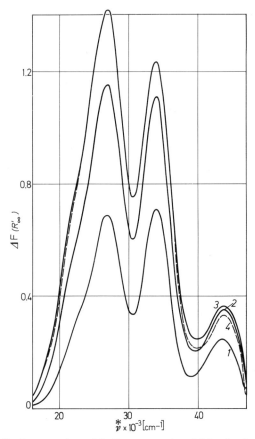

Fig. 119. Reflection spectra of iodine vapor ($p = 0.2$ torr) adsorbed on undried MgO as a function of time. *1* after 1 hour, *2* after 6.5 hours, *3* after 24 hours, *4* after 5d desorption

sorbed water. Besides we observe two intense bands at 26,800 and 34,000 cm^{-1}. They are characteristic of the I$_3^-$ complex formed in this case through the reaction of adsorbed iodine vapor with adsorbed water to give I$^-$ (I$_2$ + H$_2$O → IOH + HI) which in turn reacts with another molecule of adsorbed iodine to give I$_3^-$ (I$^-$ + I$_2$ → I$_3^-$). The weakly indicated shoulder at about 23,000 cm^{-1} indicates the absorption of hydrated iodine molecules. The correctness of these band assignments is evident from the fact that it is impossible to completely desorb the iodine even with several days' evacuation at 10^{-6} torr and a liquid nitrogen cooled cold trap. Only the band of the OH^-I_2 complex decreases slightly. This complex is thus evidently somewhat less stable than the I_3^- complex which owes its existence to a hydrolytic surface reaction.

Redox Reactions

Particularly interesting results were obtained when aromatic hydrocarbons were adsorbed on SiO_2–Al_2O_3 catalysts. These "cracking" catalysts are prepared by precipitation from silicate-aluminate solutions. Because Si^{4+} is partly exchanged by Al^{3+}, Na^+ is also incorporated to preserve the charge balance. NH_4^+ can be exchanged for the Na^+. Then, during heating, NH_3 is removed, leaving a solid with a high protonacidity. This can be shown by IR spectroscopy. Also, the electron deficiency of Al^{3+} causes the catalyst to exhibit strong Lewis-acidity.

If aromatic hydrocarbons M are adsorbed on such a catalyst under the high vacuum conditions described on p. 244, reflectance bands can be obtained which correspond either to charge transfer complexes MH^+ or to radical ions M^+. These are formed because the high electron affinity of Al^{3+} in the mixed catalyst causes a redox reaction between M and the adsorbent, and their existence has been proved by electron spin resonance measurements. In some instances, it seems that an equilibrium of the type:

$$MH^+ + K \rightleftharpoons M^+ + K^- + H^+$$

is established between the two forms[313]. In some instances the equilibrium is established in such a way that both forms can exist simultaneously. Benzene seems to be a good example. The reflectance spectrum of benzene adsorbed on a SiO_2–Al_2O_3 catalyst has been investigated by several authors[314, 315]. In addition to the band corresponding to physically adsorbed benzene at 39,300 cm^{-1}, three new bands appear at 18,000 cm^{-1}, 21,750 cm^{-1} and 31,500 cm^{-1}. The first has been assigned to the radical ion $C_6H_6^+$, and the others to the complex $C_6H_6H^+$. The same specimen shows an intense electron spin resonance signal. The first band vanishes immediately if the sample comes into contact with the air.

Numerous aromatic hydrocarbons behave similarly (see Fig. 120). Occasionally redox reactions also occur which, under the influence

---

[313] *Aalbersberg, W. I., G. J. Hoijtink, E. L. Mackor,* and *W. P. Weijland:* J. Chem. Soc. **1959**, 3049.

[314] *Kortüm, G.,* and *V. Schlichenmaier:* Z. Physik. Chem. N.F. **48**, 267 (1966).

[315] *Barachevski, V. A.,* and *A. N. Terenin:* Opt Spectr. USSR **17**, 161 (1964).

of a catalytically active surface, correspond to a strong polarization of a bond or even an electrolytic dissociation. By the adsorption of triphenylchloromethane on SiO_2, CaF_2, $MgSO_4$, or $BaSO_4$, for example, the triphenylmethyl cation is easily recognized by its characteristic bands at approximately 23,000 and 24,000 cm^{-1} [316]. The cation is stable and does not react further. After desorption with methanol, triphenylmethanol is obtained. On the other hand, when adsorbed on Al_2O_3,

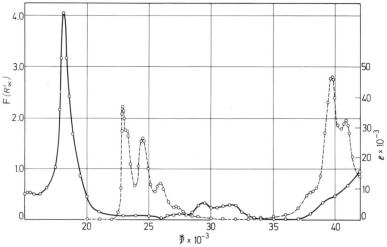

Fig. 120. Reflection spectrum of perylene adsorbed on a SiO_2-Al_2O_3 cracking catalyst under high vacuum conditions. $c = 4.5 \cdot 10^{-5}$ mole/g. —— perylene in hexane

BeO, MgO, or CaO, the initially formed cation does react further. Triphenylmethane, 9-phenylfluorene, and the 9-phenylfluorene cation, among other things, can be found on highly dried MgO. This is presumably consistent with the assumption that these adsorbents can split off free hydroxide ions.

Reversible Cleavage Reactions

Often, during the adsorption of cyclic organic compounds on ionic lattices, a reversible cleavage occurs that produces zwitterions, which have a completely different spectrum. The colorless lactone of malachite green-o-carboxylic acid (MGL)[317] has been studied in detail. *Weitz* has found that when MGL is adsorbed onto dried salts, it is

---

[316] *Kortüm, G.*, and *M. Friz:* Ber. Bunsenges. (in press).
[317] *Kortüm, G.*, and *J. Vogel:* Chem. Ber. **93**, 706 (1960).

greenish-blue. The reflectance spectra on dried NaCl at various concentrations are shown in Fig. 121 (for mole ratios $10^{-3} - 10^{-4}$), and are compared with the spectrum of MGL in methanolic solution. The bands at 16,500 cm^{-1} and 24,000 cm^{-1} are characteristic of the "typical color curve" of the adsorbed substance and are identical with the bands of the solution spectrum of malachite green[318]. Therefore it can be assumed that the lactone ring of MGL is cleaved during the adsorption, and a resonance stabilized zwitterion is formed:

The IR spectrum of the adsorbed lactone shows that this explanation is correct, because the $-COO^-$ band can be observed as well as the ring frequencies[319]. Because the lactone cannot be converted to the zwitterion by heating in solution or by melting, the ring cleavage must have a higher activation energy. This energy barrier is lowered during adsorption by the polarizing effect of the ionic lattice, so that a partial cleavage occurs even at room temperature. A plot of $F(R'_\infty)$ for the maxima of the bands at 16,500 cm^{-1} and/or 24,000 cm^{-1}, which correspond to the cleaved lactone, against the mole ratio x gives a Langmuir isotherm similar to that in Fig. 114. This suggests that the first monolayer is completely saturated.

The degree of ring cleavage, as measured by the band intensities, decreases from LiCl through to CsCl for a given surface coverage on these adsorbents. The positions of the zwitterion bands are virtually unchanged. On the contrary, exchange of the anions (F$^-$, Cl$^-$, Br$^-$, I$^-$) has no effect on the intensity of either band. Thus the bonding of the zwitterion to the adsorbent is mainly a *coulombic* interaction between the lattice cations and the carboxylate groups, because the positive charge of the zwitterion is distributed by resonance over the greater part of the molecule. Therefore, the smaller the lattice cation, the greater is its polarizing effect on the lactone ring and the more the equilibrium I ⇌ II is shifted to the right. A similar dependence of band intensity on the size of the lattice cation is found for the adsorption of MGL on dried alkaline earth sulphates and oxides. The stronger polarizing effect of

---

[318] *Lewis, G. N.*, and *H. Bigeleisen*: J. Am. Chem. Soc. **65**, 2102 (1943).
[319] *Kortüm, G.*, and *H. Delfs*: Spectrochim. Acta **20**, 405 (1964).

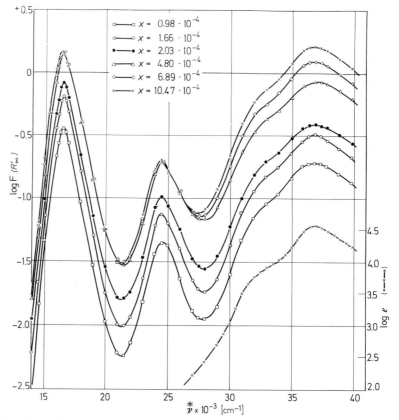

Fig. 121. Reflection spectra of MGL adsorbed on dried NaCl and transmission spectrum in methanol solution

of the doubly charged cation is shown by the greater degree of splitting of the ring, for a given surface coverage, compared with the alkali metal ions.

The ring opening reaction is reversible. After desorption, the colorless lactone is again obtained. This also occurs in the presence of water vapor (moist air). This displacement of the zwitterion by H_2O also depends on the size of the cation. When the adsorption occurs on LiCl, the displacement is very slow, and even after 12 hours exposure to the air, the hygroscopic crystals are still blue. For large cations, on the other hand, the effect of a preliminary drying of the adsorbent on the ring opening is more pronounced, i.e., the larger the cation, the smaller the polarizing effect. Fig. 122 demonstrates such behaviour on NaCl. It is just possible to observe ring opening on undried NaCl, but with increased preliminary drying, the bands of the zwitterion are developed better. The ring opening reaction cannot be observed at all on undried KCl, RbCl or CsCl.

The *spiropyranes* also give reversible ring opening under the influence of adsorptive forces. In solution these substances are thermochromic [320] and photochromic [321], i.e., the spirane ring is partially opened

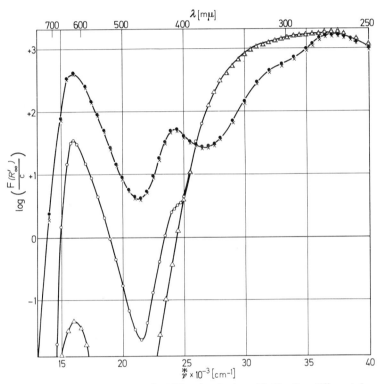

Fig. 122. Reflection spectra of MGL adsorbed on NaCl after different degrees of predrying. ●——● Dried for 1 hour at 600° C; ○——○ Dried for 30 hours at 100° C; ×——× Dried for 1 hour at 400° C; △——△ Not dried

by warming and by UV irradiation at low temperatures. The ring opening produces a zwitterion which is in resonance with a merocyanine, and is therefore colored. A specific example, 1,3,3-trimethylindolino-β-naphthospirane, is shown in Fig. 123. The equilibrium ratio of mero-

---

[320] *Dilthey, W.*, and *H. Wübken*: J. Prakt. Chem. **114**, 179 (1926). — *Löwenstein, A.*, and *W. Katz*: Chem. Ber. **59**, 1377 (1926).

[321] *Bergmann, E. D., W. Weizman*, and *E. Fischer*: J. Am. Chem. Soc. **72**, 5009 (1950). — *Hirshberg, Y.*, and *E. Fischer*: J. Chem. Soc. **1954**, 297, 3129.

cyanine that exists in diphenyl ether solution at 250° C is 1%. Yet the ratio is 50% when the spirane is adsorbed on MgO at room temperature. The reflectance spectrum is identical (except for a certain shift in the bands caused by the medium) with that of the compound produced photo- or thermo-chromically in solution. If the adsorption takes place on highly dried NaCl, another wide band at 20,500 cm^{-1} appears at small

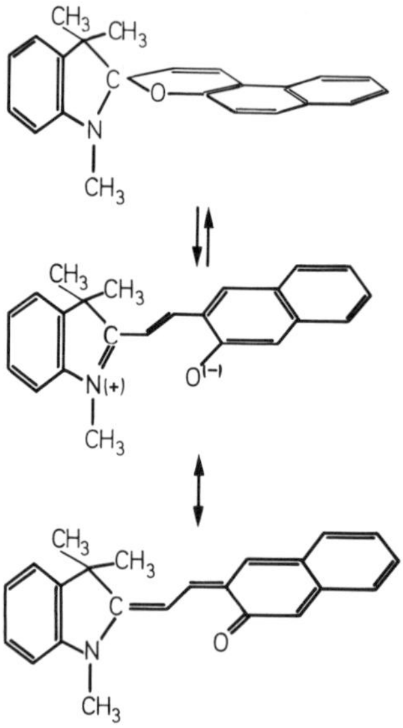

Fig. 123. Splitting-up of 1,2,3-trimethylindolino-β-naphtospirane by adsorption in a zwitterion

surface coverages. This is overlapped by the thermochromic bands. As the surface coverage increases, the new band decreases relative to the thermochromic bands (Fig. 124)[322]. The same bands can be observed in acidic, alcoholic solutions. Thus, another acid-base equilibrium can be ascertained, in which the Na$^+$ of the adsorbent acts as an acid, and OR$^-$ as a base. The spiropyrane can be adsorbed simultaneously, therefore, during ring opening, as a cation and as a zwitterion.

Surface Area Determination of Powders

The color change of selected adsorbed organic compounds arising from the acid-base interaction with the adsorbent (Fig. 113) or from the ring opening reaction (Fig. 121) can only occur in the mono-

---

[322] Kortüm, G., and G. Bayer: Z. Physik. Chem. N. F. **33**, 254 (1962).

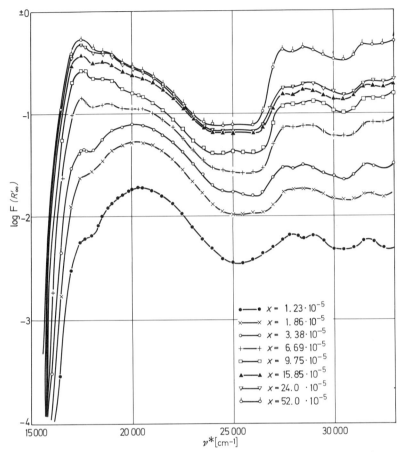

Fig. 124. Reflection spectrum of the 1,2,3-trimethylindolino-β-naphtospirane adsorbed on NaCl at various concentrations (mole fractions)

layer at the surface of the adsorbent, as it is only in this layer that chemisorption takes place. The quantitative evaluation of the Kubelka-Munk function $F(R_\infty)$ over a wide concentration range under constant conditions (temperature, particle size of the adsorbent, degree of drying, etc.) gives Langmuir adsorption isotherms (as shown in Fig. 114). These isotherms can be used for the very accurate determination of the surface areas of the adsorbents used [323]. The results are comparable with those obtained by the BET method.

Fig. 125 gives an example in which the Langmuir isotherm characteristic of the open ring form of the lactone is compared with the iso-

---

[323] Kortüm, G., and D. Oelkrug: Z. Physik. Chem. N.F. **34**, 58 (1962).

therm characterizing the physical adsorption, C (using the band at 37,000 cm^{-1}, corresponding the total of the lactone). The isotherm for the open form was obtained from Fig. 121 for the band at 16,500 cm^{-1} for two samples of H$_2$O-free NaCl of different particle size and surface area (A, B). The physical adsorption isotherm shows no saturation because higher adsorption layers can also participate. However, the Kubelka-Munk function has a linear dependence on x when x is less

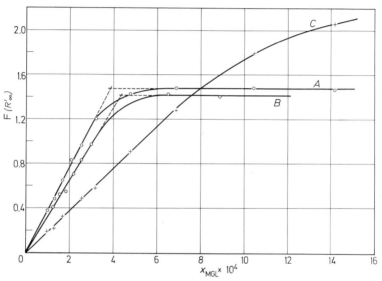

Fig. 125. Adsorption isotherms of the chemisorption of MGL on two samples intensely dried NaCl of different grain sizes (A, B) and adsorption isotherms of the physical adsorption (C) under the same conditions

than $6 \cdot 10^{-4}$. At higher concentrations, the function no longer describes the dependence, as has been shown several times previously. The Langmuir isotherms first show the expected, linear increase of $F(R'_\infty)$ with increasing x, and then reach limiting values at higher concentrations. When the linear part and the horizontal part are extrapolated, the intersection gives a value for the saturation mole fraction for chemisorption of $x_{mA} = 3.85 \cdot 10^{-4}$ and $x_{mB} = 4.33 \cdot 10^{-4}$ mole MGL/mole NaCl. For the larger surface, the values of $F(R'_\infty)$ are smaller because, with increasing surface area, i.e., with smaller average particle size, the scattering coefficient S increases (cf. p. 206). The ratio of surface areas for the two NaCl samples is $x_{mB}/x_{mA} = 1.12$. The value for the same powders, using the BET method with nitrogen adsorption, was 1.11, which is in good agreement.

The determination of surface areas of powders using the adsorption of dyes from the liquid phase has been in use for a long time. This can easily be carried out because the decrease in the concentration of the dye in the solution as a result of the adsorption can be measured photometrically. There are two criticisms of this method. First, it gives no indication whether the adsorption occurs in one or more layers. Second, part of the surface may be covered with solvent molecules. This is particularly relevant when the adsorption takes place from aqueous or other polar solutions. An example is the adsorption of MGL on LiCl, described on p. 267.

Establishment of Equilibria and Orientation at Surfaces

For problems of heterogeneous catalysis and chemical kinetics on surfaces, it is important to understand to what extent the adsorbed molecules can be considered to be mobile on the surface when sufficiently low surface concentrations are used. If such mobility exists, it is possible that association or dissociation equilibria are established at the surface. To illustrate and study these problems, the adsorption of charge transfer molecular complexes on various adsorbents has been investigated [324].

The reflectance spectra of the molecular complex of pyrene and trinitrobenzene adsorbed on water-free NaCl are reproduced in Fig. 126. Various concentrations were used, and the same NaCl was used as a reference standard. It was not possible to make the curves coincide by a parallel shift, as was possible in Fig. 124, even when very low concentrations were used. Only the short wave region of the spectrum allows overlapping. This part corresponds to the spectra of the separate components, pyrene and trinitrobenzene. The charge transfer band of the molecular complex at 22,000 cm^{-1} decreases faster with decreasing concentration than the other bands. This was originally considered to be a true deviation from the Kubelka-Munk equation, which requests a linear relationship between $F(R_\infty)$ and x. It was supposed that the behaviour was analogous to deviations from the Lambert-Beer Law in solution, and was interpreted as a partial dissociation of the molecular compound at the surface. If values of $F(R'_\infty)$ for the maximum of the pyrene band at 29,500 cm^{-1} and the charge transfer band at 22,000 cm^{-1} are plotted as a function of x, the Kubelka-Munk limiting linear relationship can be obtained only for the pyrene band. The charge transfer band gives a curve that deviates towards lower values (Fig. 127). A new study shows that this situation is similar to that shown in Fig. 63 [325]. The deviations

---

[324] *Kortüm, G.,* and *W. Braun:* Z. Physik. Chem. N.F. **18**, 242 (1958); **28**, 362 (1961).

[325] *Kortüm, G.,* and *W. Braun:* Z. Physik. Chem. N. F. **48**, 282 (1966).

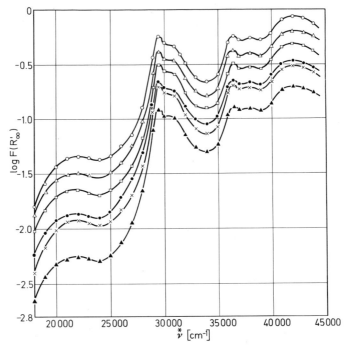

Fig. 126. Reflection spectra of the molecular compound pyrene-s-trinitrobenzene adsorbed on intensely dried NaCl at various mole fractions as follows: $x = 2.75$; 2.00; 1.48; 1.00; 0.90; $0.52 \cdot 10^{-4}$

from the Kubelka-Munk function arise from the use of relative values, R'_∞, instead of absolute values, R_∞. Thus, the very small absorption of the standard had been neglected, and it was deduced on p. 185 that especially at small absorptions this gives large errors. In fact, the curve in Fig. 127 is very well described by Eq. (V, 17). If the values of $F(R'_\infty)$ for the pyrene band are similarly corrected, only the plot against x has a different slope, but it is still linear, because the values of $F(R'_\infty)$ are an order of magnitude greater than those of the charge transfer band. This demonstrates very clearly that the measured relative values of R'_∞ should be converted into absolute R_∞ values for the evaluation of the Kubelka-Munk function, even when the reflectance of the reference standard is high (0.98 to 0.95).

In contrast to the earlier interpretation of these measurements, it cannot be assumed that dissociation of these molecular complexes takes place at the two-dimensional interface when they are adsorbed on NaCl. This is also true for NaCl that has been dried in the presence of air and thus contains some water. On the other hand, adsorption of the same molecular complex on silica gel gave indications of a dissociation effect. If the molecular complex on the adsorbent and an equivalent excess of one of the components are ground together, the absorption of the charge-transfer band increases. This observation cannot be explained by the fact that the characteristic absorption of the adsorbent was not considered.

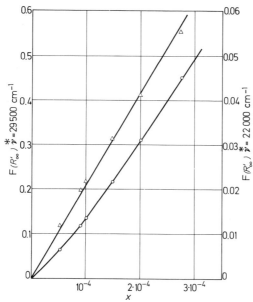

Fig. 127. $F(R'_\infty)$-values in the maximum of the pyrene band at 29,500 cm^{-1} ($\triangle$) and of the charge-transfer-band of the complex pyrene-s-trinitrobenzene at 22,000 cm^{-1} ($\bigcirc$) as function of the concentration on dry NaCl as adsorbent

In order to clarify this contradiction, measurements were made on the hexamethylbenzene-s-trinitrobenzene molecular complex adsorbed on silica gel. Freshly deposited MgO served as the comparison standard so that the absolute reflectance could be calculated [326]. In contrast to the measurements with the NaCl adsorbent, when the calculated $F(R_\infty)$ values were plotted against the mole fraction x, a considerable deviation from the expected linear course of the $F(R_\infty) - x$ function was obtained. This deviation must be interpreted as a dissociation effect. From the mass action law, it follows that

$$K_x = \frac{\alpha^2}{1-\alpha} \cdot x_0 \quad \text{or} \quad \alpha = \frac{1}{2}\left[\sqrt{\left(\frac{K_x}{x_0}\right)^2 + \frac{4K_x}{x_0}} - \frac{K_x}{x_0}\right]. \quad (1)$$

---
[326] Braun, W., and G. Kortüm: Z. Physik. Chem. N.F. **61**, 167 (1968).

Since in the region of the charge-transfer band

$$F(R_\infty) = k x_{AD} = k x_0 (1 - \alpha), \qquad (2)$$

it follows that

$$F(R_\infty) = k \left[x_0 + \frac{K_x}{2} - \frac{K_x}{2} \sqrt{1 + \frac{4 x_0}{K_x}} \right]. \qquad (3)$$

α is the degree of dissociation, x_0 the original concentration of the molecular complex, and x_{AD} the equilibrium concentration of the molecular compound. If the two components are weighed separately and not in equivalent amounts so that

$$\left.\begin{array}{l} x_A = x_{0A} - x_{AD} = x_{0A} - \dfrac{F(R_\infty)}{k} \\[6pt] x_D = x_{0D} - x_{AD} = x_{0D} - \dfrac{F(R_\infty)}{k} ,\end{array}\right\} \qquad (4)$$

the four equations yield the relationship

$$\frac{x_{0A} x_{0D}}{F(R_\infty)} + \frac{F(R_\infty)}{k^2} = \frac{K_x}{k} + \frac{1}{k}(x_{0A} + x_{0D}), \qquad (5)$$

which can be evaluated by known procedures[327]. From known mole fractions x_{0A} and x_{0D} and measured $F(R_\infty)$ values, it was found that $K_x = 8.6 \cdot 10^{-4}$ and $k = 236$. When the left side of Eq. (5) is plotted against $(x_{0A} + x_{0D})$, the expected straight line is obtained.

That the dissociation equilibrium on the surface can be described by the mass action law is qualitatively obvious from Fig. 128 which gives the spectra of the molecular complex with and without an equivalent excess of one of the components. An excess of hexamethylbenzene raises the charge-transfer band about the same amount as an equally large excess of s-trinitrobenzene.

If the two components are allowed to stand several days in the presence of a rather coarse, crystalline adsorbent, the dissociation equilibrium appears on the surface of the adsorbent. This occurs presumably partly through the gas phase but preponderantly by spreading on the surface as already observed in the example of HgI_2 adsorbed on KI (see p. 260). Therefore, the forces of interaction between the adsorbent and the individual components of the molecular complex must be of about the same order as the charge-transfer interaction forces between the components themselves. This must be the case if an equilibrium, corresponding formally to dissociation in solution, between the un-

---

[327] Compare *Kortüm, G.*, and *W. Braun*: Z. Physik. Chem. N.F. **28**, 362 (1961).

dissociated molecular complex and the chemisorbed components is to be established. If the activity of the adsorbent toward the components is small (NaCl), the components will be on the surface as a pure molecular complex and no dissociation will be observed.

Picric acid (PiOH) can form two "tautomeric complexes" with aromatic amines [328]; either salts (PiO$^-\cdots{}^+$HNAr) or charge transfer complexes (PiOH$\cdots$NAr). The conversion of the yellow salt into the red complex by heating can be proved directly from the reflectance spectrum [329].

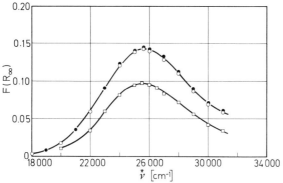

Fig. 128. Increase in the absorption of the charge-transfer band by addition of either one of the components of the complex. □–□–□ molecular complex $x = 10^{-3}$; ●–●–● hexamethylbenzene $x = 10^{-3} +$ s-trinitrobenzene $x = 2.05 \cdot 10^{-3}$; ○–○–○ s-trinitrobenzene $x = 10^{-3} +$ hexamethylbenzene $x = 2 \cdot 10^{-3}$

A number of investigations on dyestuffs has been carried out to show the orientation of adsorbed molecules [330]. If a glass or crystal surface (e.g. NaCl, CaF$_2$) is polished by rubbing always in the same direction, and the surface is then immersed in a dilute methanolic solution of methylene blue, the film of adsorbed dyestuff consists of orientated molecules. These reflect a linearly polarized beam differently according to whether the electrical vector is parallel or perpendicular to the direction of polishing. The use of such reflection measurements in the visible and IR regions shows that the adsorbed molecules lie flat on the surface, i.e., with the benzene rings parallel to the surface, and that the longest molecular axis is perpendicular to the direction of polishing. This could be related to epitaxy, in which a heterogeneous crystal phase grows in an orientated manner on a given structural plane of a crystal.

---

[328] *Hertel, E.*: Liebigs Ann. Chem. **451**, 179 (1926).

[329] *Briegleb, G.*, and *H. Delle*: Z. Physik. Chem. N.F. **24**, 359 (1960).

[330] *Demon, L.*: Ann. Phys. Paris **1**, 101 (1946). — *Anderson, S.*: J. Opt. Soc. Am. **39**, 49 (1949).

Photochemical Reactions

As was shown above for malachite green-o-carboxylic acid lactone and the spiropyranes, the activation energy of the cleavage reaction can be lowered by adsorption on suitable surfaces. In addition to this specific effect of the adsorbent there are many instances in which reactions follow new paths owing to interaction with an adsorbent. For instance, in

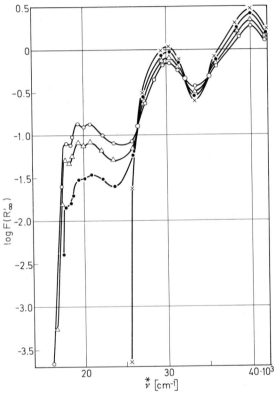

Fig. 129. Photochemical reaction of anthraquinone adsorbed on Al_2O_3 in presence of oxygen. ×—× not irradiated; ●—● irradiated for 10 min; △—△ irradiated for 20 min; ○—○ irradiated for 30 min

ultraviolet light, anthracene in solution dimerises to dianthracene, but on various adsorbents, in the presence of oxygen, it gives anthraquinone. The rate of this oxidation is strongly dependent on the kind of adsorbent [331]. Furthermore, anthraquinone adsorbed on Al_2O_3 reacts with oxygen in UV light to give other oxidation products. During this

---

[331] *Kortüm, G.*, and *W. Braun:* Liebigs Ann. Chem. **632**, 104 (1960).

process, a new band in the visible region is formed, indicating a vibrational structure, whereas the typical bands of anthraquinone diminish. The isosbestic points (Fig. 129) indicate the formation of just one compound. This product had a spectrum which was identical with that of 1,2-dihydroxyanthraquinone (quinizarine). But, depending on the properties of the Al_2O_3 used, alizarine or chrysazine can sometimes also be isolated [332]. Also, anthraquinone adsorbed on SiO_2 can be photochemically transformed into quinizarine. Nevertheless, the quantum yield is considerably less. Anthraquinone is not oxidized when it is adsorbed on KCl.

Another example of a reversible photochemical reaction of an adsorbed molecule is the tautomeric transformation of 2-(2′,4′-dinitrobenzyl)pyridine. This colorless compound becomes blue when it is irradiated, and reversibly reverts to the colorless form when placed in the dark [333]. This can be observed both for the crystalline substance and its solution. The rate of fading in solution is so large that it can be measured only at very low temperatures [334] or by a special flash technique [335]. The following reaction scheme has been discussed for the transformation:

The colorless substance can be adsorbed on an excess of a suitable adsorbent (NaCl, LiF, SiO_2), and irradiation again will turn it blue. This proves that a photochemical reaction of single molecules is occuring. However, the rate of fading is much slower than the rate observed in solution, so that the rate of change of color can be measured at room temperature by reflectance spectroscopy [336]. The reflectance spectra of

---

[332] *Voyatzakis, E.*, et al.: Compt. Rend. **251**, 2696 (1960).

[333] *Hardwick, R.*, et al.: Trans. Faraday Soc. **56**, 44 (1960); J. Chem. Phys. **32**, 1888 (1960).

[334] *Sousa, J.*, and *J. Weinstein*: J. Org. Chem. **27**, 3155 (1962).

[335] *Wettermark, G.*: J. Am. Chem. Soc. **84**, 3658 (1962).

[336] *Kortüm, G., M. Kortüm-Seiler*, and *S. D. Bailey*: J. Phys. Chem. **66**, 2439 (1962).

the non-irradiated and irradiated compounds adsorbed on NaCl are shown in Fig. 130. The spectra obtained on LiF and SiO$_2$ have similarly an additional maximum at 600 mµ. This was ascribed to the product of the photochemical reaction.

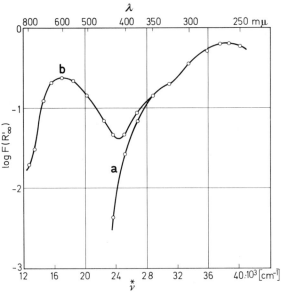

Fig. 130. Reflection spectra of the 2-(2′, 4′-dinitrobenzyl)-pyridine adsorbed on NaCl before irradiation (*a*) and after 30 min irradiation with diffuse daylight (*b*)

When MgO is used as the adsorbent, an irreversible reaction between the adsorbent and the adsorbed compound occurs during grinding in darkness. Based on polarographic measurements, the reaction was formulated as:

The quinoid structure of the reaction product explains why the reflectance is completely different, and has 2 peaks (670 and 480 mµ). This structure does not prevent the tautomerisation between I and II, so the mechanism of the tautomerisation can be described by an equilibrium between I and III. This conclusion is supported by the fact that

2(4'-nitrobenzyl)pyridine shows no photochemical change when adsorbed on NaCl or SiO_2. Thus it seems that an o-nitro group is essential for this type of tautomeric change.

c) Kinetic Measurements

Reflectance spectroscopy is particularly important for kinetic studies of reactions between solid phases or of the reactions of adsorbed substances, because no other generally applicable methods are known.

An example is the fading reaction III→I discussed above. The reaction rate of the tautomer of 2(2',4'-dinitrobenzyl)pyridine, adsorbed on SiO_2, was studied at various temperatures. The samples were fused into quartz cuvettes, which were placed in a thermostat after irradiation. After selected time intervals, the reflectance intensity at 600 mμ was measured (at the maximum of the band) against pure adsorbent as the reference standard. After the values of R'_∞ had been recalculated to give absolute values (cf. p. 150), the values of $\log F(R_\infty)$ were plotted against time for four different temperatures (Fig. 131). Two first order reactions of different rates were observed, corresponding to the fading of the neutral molecule and of the cation that is formed at the surface by the reaction with SiOH groups:

This can be deduced, on the one hand, from observations that the corresponding reaction in solution is pH-dependent[337], and on the other, that no intersections on the linear plot of $\log F(R_\infty)$-t are found for adsorption on NaCl or LiF. From the slope of both linear dependences, the corresponding rate constants k were calculated. Plots of $\log k$ against T^{-1} are linear, to a good approximation (Fig. 132). From their slopes, the activation energy of the faster reaction was calculated to be $E_1 = 15.5$ kcal/mole, whereas that for the slower reaction was $E_2 = 13.4$ kcal/mole. These activation energies are three times greater than those in solution; this is expected from the much slower fading in the adsorbed state. Because $E_1 > E_2, f_1 > f_2$ (f is the frequency factor). Thus the cation-adsorbent interaction is different from that of the neutral molecule with the adsorbent.

---

[337] *Wettermark, G.:* J. Am. Chem. Soc. **84**, 3658 (1962).

Wettermark and *King* [338] have made similar studies of benzaldehyde phenylhydrazone and cinnamaldehyde semicarbazone.

There are difficulties in applying the reflectance method at higher temperatures because a temperature gradient can occur between the sample surface and the bottom of the heated plate on which the sample is placed. Thus the temperature of the sample is not defined. Furthermore, for

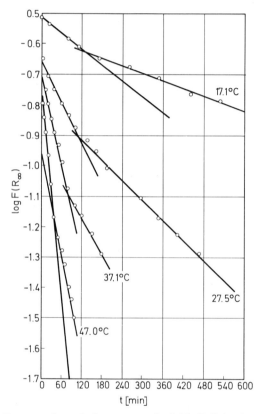

Fig. 131. Fading reaction of the tautomeric 2-(2′, 4′-dinitrobenzyl)-pyridine adsorbed on SiO_2 as function of the temperature

volatile substances, a concentration gradient may be set up, because these substances are accumulated at the surface (cf. p. 229). This error is eliminated if the reaction is separated from the measurement. This is achieved by carrying out the reaction in the mixed solid phase in a thermostat or an oven, and measuring the reflectance after quickly cooling

---

[338] *Wettermark, G.*, and *A. King*: Photochem. Photobiol. **4**, 417 (1965).

to room temperature after chosen time intervals. The reaction

$$ZnS + CdO \to ZnO + CdS$$

in the temperature range between 458 and 511° C was followed in this way [339] as was the dehydration of phenylbenzyl carbinol to stilbene at 130–150° C on Al_2O_3 as catalyst [340].

In the latter example, the reflectance at 32,500 cm^{-1} was measured as a function of time, because the carbinol does not absorb in this region.

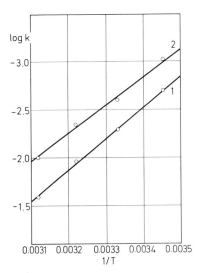

Fig. 132. Calculation of the activation energies from the rate constants belonging to the reaction of Fig. 131. Rates obtained by means of reflection spectroscopy

In order to evaluate the measurements the dependence of the $F(R_\infty)$-values of stilbene on its concentration had first to be verified. Only then could the concentration of stilbene, formed at a given time, be obtained from the curves. The dependence on concentration is shown in Fig. 133 (x = mole ratio of stilbene). For adsorption on NaCl, a linear relationship, to a good approximation, is obtained, whereas adsorption on Al_2O_3 gives a curve, with very much smaller $F(R'_\infty)$-values in the same concentration range. The latter is caused by the very much smaller particles of Al_2O_3 which have a greater scattering coefficient than the finely ground NaCl. The non-linear dependence of $F(R'_\infty)$ on x for Al_2O_3 is

---

[339] Kleykamp, H., G.-M. Schwab, and R. Sizmann: Z. Physik. Chem. N.F. **44**, 15 (1965).

[340] Schlichenmaier, V.: Diplomarbeit, Tübingen 1962.

caused by neglecting the self-absorption of Al_2O_3 in this region, which corresponds to a reflectance of only ca. 70%. Only when Eq. (V, 17) is used, is it possible to obtain a linear dependence (broken-line) after conversion into absolute R_∞-values. This conversion is not necessary if the curved $F(R'_\infty) - x$ plot is used simply as a photometric calibration curve. Using the time dependence of the measured values of $F(R'_\infty)$, the concentrations x of the stilbene formed after various time intervals can be determined. Plots of $\log x$ as a function of time give a group of linear

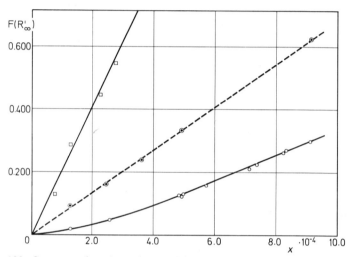

Fig. 133. Concentration dependence of the Kubelka-Munk function of stilbene, adsorbed on NaCl (□) active Al_2O_3 (○), at 32,500 cm^{-1} (band maximum). $x =$ mole fraction of the stilbene

isotherms, showing that the decomposition is first order. Likewise, the graph of $\log k$ vs. T^{-1} is linear, and an activation energy of 23.9 kcal/mole can be calculated from its slope. This value is in good agreement with the value obtained in a completely different way (manometrically)[341].

Reflectance spectroscopy was extraordinarily useful for explaining the *piezochromism* and *thermochromism* of bianthrone and bixanthylene[342]. First of all, it could be shown that a molecule monomolecularly adsorbed on certain adsorbents such as CaF_2 could be changed by warming and/or high pressure from its normal form A to a conformeric form B that was green and whose spectrum possessed an additional absorption

---

[341] *Dohse, H.*, and *W. Kälbener:* Z. Physik. Chem. B **5**, 131 (1929); **6**, 343 (1930).

[342] See *Kortüm, G.*, and *W. Zoller:* Chem. Ber. **100**, 280 (1967), and the literature cited therein.

band at 15,000 cm^{-1}. The spectra of the thermochromic form and the piezochromic form were identical. The reversible transformation B→A could likewise be followed at various temperatures by reflectance spectroscopy. The kinetic evaluation of these measurements provided the rate constants of the two transformations $A \underset{k_2}{\overset{k_1}{\rightleftarrows}} B$ and the corresponding activation energies $E_1 = 12.5$ kcal/mole and $E_2 = 1.8$ kcal/mole. From these values clues regarding the conformational isomerism of the molecular forms A and B could be extracted.

d) Spectra of Crystalline Powders

The crystal field theory attempts to calculate the splitting of the atomic energy levels of transition metal ions (in particular, those that have incompletely filled *d*-orbitals) from the symmetry properties of the electric fields in the crystal. Ligand field theory is a similar treatment that assumes that the active field depends on the ligands immediately around the particular ion rather than on the whole crystal. In this instance, *coulombic* forces participate as well as the co-valent ones. However, the original hope that the splittings would be able to be calculated on the basis of a simple *coulombic* model has only been partially fulfilled. Such a calculation gives only a rough approximation, and semi-empirical corrections, based on spectroscopic measurements, are necessary.

Reflectance spectroscopy has been particularly important in the study of crystalline powders. It is often very difficult and tedious to grow single crystals. Sometimes, as with oxides, it is virtually impossible. Reflectance spectroscopy of powders also has advantages even when compared with the investigation of solutions. They can be summarized as follows:

1. The ligands can be varied over a wide range of structures.

2. The substances that exist in a crystal are definite compounds, whereas complexes in solution can undergo exchange reactions. Thus, the crystal spectra are sometimes better defined than the solution spectra (e.g. $MnCl_2 \cdot 4H_2O$).

3. The central ion-ligand distances are known exactly in crystals.

4. The splitting of the spectral terms can be studied at various field symmetries (e.g. octahedral, tetrahedral, tetragonally or rhombically distorted octahedra).

5. Spectra can easily be measured at low temperatures.

6. The ions to be studied can be implanted in a non-absorbing host-lattice. This enables the central atom-ligand distance to be varied, and an unusual symmetry to be stabilized.

7. It is possible to study the effect of the outer coordination sphere as well as that of the closest neighbours.

Because of these advantages, the number of contributions of reflectance spectroscopic investigations to ligand field theory has increased steadily [343]. In most instances, the position of the absorption band is important in order to identify the electronic and vibrational transitions. Such measurements are simple when the scattering coefficient S does not depend greatly on the wavelength in the spectral region concerned (cf. p. 210). If the absorption is large, the sample should be diluted, using an inert standard as indicated previously. For weak bands, as ligand field bands frequently are, the pure crystalline powder can be measured against a MgO standard. Also, it is generally unnecessary to convert measured relative values into absolute values (p. 149).

Even when the radiation yield is smaller than in the corresponding transmission measurement, good resolution can be attained if an efficient spectrometer is used. Spectra at low temperatures can also be obtained without difficulty (cf. p. 230). To illustrate the efficiency of the method, a part of the reflectance spectrum of $MnCl_2 \cdot 4H_2O$ at room temperature and at 78° K has been compared [344] with an absorption spectrum of an aqueous solution (Fig. 134). The spectrum corresponds to the transitions in an octahedral field:

$$^6A_{1g}(^6S) \rightarrow {}^4A_{1g}(^4G) \quad \text{and} \quad {}^6A_{1g}(^6S) \rightarrow {}^4E_g(^4G).$$

For the spectrum obtained at room temperature, the extinction coefficient ε or $F(R'_\infty)$, measured with respect to $BaSO_4$ as standard, is plotted against $\tilde{v}$. It was necessary to reduce the intensity of the reference beam for the measurement of the low temperature spectrum because of the recording technique used. Thus, values that were only proportional to $F(R'_\infty)$ could be obtained. The bands of the solid are sharper than those of the solution. The resolution is comparable with that obtained in a single crystal spectrum [345] at 20° K. The small bands can be interpreted as combinations of electron transitions with octahedral vibrations. This is possible because a spin-orbit or spin-spin coupling can cause a splitting

---

[343] See e.g. *Assmussen, R. W.*, et al.: Acta. Chim. Scand. **11**, 745, 1097, 1223, 1331 (1957). — *Schmitz-Du-Mont, O.*, et al.: Z. Anorg. Chem. **295**, 7 (1958); **300**, 159 (1959); **312**, 121 (1961); **314**, 260 (1962); Ber. Bunsenges. **63**, 978 (1959).— *Neuhaus, A.:* Z. Krist. **113**, 195 (1960). — *Jørgensen, C. K.:* Mol. Phys. **4**, 231 (1961); **4**, 235 (1961); Acta. Chim. Scand. **17**, 1034 (1963). — *Baldwin, M. E.:* Spektrochim. Acta **19**, 319 (1963). — *Clark, R. J. H.:* J. Chem. Soc. **1964**, 417. — *Jassie, L. B.:* Spectrochim. Acta **20**, 169 (1964). — *Gans, P.*, et al.: Spectrochim. Acta **21**, 1589 (1965). — *Sintra, S. P.:* Spectrochim. Acta **22**, 57 (1966), etc.

[344] *Kortüm, G.*, and *D. Oelkrug:* Naturwiss. **53**, 600 (1966).

[345] *Pappalardo, R.:* Phil. Mag. (8) **2**, 1397 (1957).

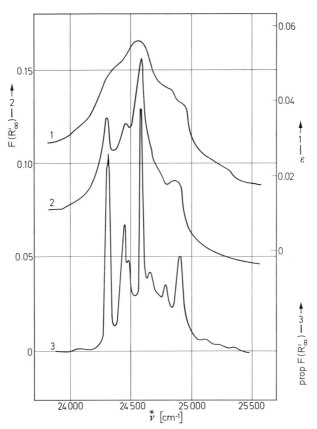

Fig. 134. Reflection spectra of $MnCl_2 \cdot 4H_2O$ at room temperature ② and at 78°K ③ in comparison with the transmission spectrum of the saturated aqueous solution ①. Measurement with the Model 14 Cary spectrometer with integrating sphere in the reflection attachment

only of the order of magnitude of 10 cm^{-1}. In the present spectra, repeating band differences of 235 cm^{-1} and 110 cm^{-1} are observed.

Another example are the spectra of manganese(II)sulphides (Fig. 135) obtained for a) the octahedrally coordinated, green modification that has a NaCl structure, and b) the tetrahedrally coordinated, red that has a Wurtzite structure[346]. The electrostatic perturbation calculation indicates that the ratio of the ligand field parameters, Δ, for tetrahedral and octahedral coordination, $\Delta_{Td}/\Delta_{Od} = -4/9$. Thus Mn^{2+}, under otherwise identical conditions (the same ligands, practically the same Mn–S distances) exhibits a greater energy difference for the splitting

---

[346] Oelkrug, D.: Ber. Bunsenges. **71**, 697 (1967).

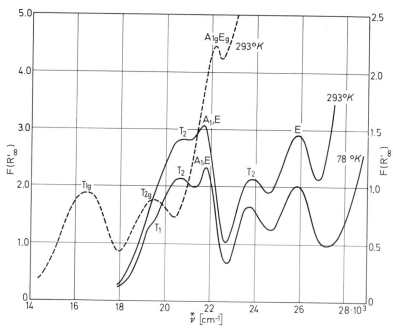

Fig. 135. Reflection spectra of green MnS (dotted curve, left-hand ordinate) and of red MnS (full-line curve, right-hand ordinate), undiluted against BaSO$_4$ as standard and with glass covers. Model 14 Cary spectrometer

terms $^4A_1$, ^4E, $^4T_1$ and $^4T_2$ derived from the ^4G term of the free Mn^{2+} ion for the octahedral structure (dotted line) than for the tetrahedral form (solid line). This has been verified experimentally. The $^4T_1$ transition of the red form can be observed as a shoulder, but only at low temperatures [347].

e) Dynamic Reflectance Spectroscopy

Reflectance spectroscopy can further be applied with advantage to detect and follow structural changes, thermal decompositions and the transformation of complexes as well as in the study of crystal structure in connection with the crystal field and ligand field theories discussed previously.

Ions and molecules in the S ground state are independent of the crystal field and show no term splitting. Nevertheless, a transition into

---

[347] Bands at 23,800 and 25,800 cm^{-1} correspond to other transitions.

an allotrope is usually accompanied by pronounced changes in the reflectance spectrum. In the opposite way, such changes in the spectra indicate a change of crystal system. Examples are the transitions of the red HgI_2 into its yellow modification [348] at 127° C or the transition of β-AgI into its α-form at 145° C. Different spectra are obtained above and below the transition temperature. This indicates a reversible transformation.

Recently, reflectance spectroscopy has been applied to the study of thermal transformation and decomposition reactions, as a complement to well-known techniques such as thermogravimetry, differential thermal analysis, pyrolysis, etc. This method is described as "dynamic" reflectance spectroscopy [349]. The reflectance of a compound is measured at a selected wavelength as a function of temperature. During the experiment, the temperature is increased linearly with time, e.g., at 2° per minute.

The transition of bis-pyridine cobalt(II)chloride from its violet, octahedral, α-form to the blue, tetrahedral, β-form was studied in this way [350]. The reflectance begins to decrease at 100°, but becomes constant again at 135° C. This agrees well with differential thermal analysis measurements, which demonstrates the transition by an endothermic peak in the same temperature range. The changes of $CuHgI_4$, Ag_2HgI_4 and AgI into modified structures with different coordination numbers were followed in the same way [351], as was the reaction of $CoCl_2 \cdot 6H_2O$ with KCl [352].

As with other dynamic methods, this method is limited by the need to increase temperatures very slowly if transition temperatures are to be determined accurately. This is caused by the delay between the transformations or reactions and the recording of temperature. The reflectance method, moreover, is affected by the temperature gradient that is established between the reflecting surface and the inner parts of the sample. This temperature gradient increases with increasing temperature. Even when, as is usual practice, the temperature is measured with a thermocouple placed immediately below the sample (cf. p. 229), the actual surface temperature is appreciably lower than this value. This explains why the $\beta \rightarrow \alpha$ transition for AgI for instance, that occurs at 145° C, apparently occurs at 155° C when this method is used [351].

---

[348] *Kortüm, G.:* Trans. Faraday Soc. **58**, 1624 (1962).

[349] *Wendlandt, W. W.:* Science **140**, 1085 (1963).

[350] *Wendlandt, W. W.:* Chemist-Analyst **53**, 71 (1964). See also *Wendlandt, W. W.:* Modern aspects of reflectance spectroscopy, p. 53 ff. New York: Plenum Press 1968.

[351] *Wendlandt, W. W.,* and *T. D. George:* Chemist-Analyst **53**, 100 (1964).

[352] *Wendlandt, W. W.,* and *R. E. Cathers:* Chemist-Analyst **53**, 110 (1964).

f) Analytical Photometric Measurements

It has repeatedly been mentioned that the value of $F(R_\infty)$ at a given wavelength for a diluted system depends linearly on the concentration of the adsorbed substance (cf. Figs. 61 and 62). This makes it possible to determine unknown concentrations in mixtures by using a linear calibration curve. The Kubelka-Munk function can therefore be applied for photometric measurements in mixtures of solids [353]. Also, at higher concentrations, where the linear relationship is no longer valid, the non-linear calibration curve can be used for determinations, provided that it is reproducible. The assumptions that must be fulfilled for such a treatment to be valid are discussed below in a special example.

Both titanium dioxide modifications – rutile and anatase – are technically very important as white pigments. Rutile has a higher refractive index and better covering properties, and so is a better clearing agent (cf. p. 112). Therefore, it is important to determine rutile in mixtures with anatase. A photometric procedure is based on the different reflectance spectra of both substances [354] (Fig. 136).

The analysis should be carried out in a way suitable for the type of instrument available. The reflectance is, for instance, measured at 26,000 cm^{-1} (Fig. 136) where the difference in the absorption of the two forms is greatest. An almost linear calibration curve of the dependence of $F(R'_\infty)$ against the mole ratio of rutile, x_{rutile}, is obtained if the sample to be analyzed is diluted with MgO. The wavelength must be controlled precisely because the measurement is carried out on a steep part of the absorption curve [355]. This is best achieved with a mercury discharge lamp, which gives a line spectrum, and interference filters are used to isolate the required line.

When, on the other hand, the reflectance of the undiluted mixed powder is measured against pure anatase (which absorbs less than rutile) as standard, an "absorption band" is obtained (Fig. 137) that corresponds to the difference in absorption between both materials that are compared. The band-heights, or, better, the areas under the band, can be plotted against the mole ratio x of the mixture to give another calibration curve. This is no longer linear, but it is reproducible.

The integration of the area under the band in Fig. 137 is obtained automatically when the reflectance of the samples if measured with a

---

[353] It is difficult to understand that in some papers [cf. e.g. *Griffiths, T. R.*: Anal. Chem. **35**, 1077 (1963)] it is assumed and verified experimentally that the extinction, $\log(1/R_\infty)$, is linearly proportional to concentration.

[354] *Kortüm, G.*, and *G. Herzog*: Z. Analyt. Chem. **190**, 239 (1962).

[355] See *Kortüm, G.*: Kolorimetrie, Photometrie und Spektrometrie, 4th ed. Berlin-Göttingen-Heidelberg: Springer 1962.

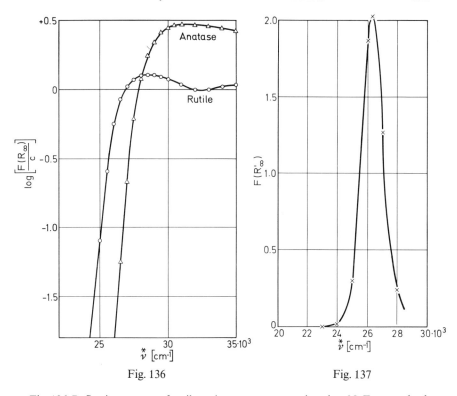

Fig. 136. Reflection spectra of rutile and anatase, measured against NaF as standard

Fig. 137. Kubelka-Munk function of a rutile-anatase mixture measured against pure anatase as standard

filter photometer. The filter chosen should cover the spectral range in which rutile absorbs more strongly than anatase, i.e., between 24,000 and 27,000 cm^{-1}.

Generally, powder mixtures composed of two (or more) components can be analysed quantitatively in this way if each component absorbs sufficiently in an accessible spectral region, and this absorption differs appreciably from the absorption of the other components in the same region. An accuracy of $\pm 2\%$ can be achieved under optimum conditions, even when the dependence between $F(R_\infty)$ and x is non-linear, as is usually the case for strongly absorbing samples. Assuming the additivity of K and S of the individual components i of a mixture, we can generally write

$$F(R'_\infty) = \frac{\Sigma c_i K_i}{\Sigma c_i S_i}, \qquad (6)$$

292   VII. Applications

where the c_i's represent the concentrations of the respective components. According to the experiences of Fig. 112, this assumption is only admissable in practice for mixtures that contain a large excess of diluent such that the scattering coefficient is completely determined by the diluent. In this case, for any given wavelength the simplification

$$F(R'_\infty) = \frac{\Sigma c_i K_i}{S} \cong \frac{\Sigma x_i K_i}{S} \tag{7}$$

is valid; i.e., we obtain an approximately linear relationship between $F(R'_\infty)$ and the concentration of the component to be determined. We measure at the same number of wavelengths (chosen to maximize the exactness of the measurements) as there are components in the mixture, so that, for a binary mixture of the components A and B, for example, we obtain [356]

$$\frac{K_{A,\lambda_1}}{K_{A,\lambda_2}} \gg \frac{K_{B,\lambda_1}}{K_{B,\lambda_2}}. \tag{8}$$

After all, this procedure corresponds throughout to that which has been usual practice in solution for some time [357] and more recently at times used for powder mixtures [358].

An analytical procedure that is similar in principle has been given by *Ringbom* [359] and *Giovanelli* [360]. It rests on the observation previously made in Fig. 55 that R_∞ itself in the region from 0.2 to 0.7 is nearly a linear function of the $\log(a/\sigma)$. If we wish to determine solely the concentration of a component A in a mixture, as is often the case especially in biochemical problems, we first combine the absorptions of all other materials present into an absorption constant a_u of the "background". By means of an iteration procedure, we can separate the reflectance spectrum of this background from that of A which must be assumed approximately known. In this manner we obtain an initial $\Delta R_{\tilde{\nu}\max}$ in the maximum of a band of the A spectrum and, by using Lorentzian dispersion curves [361], the half-width b of the band in cm^{-1}. Then for the

---

[356] Compare in this regard *Kortüm, G.*: Kolorimetrie, Photometrie und Spektrometrie, 4th ed., p. 27 ff. Berlin-Göttingen-Heidelberg: Springer 1962.

[357] Compare *Mayer, F. X.*, and *A. Luszczak*: Absorptionsspektralanal. Berlin: Walter de Gruyter & Co. 1951. — *Davidson, H. R.*, and *J. H. Godlove*: Am. Dyestuff. Rep. **39**, 628 (1950).

[358] *Everhard, M. E.*, et al.: J. Pharm. Sci. **53**, 173 (1964). — *Frei, R. W.*, et al.: Can. J. Chem. **44**, 1945 (1966).

[359] *Ringbom, A.*: Z. Analyt. Chem. **715**, 332 (1939).

[360] *Giovanelli, R. G.*: Nature **179**, 621 (1957); Australian J. Exp. Biol. Med. Sci. **35**, 143 (1957).

[361] Compare, for example, *Kortüm, G.*: Kolorimetrie, Photometrie und Spektrometrie, 4th ed. Berlin-Göttingen-Heidelberg: Springer 1962.

absorption coefficient of A,

$$a_A = \frac{kc}{1 + \left(\frac{\Delta \tilde{v}^*}{b}\right)^2} \qquad (9)$$

is valid where k represents a composite of all physical constants from the dispersion theory and $\Delta \tilde{v}^*$ the distance in wave numbers from the center of the band. If $\Delta R_{\tilde{v}\max}$ is small, the approximation

$$\frac{\Delta R_{\tilde{v}}}{\Delta R_{\tilde{v}\max}} = \frac{1}{1 + \left(\frac{\Delta \tilde{v}^*}{b}\right)^2} \qquad (10)$$

is valid in the region of the band. We test to see if the difference ΔR between the eliminated background and measured reflectance curves correspond to this relationship with sufficient exactness and must eventually repeat the calculation with an improved background curve. With $\Delta R_{\tilde{v}\max}$ thus ascertained, using Fig. 55 we can also determine a_A/σ. Here too we assume that σ is a constant.

The reflectance spectra of heavy metal complexes concentrated on ion-exchange resins have been very useful for the photometric determination of heavy metal ions [362]. Because their concentration has been increased by some 10^3 or 10^4 times, it is quite easy to determine quantitatively traces of these metals. The measurement is compared with that of the pure resin as reference standard. The absolute reflectance, R_∞, of the heavy metal complex can be obtained after correcting for the self-absorption of the resin. This is measured by comparing the reflectance of the resin with a freshly prepared MgO standard (cf. p. 148). Fig. 138 shows typical color curves of $Cu(H_2O)_4^{2+}$ ions in air-dried Dowex 50W-X8 (H-form). The concentration of the enriched complex is expressed in g. atoms of copper(II) per g equiv. of the binding group [363]. The calibration curve showing the dependence of $F(R_\infty)$ on concentration c is linear between 0.4 and 30% of the exchange capacity. This shows that here the Kubelka-Munk theory is strictly valid within the accuracy of measurement of $\pm 2\%$. Similar conclusions have been reached concerning $[Co(II)(NCS)]_4]^{2-}$ concentrated in Dowex 1–X8 anion exchange resin.

Reflectance in the near IR has been used for the quantitative determination of water in various substances. Based on the specific absorp-

---

[362] *Fujimoto, M.*, and *G. Kortüm*: Ber. Bunsenges. **68**, 488 (1964).

[363] The values of $F(R_\infty)$ first increase with time, until a steady state in the occupation of the binding sites is reached, which is governed by the diffusion of complexes inside the swollen resin particles.

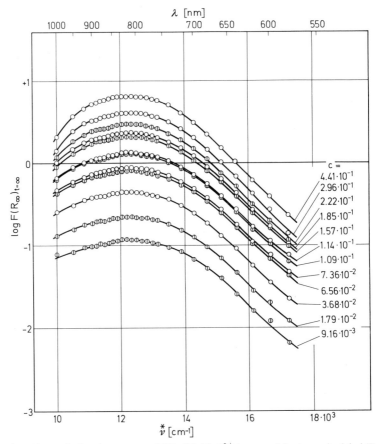

Fig. 138. Typical color curves of $[Cu(H_2O)_4]^{2+}$-ions enriched on air-dried Dowex 50W-X8(H-form) at different concentrations (gAtomCu^{2+}/g equ. Resin Sites)

tion at 1.93 and 1.7 µ, a continuous optical method can be developed for following the changes in humidity [364]. In Fig. 139, $F(R'_\infty)$ at 1.93 µ is plotted against the water content in percent of the dry weight measured gravimetrically. Hostaflon TF powder was used as the reference standard; its reflectance in this spectral region is >95%. In most instances, a linear relationship was found, indicating the validity of the Kubelka-Munk theory for this system. The slope of the linear plot is inversely proportional to the scattering coefficient of the various substances studied. With higher water content and for hydrophobic substances, the cavities are filled successively with water. This causes the change in the scatter-

---

[364] *Hoffmann, K.*: Chem. Ing. Techn. **35**, 55 (1963).

ing coefficient. These effects can be eliminated in the same way as are perturbations owing to the change in the properties of surface and in the packing density, by plotting $Q \equiv R'_{\infty 1,9\mu}/R'_{\infty 1,7\mu}$ as a function of water content (Fig. 140). These plots indicate that the measurement is particularly sensitive at low water concentrations. This is especially important for technological control of drying installations, so that maximal drying efficiency can be achieved.

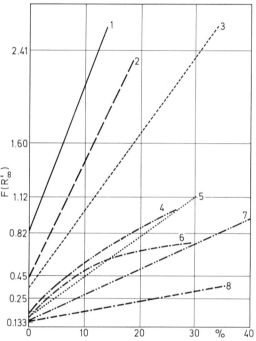

Fig. 139. Kubelka-Munk function of various materials at 1.93 μ as function of the water content in percent of the dry weight. 1 Methyl cellulose, 2 Gelatine, 3 Wool, 4 Cotton, 5 Starch, 6 Flour, 7 Paper, 8 Leather

An area of application especially rich with prospects for reflectance spectroscopy is the analytical evaluation of *thin-layer chromatograms*. It should be possible to directly analyze the separated spots in the thin layer qualitatively and eventually also quantitatively without have to cut them out and extract them. This commonly involves loss of material. At the very least, it should be possible to eliminate the extraction process by cutting out the spots, grinding them with the suitable adsorbent, preparing samples and measuring the reflectance spectra of them and/or determining the separated materials quantitatively by means of calibration curves.

For the latter process, the commercial adsorbents of thin-layer chromatography have shown themselves to be sufficiently transparent in the visible and near UV[365] that the Kubelka-Munk function for

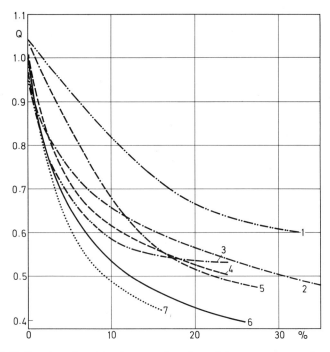

Fig. 140. Quotient $Q = R_{1.93\mu}/R_{1.7\mu}$ as function of the water content in percent of the dry weight. 1 Leather, 2 Paper, 3 Flour, 4 Cotton, 5 Starch, 6 Wool, 7 Gelatine

diluted systems using the pure adsorbent as standard can be applied. In this procedure, the spots are first localized in the usual manner through warming, fluorescence, UV-absorption, color reactions, etc. The spot is completely removed with a hollow stylus (e.g., a cork borer), placed in an agate mortar, and diluted with a determined amount of adsorbent to a given weight (e.g., 100 mg). The material is then ground until a homogeneous mixture is obtained and the contents of the sample can be determined by reflection measurements using a predetermined calibration

---

[365] Compare, for example, *Frei R. W.*, and *M. M. Frodyma:* Anal. Chim. Acta **32**, 501 (1965). — *Stahl, E.:* Proc. Soc. Anal. Chem. **1**, 121 (1964).

curve. These manipulations have been described individually for routine analysis [366]. $F(R_\infty)$ plotted against the concentration will only approach linearity if the reflectance of the adsorbent vs. MgO is corrected to absolute values (compare p. 148). Naturally, if this is not done, a bent calibration curve can also be used directly. Because the reflectance of the adsorbent varies from sample to sample and also frequently, when preserved a long time, also with time, it is recommended that the calibration curve be frequently tested for correctness.

The second process of directly analyzing the separated spots on the thin layer in situ by means of quantitative reflectance measurements [367] has many great difficulties and leads to inexact results. One reason for this is that the characteristic absorption of the adsorbents, especially in the short wavelength part of the spectrum, can no longer be neglected and must be eliminated by measuring its reflectance against MgO as a standard and subsequently establishing its absolute reflectance via procedures outlined on p. 150. If this is not done, the "typical color curve" can be considerably perturbed, for example, with regard to the positions of the absorption maxima. The second essentially more serious error source for *quantitative* determinations stems from the fact that the respective size of the spots varies and that as a rule, the substance to be determined is not homogeneously distributed in its respective spot. This can lead to extraordinarily large errors (compare p. 299). Finally, it must also be recognized that the thin layers are frequently not thick enough and the insufficient reflection of the background can perturb the results. Also, if the background is "white", layers that are too thin show considerable deviations from true R_∞-values. In spite of the greater amount of effort involved, the first process is nevertheless preferred in cases where greatest accuracy is desired.

On the other hand, the *qualitative* analysis of thin-layer chromatograms by means of reflectance and/or fluorescence measurements has extraordinarily large advantages. No color reactions or other measures are necessary for locating the adsorption zones because the dregree of remission (also in the UV) as a function of position can easily be measured and recorded. It is then possible to obtain "typical color curves" from the individual chromatogram spots by means of which it is often possible

---

[366] *Frodyma, M. M.*, et al.: J. Chromatog. **13**, 61 (1964); Anal. Chim. Acta **33**, 639 (1965). — *Frodyma, M. M.*, and *V. T. Lieu:* Modern aspects of reflectance spectroscopy, p. 88 ff. New York: Plenum Press 1968.

[367] *Frei, R. W.*, et al.: Chimia **20**, 23 (1966). — *Frodyma, M. M.*, and *V. T. Lieu:* Anal. Chem. **39**, 814 (1967). — *Lieu, V. T.*, et al.: Anal. Biochem. **19**, 454 (1967). — *Jork, H.:* Cosmo Pharma **3**, 33 (1967). — *Stahl, E.*, and *H. Jork:* Zeiss Inf. Bul. **16**, 52 (1968).

to directly identify the separated materials[368]. With the chromatogram spectrophotometer described on p. 225, such measurements can be carried out quickly and routinely.

If this apparatus is to be used for *quantitative* determinations, calibration substances must, as a rule, be allowed to follow on the sheet. Equivalent solvents and take-off zones (spot sizes) are necessary here. The reasons for this will be discussed later. In such cases reproducibilities of ± 3–4% in the measured concentrations have been reported[369].

Repeated attempts have been made[370] to apply reflectance measurements to the quantitative evaluation of paper chromatograms, and thus circumvent the need to elute the separated substances. The main difficulty here also is that the adsorbed substance is never homogeneously distributed over the whole paper sample, which is required by theory for a relationship between the reflectance and the concentration of the adsorbed substance[371]. Nevertheless, it has been shown[372] that if a dyestuff is distributed regularly over spots of comparable area, and if the measurement is carried out against a background of ten layers of the same paper (to achieve "infinite layer thickness"), reproducible calibration curves can be obtained. These curves can then be used for a quantitative determination of the given substance. In practice, the spots formed on paper chromatograms rarely have a regular size or shape so that this simple method cannot be used.

It can nevertheless be shown that consideration of the spot size and the characteristic reflectance of the chromatogram paper facilitates the derivation of an equation that represents a linear relationship between the Kubelka-Munk function of the sample and the amount of the substance of interest it contains[373]. When the reflectance of the sample is not too large, small inhomogenieties in the spots only slightly effect the exactness of the determinations.

---

[368] Compare in this regard *Stahl, E.,* and *H. Jork:* Zeiss Inf. Bull. **16**, 52 (1968), and the literature cited therein.

[369] *Jork, H.:* J. Chromatog. **33**, 297 (1968).

[370] *Vaeck, S. V.:* Nature **172**, 213 (1953); Analyt. Chim. Acta **10**, 48 (1954). — *Fischer, R. B.,* and *F. Vratny:* Anal. Chim. Acta **13**, 588 (1955). — *Korte, F.,* and *H. Weitkamp:* Angew. Chem. **70**, 434 (1958).

[371] Attempts have been made to measure the differential extinctions in transmission, obtained when the paper was soaked in paraffin oil, cut into narrow strips, and the transmittance measured separately for each strip [see, e.g. *Block, R. J.:* Science **108**, 608 (1948). — *Bull, B. H.,* et al.: J. Am. Chem. Soc. **71**, 550 (1959)]. Nevertheless, this procedure is too complicated to be used as a practical analytical method.

[372] *Kortüm, G.,* and *J. Vogel:* Angew. Chem. **71**, 451 (1959).

[373] *Braun, W.,* and *G. Kortüm:* Zeiss Mitteilg. **4**, 379 (1968); Zeiss Inf. Bul. **16**, 27 (1968).

The problem consists mainly of the fact that the measured reflectance depends on both the irregularly formed spot and the surrounding paper border. Moreover, for application of the Kubelka-Munk theory, the necessary condition of "infinite layer thickness" must be approximately realized by underlaying the chromatogram with many layers of the same paper so that the transparency is negligibly small. If the dyed or adsorbate covered fraction of the total sample surface amounts to $1/n$, the surrounding paper border is given by $(n-1)/n$, and the average measured *absolute* reflectance is given by

$$\bar{R}_\infty = \bar{R}'_\infty \varrho = \frac{n-1}{n} R'_{\infty 1} \varrho + \frac{1}{n} R'_{\infty 2} \varrho, \tag{11}$$

where $R'_{\infty 1}$ and $R'_{\infty 2}$ are the relative reflectances of the surrounding paper and the colored spot, respectively, and ϱ is the known absolute reflectance of MgO (see p. 146). From this, the desired absolute reflectance can be extracted by

$$R_{\infty 2} = n \varrho \bar{R}'_\infty - (n-1) \varrho R'_{\infty 1}. \tag{12}$$

According to the Kubelka-Munk function,

$$F(R_{\infty \text{ ads.}}) = F(R_{\infty 2}) - F(R_{\infty 1}) \tag{13}$$

is valid. If $R_{\infty 2}$ is inserted, (13) becomes

$$F(R_{\infty \text{ ads.}}) = \frac{b^2}{2(1-b)} - F(R_{\infty 1}) = \frac{K}{S} = \text{prop.}\, c. \tag{14}$$

c is the concentration in gm/unit surface area and

$$b \equiv 1 + (n-1) R'_{\infty 1} \varrho - n \bar{R}'_\infty \varrho. \tag{15}$$

Since the area covered with adsorbate amounts to $1/n$ of the total surface, the $F(R_{\infty \text{ ads.}})$ of (14) must be multiplied by $1/n$ in order to get the dependence on the adsorbed amount in grams:

$$F(R_{\infty \text{ ads.}}) \cdot \frac{1}{n} = \frac{b^2}{2n(1-b)} - \frac{F(R_{\infty 1})}{n} = \text{prop.}\, g. \tag{16}$$

If the characteristic absorption of the paper is small (e.g., $R'_{\infty 1} = 0.93$, so that $F(R_{\infty 1})/n = 0.002$), the last term in (15) can be neglected and the approximation

$$F(R_{\infty \text{ ads.}}) \cdot \frac{1}{n} = \frac{b^2}{2n(1-b)} = \text{prop.}\, g \tag{16a}$$

is valid. This is only permissable however in the long wavelength region of the visible spectrum (at 400 mµ, $R'_{\infty 1}$ already amounts to about 0.84 with chromatographic papers). If the characteristic absorption of the paper is neglected completely, i.e., $R_{\infty 1} = 1$, we obtain

$$F(R_{\infty \text{ ads.}}) \cdot \frac{1}{n} = F(\bar{R}_{\infty}) \cdot \frac{n\bar{R}_{\infty}}{n\bar{R}_{\infty} - (n-1)}. \qquad (17)$$

The following example shows, however, that a very small characteristic absorption of the paper already leads to very large errors. Let $n = 2$, $R'_{\infty 1} = 0.93$ and/or 1. Then, according to (16a), $F(R_{\infty \text{ ads.}}) \cdot \frac{1}{n} = 0.506$

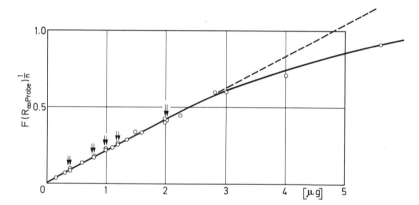

Fig. 141. Dependence of $F(R_{\infty \text{ ads.}})/n$ on the amount of material present in paper chromatograms

and, according to (17), 0.800. Thus the value obtained without considering the characteristic absorption of the paper is in error by about 58%.

To test Eq. (16a), the size n of a spot must be planimetrically determined. This can be done within about 2% by averaging the results of numerous individual determinations. $R'_{\infty 1}$ and $\bar{R}'_{\infty}$ are measured and $F(R_{\infty \text{ ads.}})/n$ is calculated from (16a). If these values are plotted against given amounts of adsorbed substance in grams, a straight line should be obtained. Measurements on malachite green at 16,00 cm^{-1} showed that the expected proportionality could actually be obtained (Fig. 141). Of particular significance was the result that samples with equivalent amounts of material distributed in spots of different sizes (arrows) gave results that corresponded very well.

A further calculation showed that relatively small inhomogenieties in the chromatogram spots ($\Delta R < 0.1$) effected the accuracy of the measurements only slightly when the total reflectance $\bar{R}_\infty \leq 0.6$. On the other hand, with very light samples, the error increased very quickly.

Numerous other literature references to application possibilities of reflectance spectroscopy to analytical problems can be found in *Wendlandt, W.,* and *H. G. Hecht*: Reflectance Spectroscopy. New York: Interscience Publ. 1966.

g) Color Measurement and Color Matching

The most important industrial application of reflectance spectroscopy is the measurement and adjustment of "bulk dyes". This is necessary, for example, in the textile, laquers, pigments and plastics industries. These applications have recently been described in numerous publications[374], and it is possible to refer only to a small portion of them.

A recording spectrophotometer has proved to be best for measuring bulk dyes, so that the reflectance R_∞ of the sample is depicted as a function of λ or $\overset{*}{\nu}$ (cf. p. 226). The knowledge of the relationship between the reflectance and the dyestuff concentration is a prerequisite for the calculation of dyeing recipes. This relationship is linear if the Kubelka-Munk function is valid, but it is valid only for "diluted systems" (cf. Fig. 57). The Kubelka-Munk theory is also limited by contributions of regular surface reflection, by the dependence of the scattering coefficient on wavelength, by non-uniform particle size, etc. These have already been discussed in detail. A further limitation is that the diffusely scattering particles must be surrounded by the same medium as the incident light; otherwise the Kubelka-Munk function is not valid, and additional reflections occur at the phase boundary (cf. p. 131). The contribution of gloss, $\overset{*}{r}_1$, and partial total reflection on the inner side of the sample, r_2, to the reflectance for perpendicular and parallel illumination, $R_{\infty\exp}$, can be expressed[375], by analogy with the equations of *Ryde*, (IV, 107) and (IV, 108), using the equation:

$$R_{\infty\exp} = \frac{(1-\overset{*}{r}_1)(1-r_2)R_\infty}{1-r_2 R_\infty}. \tag{18}$$

---

[374] Cf. especially the exhaustive description in *Judd, D. B.,* and *G. Wyszecki*: Color in business, science and industry, 2nd ed. New York-London: Wiley&Sons 1963; and also *Wright, W. D.*: The measurement of color, 2nd ed. New York: Macmillan 1958. — *Evans, R. M.*: An introduction to color. New York: Wiley & Sons 1948. — *Reule, A.*: Zeiss-Mitteilungen **3**, 266 (1964). — *Wyszecki, G. W.*, and *W. S. Stiles*: Color science. New York: Wiley & Sons 1967. Numerous papers can be found in journals such as: Die Farbe, J. Soc. Dyer's Col., Am. Dyestuff Rep., J. Oil Col. Chem. Assoc., Textilpraxis, Z. ges. Textilind. etc.

[375] *Saunderson, J. L.*: J. Opt. Soc. Am. **32**, 727 (1942).

If the sample has a refractive index of about 1.5 in the visible region (e.g. pigment bases), $\overset{*}{r}_1$ is ca. 0.04, and r_2 is ca. 0.59 [376]. Eq. (18) then simplifies to

$$R_{\infty \exp} \cong \frac{0.4 R_\infty}{1 - 0.59 R_\infty}. \qquad (19)$$

This value of $R_{\infty \exp}$ is then used in the Kubelka-Munk function (IV, 32); the corresponding values of $R_{\infty \exp}$ and $F(R_{\infty \exp})$ have been tabulated [377]. A still better approximation is presumably Eq. (IV, 113) obtained from the Fresnel equations. It is also valid for perpendicular irradiation and assumes an isotropic scattering distribution within the sample.

The simplest example in which the Kubelka-Munk function is valid is a mixture of colored pigments with an excess of a white pigment. The absorption coefficient of the white pigment can be considered to be zero and the scattering coefficient of the white pigment governs the total scattering of the mixture. The Kubelka-Munk function can then be written in a form similar to Eq. (6):

$$F(R'_\infty) = \frac{c_1 K_1 + c_2 K_2 + c_3 K_3 + \cdots}{S} = \frac{\sum_{1}^{n} c_i K_i}{S}, \qquad (20)$$

where the K_i are the absorption coefficients and the c_i are the concentrations of each colored component. This equation can also be applied, for example, to dyed textiles, because their scattering is caused almost entirely by the substrate. When the scattering coefficient S is virtually independent of the wavelength (a condition often fulfilled for not too small particles in the visible spectral range) the equation simplifies further:

$$F(R'_\infty) \sim \sum_{1}^{n} c_i K_i, \qquad (21)$$

where K_i is proportional to the molar extinction coefficient ε_i. R'_∞ is the reflectance relative to that of the white pigment or substrate. If the substrate has a measurable self-absorption, the relative values R'_∞ must be converted into absolute values R_∞, as described on p. 149. Eq. (21) then becomes:

$$F(R_\infty) - F_0(R_\infty) = \sum_{1}^{n} K_i c_i, \qquad (22)$$

where $F_0(R_\infty)$ is the Kubelka-Munk function of the white pigment or substrate. The inverse function (23) can be used to calculate R_∞:

$$R_\infty = \phi(F_0 + K_1 c_1 + K_2 c_2 + \cdots). \qquad (23)$$

---

[376] See note on p. 132.
[377] Duncan, D. R.: J. Oil Col. Chem. Ass. **45**, 300 (1962).

When values of K_i as a function of wavelength have been determined for a number of pigments, the reflectance and therefore $F(R'_\infty)$ and $F(R_\infty)$ can be calculated for an arbitrary mixture of pigments, insofar as the additivity rule is valid.

When the white pigment is not in excess, the values $F(R'_\infty)$ of the single pigments are usually not additive. This is shown by comparing Figs. 111 and 112. In such instances, attempts can be made to find an empirical function $\phi(R'_\infty)$ for each pigment. These functions should provide, at a given wavelength, an unambiguous and reproducible *linear* relationship between concentration and the degree of reflection. The values of $\phi(R'_\infty)$ will be additive, provided that there is no mutual interaction between the pigments, and that their scattering coefficients are approximately equal.

If the colored components of a pigment mixture (or the dyestuffs used for dyeing a single substrate) are known, and if the remission spectra of the mixture and all of its components (diluted) are also known, Eq. (22) can be applied for calculating dyeing recipes. This is needed in "color matching". The properties of the white pigment or the substrate as well as that of the colored pigments or dyestuffs are variable; they depend strongly on the methods of manufacture. Therefore, it usually is impossible to use the original recipe immediately for preparing a certain mixture or color. In each instance, the proper ratios of concentrations of the components used must be determined to achieve and reproduce a given shade of color. This color matching, which is very important in technology and industry, used to be made by trial and error. This treatment is made easier by applying Eq. (22). Nevertheless, this application is limited to "diluted systems", for which the Kubelka-Munk equation is approximately valid, and the values of $F(R_\infty)$ are therefore additive.

A system of linear equations for a series of selected wavelengths can be obtained from Eq. (22):

$$F(R_\infty)_{\lambda_1} - F_0(R_\infty)_{\lambda_1} = K_1(\lambda_1)c_1 + K_2(\lambda_1)c_2 + \cdots$$
$$F(R_\infty)_{\lambda_2} - F_0(R_\infty)_{\lambda_2} = K_1(\lambda_2)c_1 + K_2(\lambda_2)c_2 + \cdots \quad (24)$$
$$F(R_\infty)_{\lambda_3} - F_0(R_\infty)_{\lambda_3} = K_1(\lambda_3)c_1 + K_2(\lambda_3)c_2 + \cdots$$
....

Measurements must be made at at least as many wavelengths as there are pigments or dyestuffs in the mixture. Values of $F(R_\infty)$ and $F_0(R_\infty)$ can be obtained from the measured remission spectra of the mixture and of the white pigment or substrate. Values of K can be obtained from the remission spectra of the individual, diluted components of the mixture. The system of equations can be solved for the required concentration values by the use of analog computers specially designed for this problem [378]. The choice of wavelength is made so as to reach a high absorption for each pigment or each dyestuff at one of the selected wavelengths. The other pigments should have as low an absorption as possible at this wavelength. Under these conditions, the elements on the main diagonal of the coefficient matrix $K_i(\lambda_k)$ are determinant. The first, approximate concentration values can be calculated from these elements. These values are inserted into the system of equations, and second approximations for the concentrations on the main diagonal are obtained. This iteration procedure is continued until the concentration values become constant. In this way, the dyeing recipe can be calculated in advance. In a similar way, an existing dyeing recipe can be corrected if the non-constant properties of the components or substrate make it necessary. Instead of $F_0(R_\infty)$, the preliminary value is substituted, and, in place of the concentrations c_i to be determined, the corresponding differences Δc_i between the preliminary concentrations and those sought are used.

It is important to understand that, with the above procedure, it is possible to attain full agreement between the required value $F(R_\infty)$ and the adjusted value only at as many wavelengths as there are dystuffs present in the mixture. Thus a color, for which the remission spectra of the model and adjusted mixtures are completely identical, is rather improbable, because of the irreproducibility of the behavior of the components. This is particularly true when white pigments or substrates of various materials and of different origins are to be colored in the same way. Colorations are then obtained that look the same under a given illumination, but have different remission spectra. Such colorations are described as *metameric*, and they create a different impression on the eye under different illumination. The "metameric match" therefore should be made for the illumination under which the object will normally be viewed.

After attaining approximately similar remission spectra for the model and matched mixtures, using this procedure, a complete matching of the color impression on the eye can be made for a given type of illumination. To do this, the normalized tristimulus values X, Y, Z must be obtained from the reflectance spectra. Two colors make the same impression on the eye (for a given intensity of illumination S_λ, showing the same energy distribution) if these normalized tristimulus values are identical. If these values *for the equal-energy spectrum* are denoted as $\bar{x}_\lambda, \bar{y}_\lambda, \bar{z}_\lambda$ (the so-called normal spectral values)[379], then the normalized tristimulus

---

[378] *Davidson, H. R., H. Hemmendinger,* and *I. L. R. Laudry:* J. Soc. Dyers Col. **79**, 577 (1963).

[379] These values are tabulated in text books on colorimetry as are the values of energy distribution of the so-called standard radiation source of the Commission Internationale de l'Eclairage (CIE).

values are defined by Eq. (25):

$$X = \int_0^\infty S_\lambda \bar{x}_\lambda R_{\infty\lambda} d\lambda,$$
$$Y = \int_0^\infty S_\lambda \bar{y}_\lambda R_{\infty\lambda} d\lambda, \qquad (25)$$
$$Z = \int_0^\infty S_\lambda \bar{z}_\lambda R_{\infty\lambda} d\lambda.$$

In these equations, $S_\lambda \cdot R_{\infty\lambda}$ is the energy of the reflected radiation flux at a given wavelength. The color impressions of two samples a and b are therefore identical to the eye for a given type of illumination, if the conditions given in Eq. (26) are fulfilled.

$$X_a = X_b; \quad Y_a = Y_b; \quad Z_a = Z_b. \qquad (26)$$

When $(R_{\infty\lambda})_a = (R_{\infty\lambda})_b$, both spectra are completely identical, and the same impression will be made on the eye in all kinds of illumination (isomeric matched coloration). As mentioned above, this condition is hardly ever found in practice. Thus metamery has to be accepted (although it may be small) in which the color of the model and matched mixtures are the same at least for a certain kind of illumination.

The three conditions formulated in Eq. (25) for a given kind of illumination indicate that it should be possible to reproduce a certain color by using three suitable dyestuffs. In this instance, the normalized color values for the model (X, Y, Z) are considered as given, and the concentrations $c_1, c_2,$ and c_3 in the matched mixture as variables that should be able to be calculated by solving the system of equations. As the latter cannot be solved explicitly, it is necessary to substitute as an approximation the integrals by sums over the same visible spectral ranges, as shown in Eq. (27):

$$X = k \cdot \sum_{\lambda=380}^{770} S_\lambda \bar{x}_\lambda R_{\infty\lambda} \Delta\lambda; \quad Y = k \cdot \sum_{380}^{770} S_\lambda \bar{y}_\lambda R_{\infty\lambda} \Delta\lambda; \qquad (27)$$
$$Z = k \cdot \sum_{380}^{770} S_\lambda \bar{z}_\lambda R_{\infty\lambda} \Delta\lambda.$$

Here, k is a normalizing factor defined as

$$k = \frac{100}{\sum_{380}^{770} S_\lambda \bar{y}_\lambda \Delta\lambda}. \qquad (28)$$

Therefore, Y gives directly the reflectance in percent, relative to an ideally reflecting standard at the same illumination. According to recent measurements with a General Electric spectrophotometer in 15 different laboratories using plastic standards and C illumination (daylight), Y could be exactly determined to within $\pm 1.5\%$. When the measurements were repeated within a short time, the scatter in the measurements amounted to only $\pm 0.09\%$ and only $\pm 0.62\%$ after 14 months [379a]. The quantities $S_\lambda \bar{x}_\lambda$, $S_\lambda \bar{y}_\lambda$ and $S_\lambda \bar{z}_\lambda$ for frequently used sources of illumination are tabulated for wavelength intervals of 5μ [380] and 10μ [381]. It is possible to calculate, therefore, the summations in Eq. (27) from reflection measurements, and then the normalized tristimulus values X, Y and Z of a given model and a matched mixture prepared according to a recipe calculated with the aid of Eq. (24) can be calculated. If these values differ by small amounts ΔX, ΔY and ΔZ, they can be written in the form:

$$\Delta X = \frac{\partial X}{\partial c_1} \Delta c_1 + \frac{\partial X}{\partial c_2} \Delta c_2 + \frac{\partial X}{\partial c_3} \Delta c_3,$$
$$\Delta Y = \cdots,$$
$$\Delta Z = \cdots. \tag{29}$$

Total agreement can be attained by varying the concentrations by trial and error. However, Δc can be calculated from this linear system when the coefficients $\partial X / \partial c_i \ldots$ are known. These are obtained by partial differentiation of Eq. (25), after substituting concentrations [382] according to Eq. (23) in place of $R_{\infty \lambda}$. The original literature should be consulted for details of the mathematical treatment. This calculation can also be carried out using analog computers attached directly to recording spectrophotometers (p. 231). This application is based on the assumption, naturally, that the remission function is linear with respect to the concentrations of colored components, for "diluted systems" this is the Kubelka-Munk function.

The examples examined for the application of diffuse reflection for the investigation of definite problems have been arbitrarily chosen and are by no means exhaustive. As already mentioned, the application possibilities are almost unlimited. Several new directions of application may be given:

Critical investigations of biological materials are in beginning stages. Worth mentioning perhaps are measurements of the saturation of blood with oxygen (reflectance oximetry) [383], in which the extinction coefficients of oxygenated and reduced hemoglobin come in at two wavelengths. A linear relationship is found between the degree of oxygen saturation

---

[379a] See Billmeyer, F. W.: J. Opt. Soc. Am. **55**, 707 (1965).

[380] Smith, T., and J. Guild: Trans. Opt. Soc. (London) **33**, 73 (1931).

[381] Smith, T.: Proc. Phys. Soc. (London) **46**, 372 (1934).

[382] See Park, R. H., and E. I. Stearns: J. Opt. Soc. Am. **34**, 112 (1944).

[383] Polanyi, M. L., and R. M. Hehir: Rev. Sci. Instr. **31**, 401 (1960).

and the reflectance at 805 mμ and 660 mμ. Reflectance measurements on green leaves [384] in the IR showed, as a rule, reflectances of less than 5–11% in the 3–14 μ range while generally higher values were found in the 2–3 μ range. Minima at 2, 3, 6, and 15 μ were attributed to water absorption. Lengthy drying increased the reflectance throughout the spectrum although the absorption bands remained marked. Leaves reflect partly regularly and partly diffusely as the measured angular distribution of the remitted radiation showed. The regular components frequently become more prominent with increasing wavelength.

In more recent work with scattering transmission, the absorption spectrum of monocellular algae in aqueous suspension was measured [385]. This work established previously supposed theoretical recognition: namely, that in the neighborhood of an absorption maximum, the scattering coefficient has a dispersion similar to that of the refractive index and is thus very strongly wavelength dependent [386]. This overlapping of absorption and scattering causes the absorption maximum to appear at longer wavelengths in transmittance measurements and at shorter wavelengths in reflectance measurements. A procedure was worked out to calculate absorption and scattering from several measured values. In this way, the exact position of the absorption maximum of chlorophyll in the living cells was determined.

Another very promising area of application for diffuse reflectance measurements is the investigation of the durability and/or reactivity of powdered *pharmaceutical preparations* under the influence of light as a function of time. Here too a series of investigations have already be undertaken [387].

Geophysical problems likewise have already be investigated with the help of diffuse reflectance measurements. Desert areas of various types, for example, have been measured in the 400–650 mμ range from heights of 1.5–200 m above the earth's surface [388]. The reflectance more than doubles with increasing wavelength in this spectral region with the exception of basalt lava. The absolute values vary between 3 and 74%. The Lambert Cosine Law is not valid which means that here too, as

---

[384] *Wong, C. L.*, and *W. R. Blevin*: Australian J. Biol. Sci. **20**, 501 (1967), and the literature cited therein.

[385] *Hagemeister, V.*: Thesis, Tübingen 1968.

[386] Compare, for example, *Charney, E.*, and *F. S. Brackett*: Arch. Biochem. Biophys. **92**, 1 (1961). — *Latimer, P.*: Plant Physiol. **34**, 193 (1959).

[387] Compare, for example, *Everhard, M. E.*, and *F. W. Goodhart*: J. Pharm. Sci. **52**, 281 (1963). — *Lochmann, L.*, et al.: Am. Pharm. Ass. J. **49**, 163 (1960); **50**, 141, 145 (1961).

[388] Compare *Ashburn, E. V.*, and *R. G. Weldon*: J. Opt. Soc. Am. **46**, 583 (1956) and the literature cited therein.

expected, regularly reflected light is also measured. The same is true for snow in various states of settlement [389]. At large angles of incident irradiation, the spectrum of snow shows large portions of regular reflection. *Bloch*[390] has proposed a theory regarding the influence of vulcanic dust on the reflectance of snow in Antarctica. This is based on historical changes in sea level as a result of volcanic activity. The IR absorption of dust particles should cause local warming and thereby a recrystallization to large ice crystals. This increases the depth of penetration of incident radiation which again leads to stronger absorption of the IR bands of water and thus brings forth a self accelerating melting process.

[389] *Middleton, W. E. K.*, and *A. G. Mungall:* J. Opt. Soc. Am. **42**, 572 (1952).
[390] *Bloch, M. R.:* Palaeogeogr. Palaeoclimatol. Palaeoecol. **1**, 127 (1965). — *Lamb, H. H.:* Palaeogeogr. Palaeoclimatol. Palaeoecol. **4**, 219 (1968).

Chapter VIII. Reflectance Spectra Obtained by Attenuated Total Reflection

a) Determination of the Optical Constants n and $\varkappa$

As was indicated in Chapt. II, c, it is possible in principle to determine the optical constants n and $\varkappa$ of materials as a function of λ from regular reflection measurements, i.e. absorption spectra can be obtained. Two measurements are always necessary, e.g., $R_\perp$ at two different angles of incidence or $R_{reg(\alpha=0)}$ and the phase shift δ, and tedious mathematical or graphical processes are then needed for the evaluation of n and $\varkappa$. The accuracy of the values obtained in this way is not very satisfactory. This is even true for metals, which have very large $\varkappa$-values (Table 2). For most substances, especially organic materials, the $\varkappa$-values are considerably smaller even in the IR region, and the refractive indices lie between 1 and 2. Only when $\varkappa > 0.2$ can reliable $\varkappa$-values be obtained in this way from such reflection measurements. This has been shown by *Fahrenfort*[391]. When $\varkappa = 0.2$, which already corresponds to very high extinction coefficients (cf. note 14 on p. 21), optical constants of most substances cannot be obtained by this method.

As *Fahrenfort* has indicated, better results can nevertheless be obtained if the interface between a dielectric of large refractive index n_1 and the test substance n_2 is used for the reflection instead of an air-test sample interface. As was shown on p. 13, radiation coming from the more dense medium at angles of incidence $\alpha > \alpha_g$ is totally reflected. If the sample does not absorb ($\varkappa = 0$), on average no radiation energy is transferred in the less dense medium, although a transversally damped wave travels along the surface because of diffraction effects; but the energy flux through the phase boundary is equally large in both directions. When, on the contrary, $\varkappa \neq 0$, this energy flux is no longer equal, and some of the transferred radiation energy will be absorbed, i.e., the reflection is no longer total. The absorption is greatest in the immediate vicinity of the critical angle, α_g, and depends, of course, on the magnitude of the absorption index $\varkappa$. It appears that this *attenuated total reflection* (ATR) is already measurable at very small $\varkappa$-values (0.1 to 0.0001). Thus, this presents another, independent method of recording optical constants by reflectance measurements. It has proved to be exceedingly useful mainly in the infrared.

---

[391] *Fahrenfort, J.:* Spectrochim. Acta **17**, 698 (1961).

Using Eqs. (II, 30 and 31) substituted with the complex refractive index (Eq. II, 43), the reflectance $R_\perp$ and $R_\parallel$ can be calculated as a function of $\varkappa$. By this means the Fresnel equations become very complicated and can be handled easily only with an electronic calculator. For thick samples of medium 2 and incident radiation flux $I_0 = 1$, the reflectances are given by

$$R_\perp = \left[\frac{n_1 \cos\alpha - [(n_2 - i\varkappa_2)^2 - n_1^2 \sin^2\alpha]^{1/2}}{n_1 \cos\alpha + [(n_2 - i\varkappa_2)^2 - n_1^2 \sin^2\alpha]^{1/2}}\right]^2 \quad (1)$$

$$R_\parallel = \left[\frac{(n_2 - i\varkappa_2)^2 \cos\alpha - n_1[(n_2 - i\varkappa_2)^2 - n_1^2 \sin^2\alpha]^{1/2}}{(n_2 - i\varkappa_2)^2 \cos\alpha + n_1[(n_2 - i\varkappa_2)^2 - n_1^2 \sin^2\alpha]^{1/2}}\right]^2. \quad (2)$$

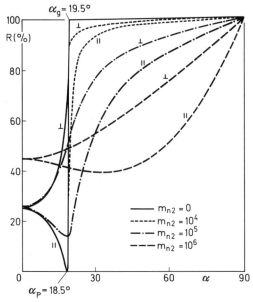

Fig. 142. Internal reflectance of a phase boundary with $n_2/n_1 = 0.333$ as a function of the angle of incidence α at $\lambda = 400$ mµ and various values of the extinction modulus m_{n2} as parameter

Separation of the real and imaginary parts of either (1) or (2) gives two equations by means of which n_2 and $\varkappa_2$, respectively, can be calculated from measurements, for example, of $R_\perp$ at two different angles α. Fig. 142 illustrates the calculated reflectance $R_\perp$ and/or $R_\parallel$ as a function of angle of incidence α at various values of the extinction modulus $m_{n,2} = 4\pi n\varkappa/\lambda_0$ as parameter (compare note 14, p. 21)[392]. It can be seen that at perpendic-

---

[392] Harrick, N. J.: Ann. N. Y. Acad. Sci. **101**, 928 (1963).

ular incidence ($\alpha = 0$) even an $m_{n,2} = 10^4$ or $n_2 \varkappa = 0.25$ scarcely effects the external reflectance at all. However, in the region of total reflection and especially in the neighborhood of the critical angle α_g the reflection losses become very large, and, to be sure, more so for $R_{||}$ than for $R_\perp$. Furthermore, the critical angle α_g losses its meaning and the $R-\alpha$-curves climb ever more slowly the greater $m_{n,2}$ becomes.

The sensitivity of this "attentuated total reflection" to changes in $\varkappa$ is illustrated in Fig. 143[393]. This figure again gives the values of $R_\perp$

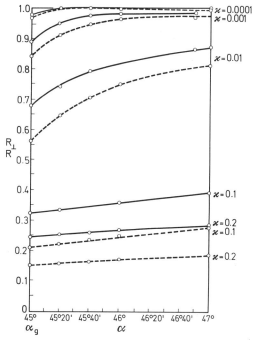

Fig. 143. Attenuation of the total reflection on the phase boundary densser medium/rarer medium with $n_2/n_1 = 0.707$ as a function of the angle of incidence with different absorption index $\varkappa$ of the rarer medium as parameter $R_\perp$ ———; R -----; $\alpha_g = 45°$

for irradiation polarized perpendicular to the plane of incidence and of $R = \frac{1}{2}(R_\perp + R_{||})$ for unpolarized light as functions of the angle of incidence in the neighborhood of the limiting angle α_g (45°) for different values of $\varkappa$.

To determine n and $\varkappa$ as a function of λ, two measurements of attenuated total reflection $R_\perp$ or $R_{||}$ must be made at two different angles of incidence. The constants can than be determined graphically. It turned out to be necessary to calculate groups of curves, that represent R

---

[393] According to *Fahrenfort, J.*: l.c.

as a function of n at different angles of incidence in the vicinity of α_g and at various values of $\varkappa$ as parameter. A measured R value at a given angle α_1 or α_2 produces a series of possible $n-\varkappa$ combinations in the form of a curve in the $n-\varkappa$-plane. The intersection of the two curves gives the desired values of n and $\varkappa$ at the wavelength used. As this method is very tedious, formulas have recently been developed[394] that give an

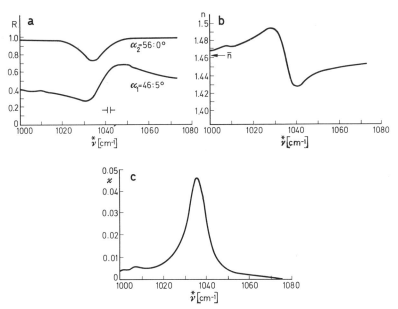

Fig. 144. Spectrum (a) from the attenuated total reflection of the 1035 cm^{-1} band of benzene and the optical constants n(b) and $\varkappa$(c) calculated from it

explicit solution to the equation for n and $\varkappa$. Thereby, $n = n_2/n_1$, where n_1 is the refractive index of the more dense medium, the dispersion of which must be known. Fig. 144 shows the experimental reflectance spectrum (a) and the calculated values of n (b) and $\varkappa$ (c) for the 1035 cm^{-1} band of liquid benzene[395].

---

[394] *Fahrenfort, J.*, and *W. M. Visser:* Spectrochim. Acta **18**, 1103 (1962). — *Hansen, W. N.:* Spectrochim. Acta **21**, 209, 815 (1965). — *Fahrenfort, J.*, and *W. M. Visser:* Spectrochim. Acta **21**, 1433 (1965).

[395] Further measurements for determining the optical constants n and $\varkappa$ can be found, for example in *Clifford, A. A.*, and *B. Crawford Jr.:* J. Phys. Chem. **70**, 1536 (1966). — *Gilby, A.*, et al.: J. Phys. Chem. **70**, 1525 (1966). — *Hansen, W. N.:* Spectrochim. Acta **21**, 209 (1965); ISA Trans. **4**, 263 (1965); Anal. Chem. **37**, 1142 (1965).

We see that reflectance measurements made considerably above the limiting angle α_g are similar to the $n\varkappa$-curves while measurements directly beneath the limiting angle approximately equal the mirror image of the dispersion curves. This result, which at first appears remarkable, could be explained by *Fahrenfort* and *Visser* by means of a detailed error discussion. The most exact $\varkappa$-values are obtained from measurements at two α-angles on both sides of α_g in the region $0.1 > \varkappa > 0.02$ with $\Delta\alpha = 10°$. For a relative error in R of ± 0.005, the relative error in $\varkappa$ amounts to only about $\pm 5\%$ over a relatively large α-range. Under the same conditions, the error in n comes to only 0.04–0.6%.

The method also lends itself advantageously to the determination of the optical constants of *metals*. The very high reflectance of the metal is reduced by the dielectric and for that reason, the difference to 100% of the total reflection can be more exactly measured. In the first approximation, the reflectance of the dielectric-metal phase boundary is given by [396]

$$R = 100 - 2.1 \cdot 10^{-4} \left(\frac{\varepsilon}{\lambda \sigma}\right)^{1/2}, \qquad (3)$$

where ε is the dielectric constant of the dielectric and σ is the specific conductivity of the metal. According to measurements by *Hansen*[397], values determined in this way agree very well with those obtained earlier by worrisome measurements according to classical methods.

b) Internal Reflection Spectroscopy

In the majority of cases of application, the ATR-method has been restricted to plotting or directly recording the apparent extinctions and/or transmittances of the investigated sample as a function of the wavelength without undertaking the somewhat difficult determination of the optical constants n and $\varkappa$ from the measured values and without consideration of the state of polarization of the incident irradiation. This method, also known as "internal reflection spectroscopy" (ITR), has recently been exhaustively described in a monograph[398] so that here we can restrict ourselves to the essentials. Fig. 142 illustrates a very instructive proof for the usefulness of this method. The figure compares the spectra of liquid dibutylphthlate under otherwise equivalent conditions measured by (A) transmission, (B) normal reflection, and (C) attenuated

---

[396] *Harrick, N. J.*: J. Opt. Soc. Am. **49**, 376 (1959).

[397] *Hansen, W. N.*: ISA Trans. **5**, 263 (1965).

[398] *Harrick, N. J.*: Internal reflection spectroscopy. New York: Interscience Publ. 1967; see also *Wendlandt, W. W.* (Ed.): Modern aspects of reflectance spectroscopy. New York: Plenum Press 1968.

314   VIII. Reflectance Spectra Obtained by Attenuated Total Reflection

total reflection using AgCl as the more dense phase [399]. The resolution and intensity of the bands are extraordinarily better in case (C) than in case (B) where normal reflection was used. The spectra in cases (A) and (C) are very similar.

The question that now arises is: how completely does the spectrum obtained in this way agree with the true spectrum; i.e., the question

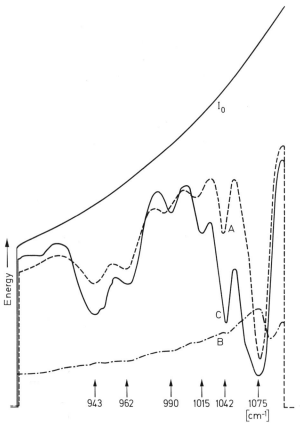

Fig. 145. Comparison of the infrared spectra of liquid dibutyl phthalate under the same external conditions. (*A*) in transmission, (*B*) in normal reflection, (*C*) in attenuated total reflection with AgCl as the more dense phase

concerns the correspondence of the measured apparent extinction with the absorption coefficients and with the depth of penetration of the radiation in the optically less dense medium. As already shown by Fig. 142, R depends on the angle of incidence and is smaller for ∥-polarization

---

[399] According to *Fahrenfort, J.*: l.c.

than for ⊥-polarization. Consequently, the absorption parameter a, defined by

$$a = (100 - R) \%, \quad (4)$$

must also be a function of these quantities. If R or a, calculated from the Fresnel equations (1) and (2), are plotted as a function of the extinction modulus m_n, the complicated curves of Fig. 146 are obtained. The ab-

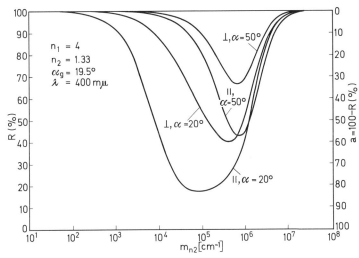

Fig. 146. Calculated dependence of the internal reflectance R or of absorption parameter $a = (100\text{-}R)$ from the extinction modulus of the optically rarer medium at two different angles of incidence. $\lambda = 400$ mµ; $n_2/n_1 = 0.333$

sorption parameter first increases from zero for non-absorbing media, goes through a maximum, and finally again falls to zero for very strongly absorbing materials (metals). A complete absorption with a single reflection cannot thus generally be attained even though m_n may be arbitrarily large. This is valid for a semi-infinite medium in which the absorbance can be changed by adjusting the concentration of the absorbing material.

According to the Maxwell theory, standing waves that penetrate in the optically thin, non-absorbing medium are formed perpendicularly to the totally reflecting surface. The amplitude of the electric field falls off exponentially with the distance from the phase boundary according to [400]

$$\vec{E} = \vec{E}_0 \cdot \exp\left[\frac{-x}{d_p}\right]. \quad (5)$$

---

[400] Harrick, N. J.: J. Opt. Soc. Am. **55**, 851 (1965).

The depth of penetration d_p, defined by the distance within which $\vec{E}$ decreases to $1/e$, can be calculated[401] by

$$d_p = \frac{\lambda_1}{2\pi(\sin^2\alpha - (n_2/n_1)^2)^{1/2}}; \quad \alpha > \alpha_g, \tag{6}$$

It is proportional to the wavelength λ_1 in the optically denser medium. If the rarer medium exhibits simultaneous absorption, the penetrating wave becomes attenuated. In the case of transmittance at very weak absorption ($m_n < 0.1$),

$$\frac{I}{I_0} = e^{-m_n d} \cong 1 - m_n d. \tag{7}$$

is valid. In a similar manner, we can write for the reflectance at weak absorption

$$R \cong 1 - m_n d_e, \tag{8}$$

where d_e is the "*effective layer thickness*". It is connected with the absorption parameter a, defined by (4), according to the relationship $d_e = a/m_n$ with a single reflection. With multiple reflections (N-fold), the corresponding expression is

$$R^N = (1 - m_n d_e)^N. \tag{9}$$

The "effective layer thickness" thus represents that layer thickness which would be necessary in transmittance measurements to obtain the same extinction as a single reflection at the phase boundary of a semi-infinite optically rarer medium[402]. d_e is generally a complicated function of the various possible variables but is independent of m_n for weak absorption. Using (5) for the change of amplitude of the electric field in the rarer medium, d_e can be calculated as

$$d_e = \frac{n_2/n_1}{\cos\alpha} \int_0^\infty E^2 \, dx = \frac{(n_2/n_1)E_0^2 d_p}{2\cos\alpha}. \tag{10}$$

This expression assumes that the field amplitude of the incident wave has a value of 1 in the more dense medium. The relative effective layer thickness accordingly depends on the penetration depth d_p, which in its turn decreases with increasing incidence angle α and is equal for ∥- and ⊥-polarization. It depends also on the sample surface, proportionally to $1/\cos\alpha$, on the ration n_2/n_1 which is independent of α, and, finally,

---

[401] Harrick, N. J.: Ann. N. Y. Acad. Sci. **101**, 928 (1963).
[402] Harrick, N. J., and F. K. du Pré: Appl. Opt. **5**, 1739 (1966).

on E_0^2 which is greater for ∥-polarization than for ⊥-polarization. The amplitude $\vec{E}_0$ of the electric field in the optically rarer medium at the phase boundary was calculated by *Harrick*[403]. In the most simple case of perpendicular polarization it has the value

$$\vec{E}_{0\perp} = \frac{2\cos\alpha}{[1-(n_2/n_1)^2]^{1/2}}. \tag{11}$$

If (6) and (11) are inserted into (10), we obtain

$$\frac{d_{e\perp}}{\lambda_1} = \frac{(n_2/n_1)\cos\alpha}{\pi(1-(n_2/n_1)^2)(\sin^2\alpha-(n_2/n_1)^2)^{1/2}}. \tag{12}$$

Correspondingly, for ∥-polarization,

$$\frac{d_{e\|}}{\lambda_1} = \frac{(n_2/n_1)\cos\alpha(2\sin^2\alpha-(n_2/n_1)^2)}{\pi(1-(n_2/n_1)^2)\left[(1-(n_2/n_1)^2)\sin^2\alpha-(n_2/n_1)^2\right](\sin^2\alpha-(n_2/n_1)^2)^{1/2}}. \tag{13}$$

All these factors together account for the interaction of the penetrating wave with the absorbing medium and thus, d_e also decreases with increasing α with, to be sure, various intensities for ∥- and for ⊥-polarization. Fig. 147 gives the measured relative penetration depths d_p, in units of λ_1, and the relative effective layer thicknesses $d_{e\|}$ and $d_{e\perp}$ as functions of α for $n_2/n_1 = 0.423$ and $\alpha_g = 25°$. For angles of incidence near the critical angle α_g, d_e becomes indeterminantly large. At grazing incidence it approaches zero because E_0 vanishes at $\alpha = 90°$. At $\alpha = 45°$, the average effective thickness $(d_{e\perp}+d_{e\|})/2$ is approximately equal to the penetration depth d_p.

Since d_p is proportional to the wavelength, it also increases with λ. This is the reason why, in the internal reflectance spectra of semi-infinite media, the longer wavelength bands are relatively stronger than in transmission spectra. Additionally, the stronger absorption on the long wavelength side of a band also causes a deformation of the band compared to its form in transmission.

Eq. (10) is rigorously valid only for low absorption. Depending on the magnitude of α, however, the equation is applicable over a wide extinction modulus range to a good approximation. For example, at $\alpha = 50°$ and $m_n = 10^5$, the equation is still applicable.

If, instead of a semi-infinite medium, *thin films* of thickness $d < d_e$ are used as the optically rarer medium, the electric field can be considered constant over the film thickness and the "effective layer thickness" is given by

$$d_e = \frac{(n_2/n_1)E_0^2 d}{\cos\alpha}, \tag{14}$$

---

[403] *Harrick*, N. J., and F. K. du Pré: Appl. Opt. **5**, 1739 (1966).

where, for perpendicular polarization, the amplitude E_0 within the film is now given by

$$E_{0\perp} = \frac{2\cos\alpha}{[1-(n_3/n_1)^2]^{1/2}}. \qquad (15)$$

Here the field strength thus depends also on the refractive index n_3 of the background of the film. The "effective layer thickness" becomes

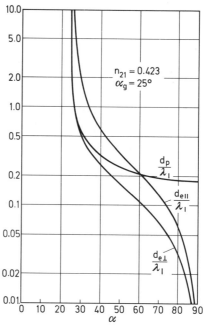

Fig. 147. Relative depth of penetration and effective layer thickness of the waves penetrating into the optically rarer medium as a function of the angle of incidence α at the phase boundary

independent of the depth of penetration, and thereby also of the wavelength, and proportional to the thickness of the film. Further advantages are that d_e does not become indefinitely large if α approaches the limiting angle α_g because now $\sin\alpha_g = n_3/n_1$, and that, because of the wavelength independence, as well the relative intensities as the forms of various bands are related to one another similarly as in transmission measurements. This is not the case when a semi-infinite medium is used. For this reason, measurements on thin films instead of semi-infinite media are recommended. An even better recommendation is the larger usuable range of α which is available, because the refractive index of the thin film no longer prescribes the angular region, but the limiting angle α_g is given by the ratio n_3/n_1. Even if $n_2 > n_1$, total reflection can be obtained so long as $n_3 < n_1$.

The equations derived for d_e are only valid for low absorption. Their region of usefulness can be extended however, if a sufficiently large angle of incidence α is chosen (in which case the effective absorption is low) and, for compensation, multiple reflections, so that an adequate contrast in the spectrum is obtained. It is, furthermore, necessary to have a good optical contact between the two media because otherwise band intensities and band profiles are not comparable with those obtained in transmission. This condition frequently limits obtaining reliable ATR-spectra of solid or powdered materials.

Finally, another source of error in measurements of internal reflectance spectra is the *dispersion* of the optically rarer medium which is "anomalous" just within absorption bands and can lead to very large deviations of n_2 with narrow and intense bands. While the changes conditioned by reflection at the phase boundaries in transmission measurements can be largely eliminated using two layer thicknesses so that "true" transparencies can be recorded, it is found that in internal reflection spectra a deformation and a shift of the bands to longer wavelengths (compared with those measured in transmission) occur in the neighborhood of the critical angle and that these changes are conditional upon the anomalous dispersion of n_2 [404]. In order to minimize this source of error, the angle of incident irradiation should be far from the limiting angle. This, on the other hand, generally means a loss of contrast (low absorption) in the spectrum. Also in this case, the errors are smaller when measuring with thin films than with semi-infinite media and likewise smaller for $\perp$-polarization than for $\|$-polarization.

c) Methods

As shown in the last section, the "effective layer thickness" d_e is determinant for the strength of the absorption and thereby the contrast in an "internal reflection spectrum". d_e becomes larger the smaller the permissable refractive index n_1 of the totally reflecting element and the smaller the angle of incidence α on the totally reflecting surface. There

---

[404] The question of whether the band positions measured in transmission or by means of internal reflection correspond to the true frequency of the oscillator has been discussed by *Clifford, A. A.*, and *B. Crawford Jr.*: J. Phys. Chem. **70**, 1536 (1966), as well as by *Young, E. F.*, and *R. W. Hannah*: Modern aspects of reflectance spectroscopy, p. 218 ff. New York: Plenum Press 1968. The latter authors point out that also in transmission measurements the band positions can be influenced by the materials in the sample holder windows.

320   VIII. Reflectance Spectra Obtained by Attenuated Total Reflection

is, nevertheless, an upper limit for d_e because of the distorting band deformations and band shifts. In cases where d_e is to small, however, the contrast of the spectrum can be increased by multiple reflection (Eq. 9). We distinguish therefore between single and multiple reflecting elements and again, in either case, between fixed and variable incidence angle.

To determine the optical constants n and $\varkappa$, it is necessary to measure at two different angles of incidence. These angles must be measured

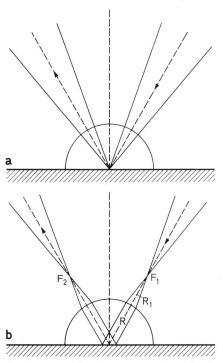

Fig. 148. Half cylinder for measuring the attenuated total reflection, according to Fahrenfort

and/or defined as exactly as possible (at least within a few minutes). Fahrenfort[405] used AgCl and KRS-5 crystals in the form of half cylinders. The horizontal plane served as the totally reflecting phase boundary while the radiation entered and emerged through the curved surface.

When the beam is focussed at the center of the cylinder (Fig. 148a) it passes straight through the cylinder. But as the angle of incidence varies in this case with the cross section, the beam is focussed at a distance $R_1 = R/(n-1)$ beyond the cylinder (Fig. 148b), so that nearly parallel

---

[405] Fahrenfort, J.: Spectrochim. Acta **17**, 698 (1961).

radiation with a better defined angle of incidence is obtained. This arrangement has the advantage that the angle of incidence can be varied between 15° and 85°, and, with the help of angle measuring equipment, can be read with an accuracy of 3'. Fig. 149 shows the modified form of a half cylinder designed by *Fahrenfort* for a five-fold reflection. The number of reflections depends on the ratio b/h. Thus, for an integer m, the number of reflections is $2m+1$. It is independent of the angle of

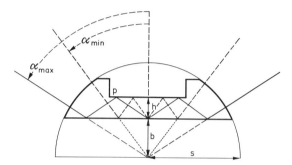

Fig. 149. Half cylinder for 5-fold reflection with variable angle of incidence, according to *Fahrenfort*

incidence, within defined limits α_{max} and α_{min}, if the incident beam is directed toward the axis of the cylinder.

Fig. 150 shows a subsequent development of this principle by *Harrick*[406], consisting of a plate with two attached quarter cylinders which can be used as single path or double path element. The considerations of Fig. 148 are valid for focussing the beam of incident radiation.

For measuring internal reflection spectra (without separate determination of n and $\varkappa$) we frequently content ourselves with reflection elements of fixed angles of incidence because they are easier to prepare. Numerous types have been described[407] for single and multiple reflection as well as single-path and double-path elements, ranging from simple equilateral to complicated achromatic prisms. The most frequently used are trapazoid or parallelogram shaped plates (Fig. 152) of various thicknesses whose sloping sides serve as the aperture. The best situation is when a parallel radiation beam enters and exits perpendicularly through the aperture so that no change of the state of polarization occurs and the angle of incidence is well defined. Weakly converging radiation is also

---

[406] *Harrick, N. J.*: Anal. Chem. **36**, 188 (1964).

[407] See, for example, *Hansen, W. N.*, and *J. A. Horton*: Anal. Chem. **36**, 783 (1964). — *Hansen, W. N.*: Anal. Chem. **35**, 765 (1963). See also the monograph by *Harrick, N. J.*, cited on p. 313.

frequently used. It is focussed on the middle of the plate so the emerging radiation is divergent. With these arrangements, the region of the angle of incidence is more or less widely fixed and cannot be arbitrarily varied as is necessary for changing the "effective layer thickness".

Double-pass elements have several advantages. The number of reflections is doubled compared to an one-path element for a given length, and it is easier to insert into the radiation path of a spectrometer. For the

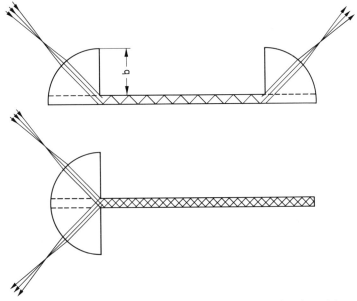

Fig. 150. Plates according to *Harrick* for multiple internal reflection with variable angle of incidence (single and double paths)

study of liquids or powders, the intensity of the absorption can be varied simply by changing the immersion depth of the element in the studied sample. It has a disadvantage, however, in that the angle of incidence at the surface at the flat end of the element is $(90-\alpha)$ except when the angle of incidence $\alpha = 45°$. This can be avoided if the bottom of the element is coated with an evaporated metal layer. Whereas in an one-path element, the angle of incidence can vary between α_g and 90°, in this case only a variation between α_g and $(90-\alpha_g)$ is possible. For instance, for the Ge–H$_2$O interface, this range lies between 18.5° and 71.5°.

A number of special constructions have been used. One is the V-shaped element [408] which has the advantage of not unfocussing the

---

[408] *Harrick, N. J.*: Phys. Rev. Letters **4**, 224 (1960); Anal. Chem. **36**, 188 (1964).

radiation beam at a suitable choice of angles of incidence and dimensions and/or inclinations of the two arms of the element. Furthermore, the entering and exiting beams are coaxial (Fig. 151) so that the element can simply be placed in the radiation path of a spectrophotometer. The *rosette*[409] is an element with whose help the radiation beam, by multiple mirror reflections, is consecutively directed to the same point on the sample surface. With this device very small samples (e.g., microcrystalls) can also be investigated.

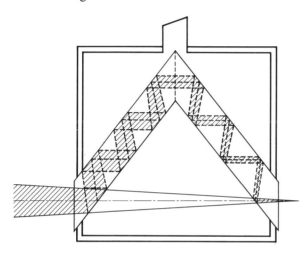

Fig. 151. V-form element for internal reflection according to *Harrick* that does not unfocus or displace the beam

For still smaller samples (e.g., adsorbed amounts of the order of 10^{-9} g), the advantages of the so-called "*interference spectroscopy*" can be utilized as, for example, in the Fabry-Perot-Etalon[410]. The principle of the method rests on the following: If a parallel radiation beam falls on a very thin film above the critical angle, a large number of internal reflections are obtained already in the range of the cross section of the beam because, as already shown, the radiation also penetrates the film and produces there a laterally damped electric field. Under suitable conditions, multiple reflection causes interference. If the phase difference

---

[409] *Harrick, N. J.*: l.c. — *Hirschfeld, T.*: Appl. Opt. **6**, 715 (1967).

[410] Compare *Kortüm, G.*: Kolorimetrie, Photometrie und Spektrometrie, 4th ed., p. 80, 117. Berlin-Göttingen-Heidelberg: Springer 1962. — *Harrick, N. J.*: Internal reflection spectroscopy. New York: Interscience Publ. 1967. — *Berz, F.*: Brit. J. Appl. Phys. **16**, 1733 (1965). — *Harrick, N. J.*, et al.: J. Opt. Soc. Am. **56**, 553 A (1966).

between the individual reflected components is given by

$$\delta = \frac{4\pi n d \cos\alpha}{\lambda_0} = 2\pi m, \tag{16}$$

where d is the film thickness, n its refractive index, λ_0 the vacuum wavelength, and m an integer, the reflectance of such a film in the ideal case is equal to zero, that is, the entire beam is transmitted. If the film and/or a medium behind the film has a weak absorbance, a large field amplitude and thereby a high amplification of the absorption will nevertheless occur because of the resonance within the film.

The demands on the elements for internal reflection are naturally extraordinarily high regarding the material of construction (transparency, refractive index and its dispersion, hardness, chemical resistance, purity, etc.) as well as the geometric form (precision of the desired dimensions and angles, surface properties, etc.). Since the method has proven useful preponderantly for the IR spectral region, elements chiefly with great transparency in the IR have been produced. The materials stretch from diamond to CsI, but the most frequently used have been KRS-5 ($n = 2.4$), Si ($n = 3.4$), Ge ($n = 4$), AgCl ($n = 2$), for the IR and quartz ($n = 1.5$), NaCl ($n = 1.5$), flint glass ($n = 1.7$), Al_2O_3 ($n = 1.8$), and MgO ($n = 1.8$) for the visible and UV. Transmittances[411] and dispersions[412] have been given in tables or graphically.

Polishing of the optical surfaces requires special care so that all diffuse reflection is avoided. A number of reflections at a surface would very quickly reduce the radiation flux. At best, single crystals are used that are cut or split so that the totally reflecting surfaces coincide with definite crystal planes. Soft materials such as CsI, KRS-5, AgCl, etc., can be pressed into the desired forms. For the most widely used plates, the ratio l/d of length to thickness determines the number of reflections at a given angle of incidence. This ratio must be chosen so that the radiation beam exits completely through the exit aperture and is not split up. Two such cases are shown in Fig. 152 for parallel and focussed radiation. Likewise, the angles of the aperture surfaces and the parallel positions of the totally reflecting surfaces must be very exactly checked so that the exiting radiation beam is not deflected sideways[413].

---

[411] *McCarthy, K. A.*, et al.: Am. Inst. of Physics Handbook, 2nd ed. New York: McGraw-Hill 1963. — *Landolt-Börnstein:* 6th ed., Vol. 1, p. 4. Berlin-Göttingen-Heidelberg: Springer 1955.

[412] *Wolfe, W. L.*, et al.: Am. Inst. of Physics Handbook, 2nd ed. New York: MacGraw-Hill 1963. — *Landolt-Börnstein:* 6th ed., Vol. 8, p. 2. Berlin-Göttingen-Heidelberg: Springer 1962.

[413] For more particulars see *Harrick, N. J.:* Internal reflection spectroscopy. New York: Interscience Publ. 1967.

To obtain spectra of high contrast and reproducibility in the investigation of solid materials (crystals, powders, etc.), optical contact between the sample and the reflection element must be as good as possible. To do this, it is necessary to press the samples against the element. Special hydraulic presses that distribute the pressure evenly when rubber is placed between the press and the sample have been developed. It is frequently more simple to establish optical contact through a non-

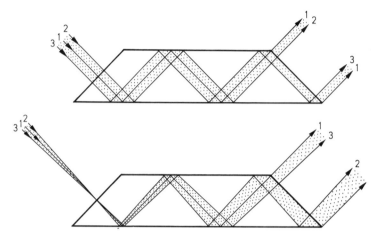

Fig. 152. Importance of the ratio l/d (lenght to thickness) of a plate for internal reflection for parallel or focussed radiation

absorbing fluid with a sufficiently high refractive index. In the IR, CS_2 ($n_D = 1.627$) is useful for this purpose. Except for individual bands, it is transparent in the 0.35–13 μ region. In the visible and near UV, s-tetrabromoethane ($n_D = 1.638$) is useful.

To determine the optical constants n and $\varkappa$, the use of linearly polarized radiation is recommended because then the calculation with the help of the Fresnel equations is easier. Disregarding this recommendation, the radiation in spectrometers, especially grating spectrometers, is always partly polarized, so that for quantitative measurements, the state of polarization must be established beforehand. It is therefore prudent to use completely polarized light throughout. Polarizers for the visible and UV as well as the IR and the principles by which they operate have been extensively described in the literatur[414].

---

[414] *Kortüm*, G.: Kolorimetrie, Photometrie und Spektrometrie, 4th ed. Berlin-Göttingen-Heidelberg: Springer 1962. — *Shurcliff*, W. A.: Polarized light. Cambridge, Mass.: Harvard Univ. Press 1962.

Also, for *building reflection elements into a spectrophotometer*, the question whether to determine the optical constants n and ϰ or to be restricted to the measurement of internal reflection spectra is crucial. The first case requires on one hand the strongest possible parallel radiation in order to establish the angle of incidence as well as possible. On the

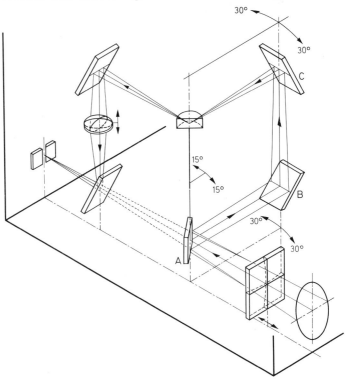

Fig. 153. Attachment for attenuated total reflection according to *Fahrenfort* for determining the optical constants n and ϰ

other hand, it requires a very exact measuring device which should be able to exactly determine these angles to within several minutes. At present there are apparently no attachments for commercial spectrophotometers that fulfill these requirements and only very few constructions have been described in which these requirements were considered.

Fig. 153 schematically represents the radiation path in an arrangement given by *Fahrenfort* and *Visser*[415]. A half cylinder serves as the reflecting element for single or multiple reflections. The mirror A and the half cylinder can be rotated about a common fixed axis so that their reflecting

---

[415] *Fahrenfort, J.*, and *W. M. Visser*: Spectrochim. Acta **18**, 1103 (1962).

surfaces always stay perpendicular to one another. Also the two solidly connected mirrors B and C can be rotated together about the same axis. A and the reflection element are mechanically coupled with B and C so that the angle of incidence can be varied between 30° and 63°. The precision of the angular position is about 3'. A mechanical system automatically maintains the focussing of the beam when α is changed. The half cylinder is fixed in a plate of epoxy resin, the underside of which is flush with the base of the cylinder. For the measurement of liquid samples, there is a space in the block situated beneath which can be filled with a capillary, and is then covered by the cylinder. Solids and films must be placed on a movable plate that can be brought up until optical contact is made. Optical contact can be improved by using a thin layer of a highly refracting, non-absorbing liquid. The entire apparatus can be built into the cell holder of a double beam spectrophotometer (Beckman or Perkin-Elmer).

The apparatus given by *Gilby* and coworkers[416] also uses a half cylinder as the reflection element. The radiation beam is made as parallel as possible here too although the spread of the beam nevertheless amounts to about 1°. The error this produces in the results of the measurements is calculated. The reproducibility of the angle adjustment is about 0.5'; the accuracy, by calibration, reaches 5'. The adjustable range of the angle of incidence varies from 11–75°.

Attachments for commercial spectrophotometers for measuring internal reflection spectra have been developed and are now sold by many firms[417]. The most simple of these consists solely of a suitable reflection element that is, without supplimentary optics, merely placed in the beam of a spectrophotometer in place of a cuvette. For parallel beam paths, this can be for instance an achromatic prism; for convergent radiation, a *V*-formed cell (Fig. 151). To be sure, the possibilities for varying the angle of incidence must be discarded in this case.

For the case where angle of incidence variations are necessary but it is desirable, nevertheless, to use the cuvette space of a commercial spectrophotometer for the installation of a reflection element, an optical attachment must be developed such as the one of *Fahrenfort* and *Visser* (Fig. 152) that has already been described. In this case there are many-sided variations of the possibilities for the combination of different reflection elements with optical and mechanical devices for radiation conduction. This has led to the development of a large number of attachments for common spectrophotometers. These are separated into

---

[416] *Gilby, A. C.*, et al.: J. Phys. Chem. **70**, 1520, 1525 (1966).

[417] For example: Barnes Engineering, Stamford, Conn.; Beckman Instruments, Fullerton, Calif.; Perkin-Elmer Corp., Norwalk, Conn.; Research and Industrial Instrument Co., London; Wilks Scientific Corp., South Norwalk, Conn.

two groups depending upon whether or not the angles of incidence are fixed or variable. These groups are again subdivided into systems for single and multiple reflection elements. Since these attachments as a rule are not designed for the determination of optical constants, the definition of the angle of incidence as well as its measurement are generally not too exact, but nevertheless completely adequate for the purposes of recording internal reflection spectra. As an example, Fig. 154 schematically illustrates the beam path in the attachment Model 9 of the Wilks Scientific Corp. for multiple reflections at fixed angle of incidence.

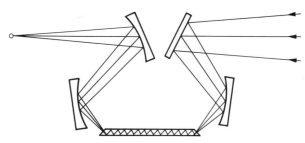

Fig. 154. Attachment model 9 of the Wilks Scientific Corp. for incorporation into the sample space of a spectrophotometer. Trapezoid shaped element for multiple reflection and fixed angle of incidence

For *special investigations*, e.g., at high or low temperatures, the sample space of common spectrophotometers is not sufficient for the incorporation of the reflection element as well as the attendant optics, ovens, dewar flasks, etc. In such cases a new construction of the optical arrangement is necessary either before entrance into the monochromator or after exit from it. Additionally, such an arrangement has the advantage that the surface of the reflection element is more easily accessible. This is useful for the investigation of samples of different sorts (films, fluids, powders, etc.). Various examples for such arrangements have been given [418].

When possible, a *double-beam spectrophotometer* should be used, and among these, those which operate on the single cell chopper process with optical balance are preferable since they deliver the most certain results [419]. For compensation of radiation loss in double-beam methods, two reflection elements that are as nearly equal as possible are required for the sample and reference beams; the same as the two equivalent cuvettes that are necessary in transmission measurements.

---

[418] *Harrick, N. J.:* Appl. Opt. **4**, 1664 (1965); Anal. Chem. **36**, 188 (1964). — *Becker, G. E., and G. W. Gobeli:* J. Chem. Phys. **38**, 2942 (1963).

[419] Compare, in this regard, *Kortüm, G.:* Kolorimetrie, Photometrie und Spektrometrie, 4th ed. Berlin-Göttingen-Heidelberg: Springer 1962.

d) Applications

The possibilities for applying this method, just like diffuse reflectance, are almost unlimited. In contrast to transmittance measurements, it often offers special advantages in the investigation of solid substances and turbid samples. As the penetration depth of the radiation is very small, very thin layers can also be investigated. When the thickness of the layer exceeds 5 μ, the spectrum obtained does not depend on the layer thickness, as it does in transmittance measurements. Water-soluble or water-containing substances, that are difficult to measure by transmittance in the IR can easily be investigated, just as two-phase systems such as suspensions, emulsions or colloids, etc., present no difficulty, so that this method is especially applicable to the investigation of biochemical problems (spectra of amino acids, polypeptides, etc.).

Numerous examples for the many-sided applicability of the method are given in the monograph by *Harrick* so that we can restrict ourselves here to examples in which the advantages of the method compared with transmission measurements are especially easily recognized.

As with diffuse reflection, internal reflection is especially useful for investigating *solid materials*. The production of samples does not require great preparations. With plastic materials such as many resins or rubber, sufficient contact with the reflecting surface can be obtained by merely pressing the sample and surface together. With hard materials the optical contact must often ultimately be improved with a few drops of a liquid of high refractive index such as CS_2. Even if this is later evaporated, the optical contact is largely maintained. As an example, Fig. 155 gives the superimposed spectra of a solid epoxy resin as obtained in transmission by the KBr method and also by the internal reflection method[420]. We recognize the band shifts and deformations discussed earlier (p. 319). The spectra are, however, largely similar to one another so that IR spectra compendia (Sadtler collection) can be used in analogous manners to identify group vibrations. For qualitative measurements, even rough surfaces provide sufficient contact with multiple reflections in any case so that well defined spectra can be obtained.

Even the spectra of *powdered materials* can be obtained with this method[421]. As would be expected, the measured extinctions, i.e., the contrast of the spectra, depend on the particle size of the powders[422]. The extinction decreases with increasing particle diameter due to the variable packing density. On the other hand, no scattering losses of the radiation are supposed to occur. It does not appear unequivocally

---

[420] *Fahrenfort, J.:* Spectrochim. Acta **17**, 698 (1961).

[421] Compare *Lyon, R. J. P.:* Infrared analysis. New York: Encycl. Earth Sci. 1967.

[422] *Harrick, N. J.*, and *B. H. Riederman:* Spectrochim. Acta **21**, 2135 (1965).

clear whether or not the shape and intensity of the bands are independent of particle size.

Because the direction of the electric vector of the penetrating wave can be set by choice of polarization and angle of incidence, the method offers special advantages for the investigation of *anisotropic materials*. In this manner, the extinction modulus in the various directions in space and/or the polarization dependence of individual bands can be directly measured [423].

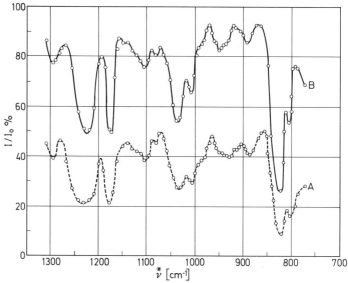

Fig. 155. Comparison between (*A*) a transmission spectrum according to the KBr method and (*B*) a reflection spectrum derived from measurements of the attenuated total reflection of a solid epoxide resin

The method of internal reflection is better suited than that of diffuse reflectance for the investigation of *surface adsorption* if the problem involves adsorption on a particularly oriented crystal surface. Even if the problem concerns monomolecular layers or only partially covered surfaces, multiple reflection provides sufficiently large extinctions so that a qualitative analysis of the surface layer can be obtained. As an example, Fig. 156 gives the surface spectrum of a polished silicon plate that was exposed to the atmosphere. At $\alpha = 45°$ with a 165-fold reflection, two bands appear at 2.9 and 3.4 μ. These can be ascribed to OH- or CH-valence vibrations of adsorbed molecules. For quantitative measure-

---

[423] *Flournoy, P. A.*, and *W. J. Schaffers:* Spectrochim. Acta **22**, 5, 15 (1966). — *Fraser, R. D. B.:* J. chem. Phys. **21**, 1511 (1953); **24**, 89 (1956).

ments, perhaps to prove a Langmuir adsorption isotherm, the diffuse reflectance method is to be preferred (compare p. 272). Because of the relatively simple adsorbent preparation by heating in vacuum (removal of adsorbed water, recrystallization, etc.), diffuse reflectance also offers additional possibilities for investigation.

Just the possibility of studying very thin surface layers with internal reflection makes this method suitable for pursuing *surface reactions* such as oxidations, reductions, the formation of intermediates (catalysis),

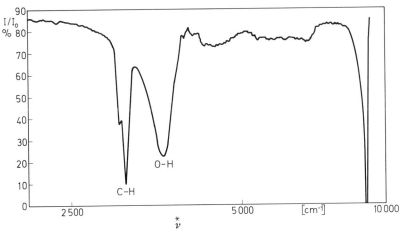

Fig. 156. Surface spectrum of a polished silicon plate by means of internal reflection. $N = 165$; $\alpha = 45°$. OH- and CH-valence vibrations of adsorbed molecules

electrode surface reactions[424], etc. Semiconductors are especially suitable here since they are largely transparent to IR radiation and additionally have high refractive indices. Furthermore, the free electrons or holes in the semiconductor barriers can absorb IR radiation. Use of this property under suitable conditions facilitates the measurement of the number of these free carriers and its change by irradiation as a function of wavelength[425].

Recently internal reflection spectroscopy has been called upon for investigating *gas chromatograph fractions*[426]. The hot gas is conducted directly through a capillary cell cooled from the side by means of the

---

[424] *Hansen, W. N.:* Modern aspects of reflectance spectroscopy, p. 182ff. New York: Plenum Press 1968.

[425] Compare *Harrick, N. J.:* Phys. Rev. **125**, 1165 (1962). — *Beckmann, K. H.:* Angew. Chem. **80**, 213 (1968).

[426] *Wilks, P. A.:* Modern aspects of reflectance spectroscopy, p. 192 ff. New York: Plenum Press 1968.

Peltier effect so that the fraction concerned deposits on the surface of the reflection element. In a similar manner, in the decomposition of samples by *pyrolysis*, the fugative components can be trapped and by reflection spectrophotometrically analyzed [426].

The measurements of *Hansen* [427] on aqueous solutions of Eosin-Y in the 430–660 mµ region have shown that the method of attenuated total reflection can be extended to the visible and UV spectral regions with good success. This work used a simple glass prism for a single internal reflection. Fig. 157a gives the ATR spectra of 5 different solutions (relative concentration unit = 50 g/l) at $\alpha = 55°$ and polarization parallel to the plane of incidence. The apparent extinction $E = \log(I_0/I)$ is plotted as the ordinate. The refractive index of the prism n_1 varied from 1.816 to 1.781 in the given spectral range. This "internal reflection spectrum" at an angle of incidence above the critical limiting angle α_g shows the characteristic band shift compared with the transmission spectrum. Mainly, this is conditioned by the anomalous dispersion of the refractive index n_2 of the solutions in the absorbance region. The relative intensity changes of the two bands with increasing concentration is also traced to this anomaly.

If the angle of incidence α is chosen less than α_g, the curves of Fig. 157c are obtained. These curves are similar to the dispersion curves. In this figure, the apparent extinction is plotted against the wavelength. The angle of incidence was 45° and thus substantially lower than α_g such that the reflectance no longer changed so strongly with α. This was necessary to avoid incorporating errors because of uncertainty in α. Figs. 157b and 157d give the measurements with ‖-polarization at two angles of incidence from which the optical constants n_2 and m_n of the Eosin solutions were calculated with the help of the Fresnel equations.

Through development of the Fresnel formulas in series, *Hansen* [428] has derived approximate equations for $R_\perp$ and $R_\|$. These equations give the extinction as a function of $\varkappa$ or the extinction modulus m_n, and thereby also of the concentration of the solutions, when $\alpha > \alpha_g$ and $\varkappa < 1$. Since the extinction is measured directly, experiment and theory can be easily compared with one another in this manner. The Hansen equations are

$$E_\perp \equiv \log \frac{1}{R_\perp} = \frac{2P}{2{,}303} \varkappa \\ - \frac{2P}{2{,}303}\left[\frac{n^2}{2\beta}\left(1 + \frac{n^2}{\beta}\right) + \frac{n^2}{1-n^2}\left(1 + \frac{2n^2}{\beta}\right) - \frac{1}{3}P^2 \right]\varkappa^3 + \cdots \quad (17)$$

---
[427] Hansen, W. N.: Anal. Chem. **37**, 1142 (1965).
[428] Hansen, W. N.: Spectrochim. Acta **21**, 815 (1965).

$$E_\| \equiv \log \frac{1}{R_\|} = \frac{2Q}{2{,}303}\left(1 + \frac{2\beta}{n^2}\right)\varkappa \tag{18}$$
$$- \frac{2}{2{,}303}\left\{\frac{n^2}{2\beta}\left(\frac{n^2}{\beta} - 1\right) + \frac{n^2 + 2\beta}{n^4 \cos^2\alpha + \beta}\left[2n^2\left(\cos^2\alpha + \frac{1}{\beta}\right) + 1\right]\right\}\varkappa^3 + \cdots$$

where
$$n \equiv n_2/n_1; \quad \beta \equiv \sin^2\alpha - n^2;$$
$$P \equiv \frac{2n^2 \cos\alpha}{(1-n^2)\beta^{1/2}}; \quad Q \equiv \frac{2n^4 \cos\alpha}{\beta^{1/2}(n^4 \cos^2\alpha + \beta)}.$$

The developed equations contain only terms with uneven powers of $\varkappa$ and converge very quickly if $\varkappa/(\sin^2\alpha - n^2) \ll 1$. Frequently it is possible to consider only the first term of these equations, which means that $E_\perp$ or $E_\|$, respectively, should be a linear function of $\varkappa$ in this region. Fig. 158 shows this for $\perp$-polarization. Since $\varkappa$ is proportional to the decadic extinction modulus m (compare p. 21),

$$\varkappa = \frac{2{,}303\,\lambda_0\,\varepsilon c}{4\pi\,n_2}, \tag{19}$$

use of the first term of (17) or (18) gives

$$E_\perp = \varepsilon b_\perp c \quad \text{and} \quad E_\| = \varepsilon b_\| c, \tag{20}$$

i.e., the Bouguer-Beer Law, where

$$b_\perp \equiv \frac{P\lambda_0}{2\pi n_2} \quad \text{and} \quad b_\| \equiv \frac{Q(1 + 2\beta/n^2)\lambda_0}{2\pi n_2} \tag{21}$$

can be considered as the "effective layer thicknesses" for the ATR element. This is valid for a single reflection, but because extinctions are additive,

$$E_\perp = N\varepsilon b_\perp c \quad \text{and} \quad E_\| = N\varepsilon b_\| c \tag{22}$$

are valid for N-fold reflections. Fig. 159 shows the extinctions $E_\perp$ and $E_\|$ for Eosin-B solutions of different concentrations (concentration unit $= 4.9$ g/l) at $\lambda = 4950$ Å and an 11-fold reflection. n_2 is independent of concentration for these dilute solutions. The Bouguer-Beer Law is thus well fulfilled. On the other hand, at higher concentrations (Fig. 160), noticeable deviations appear; i.e., the first terms of Eqs. (17) and (18) are no longer sufficient. For values of $\varkappa < 0.05$, the higher terms generally do not need to be considered, but this limit also depends on n_1 and α. Hansen has also carried through calculations for cases where the radiation is natural or only partly polarized as is often the case in spectrophotometers. Since, however, the degree of polarization varies with the wavelength, it is prudent to work throughout with $\perp$- or $\|$-polarization. For

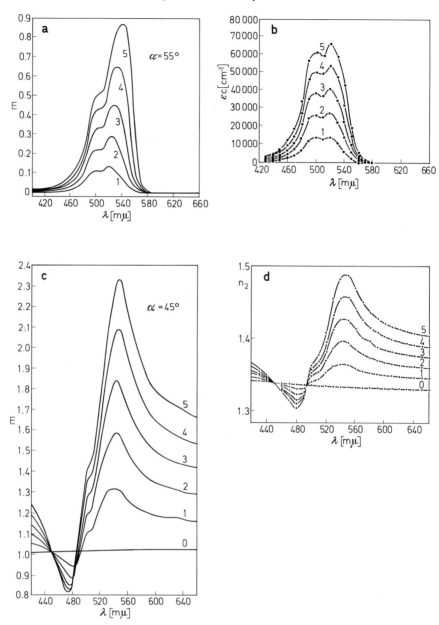

Fig. 157. ATR spectra of aqueous Eosin-Y solutions (concentration unit = 50 g/l) at $\alpha > \alpha_g$ and $\alpha < \alpha_g$ and $\|$-polarization. n- and $\varkappa$- values calculated from the spectra

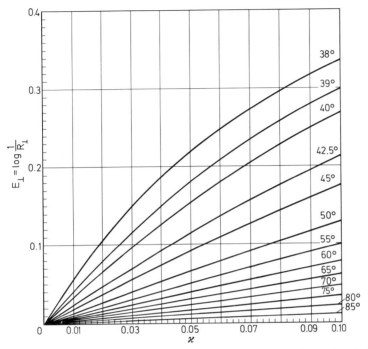

Fig. 158. The extinction $E_\perp \equiv \log(1/R_\perp)$ as a function of the absorption index $\varkappa$ at various angles of incidence. $n_2/n_1 = 0.6$

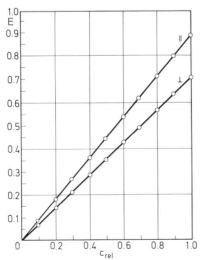

Fig. 159. Extinctions $E_\perp$ and $E_\parallel$ of an aqueous solution of Eosin B as a function of concentration. Concentration unit $= 4.9$ g/l; $n_{1D} = 1.516$; $\alpha = 72.8°$; $N = 11$

natural radiation an E_{nat} is obtained that ascends with $\varkappa$ in an approximately equal manner as shown in Fig. 158.

The derived approximate equations can, within their regions of validity, naturally be used in place of the exact equations for the cal-

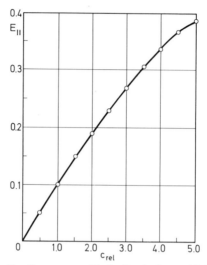

Fig. 160. Extinction $E_{\|}$ of an aquous Eosin B solution at higher concentrations as a function of c. $N = 1$

culation of the optical constants n and $\varkappa$. Disregarding higher terms, it follows from (17) and (18) that

$$\frac{E_{\|}}{E_{\perp}} = \frac{2\sin^2\alpha - n^2}{(1+n^2)\sin^2\alpha - n^2}, \qquad (23)$$

which, when solved for n, leads to

$$n \equiv \frac{n_2}{n_1} = \left[\frac{(2 - E_{\|}/E_{\perp})\sin^2\alpha}{1 - \cos^2\alpha(E_{\|}/E_{\perp})}\right]. \qquad (24)$$

If, for example, n is inserted in (17), $\varkappa$ is given by

$$\varkappa = \frac{2.303}{2P} \cdot E_{\perp}. \qquad (25)$$

The exactness of the constants calculated in this manner can be improved even more, if the values calculated according to (24) and (25) are used to compute the $\varkappa^3$-terms. These are then substracted from E and the corrected $E_{\perp}$- and $E_{\|}$-values used in (24) and (25) in order to obtain corrected values for n and $\varkappa$.

Appendix

I. Tables of the Kubelka-Munk-Function

R_∞	$F(R_\infty)$	$\log F(R_\infty)$	R_∞	$F(R_\infty)$	$\log F(R_\infty)$
0,0010	499,00	+2,6981	0,0020	249,00	+2,3962
0,0030	165,67	+2,2192	0,0040	124,00	+2,0934
0,0050	99,003	+1,9956	0,0060	82,336	+1,9156
0,0070	70,432	+1,8478	0,0080	61,504	+1,7889
0,0090	54,560	+1,7369	0,0100	49,005	+1,6902
0,0110	44,460	+1,6480	0,0120	40,673	+1,6093
0,0130	37,468	+1,5737	0,0140	34,721	+1,5406
0,0150	32,341	+1,5098	0,0160	30,258	+1,4808
0,0170	28,420	+1,4536	0,0180	26,787	+1,4279
0,0190	25,325	+1,4036	0,0200	24,010	+1,3804
0,0210	22,820	+1,3583	0,0220	21,738	+1,3372
0,0230	20,751	+1,3170	0,0240	19,845	+1,2977
0,0250	19,013	+1,2790	0,0260	18,244	+1,2611
0,0270	17,532	+1,2438	0,0280	16,871	+1,2271
0,0290	16,256	+1,2110	0,0300	15,682	+1,1954
0,0310	15,145	+1,1803	0,0320	14,641	+1,1656
0,0330	14,168	+1,1513	0,0340	13,723	+1,1374
0,0350	13,303	+1,1240	0,0360	12,907	+1,1108
0,0370	12,532	+1,0980	0,0380	12,177	+1,0855
0,0390	11,840	+1,0734	0,0400	11,520	+1,0615
0,0410	11,216	+1,0498	0,0420	10,926	+1,0385
0,0430	10,649	+1,0273	0,0440	10,386	+1,0164
0,0450	10,134	+1,0058	0,0460	9,8926	+0,9953
0,0470	9,6618	+0,9851	0,0480	9,4407	+0,9750
0,0490	9,2286	+0,9651	0,0500	9,0250	+0,9554
0,0510	8,8294	+0,9459	0,0520	8,6414	+0,9366
0,0530	8,4605	+0,9274	0,0540	8,2863	+0,9184
0,0550	8,1184	+0,9095	0,0560	7,9566	+0,9007
0,0570	7,8004	+0,8921	0,0580	7,6497	+0,8836
0,0590	7,5041	+0,8753	0,0600	7,3633	+0,8671

I. (continued)

R_∞	$F(R_\infty)$	$\log F(R_\infty)$	R_∞	$F(R_\infty)$	$\log F(R_\infty)$
0,0610	7,2272	+0,8590	0,0620	7,0955	+0,8510
0,0630	6,9680	+0,8431	0,0640	6,8445	+0,8353
0,0650	6,7248	+0,8277	0,0660	6,6088	+0,8201
0,0670	6,4962	+0,8127	0,0680	6,3869	+0,8053
0,0690	6,2809	+0,7980	0,0700	6,1779	+0,7908
0,0710	6,0778	+0,7837	0,0720	5,9804	+0,7767
0,0730	5,8858	+0,7698	0,0740	5,7938	+0,7630
0,0750	5,7042	+0,7562	0,0760	5,6169	+0,7495
0,0770	5,5320	+0,7429	0,0780	5,4493	+0,7363
0,0790	5,3686	+0,7299	0,0800	5,2900	+0,7235
0,0810	5,2133	+0,7171	0,0820	5,1386	+0,7108
0,0830	5,0656	+0,7046	0,0840	4,9944	+0,6985
0,0850	4,9249	+0,6924	0,0860	4,8570	+0,6864
0,0870	4,7906	+0,6804	0,0880	4,7258	+0,6745
0,0890	4,6625	+0,6686	0,0900	4,6006	+0,6628
0,0910	4,5400	+0,6571	0,0920	4,4808	+0,6514
0,0930	4,4228	+0,6457	0,0940	4,3661	+0,6401
0,0950	4,3107	+0,6345	0,0960	4,2563	+0,6290
0,0970	4,2031	+0,6236	0,0980	4,1510	+0,6182
0,0990	4,1000	+0,6128	0,1000	4,0500	+0,6075
0,1010	4,0010	+0,6022	0,1020	3,9530	+0,5969
0,1030	3,9059	+0,5917	0,1040	3,8597	+0,5866
0,1050	3,8144	+0,5814	0,1060	3,7700	+0,5763
0,1070	3,7264	+0,5713	0,1080	3,6836	+0,5663
0,1090	3,6417	+0,5613	0,1100	3,6005	+0,5564
0,1110	3,5600	+0,5515	0,1120	3,5203	+0,5466
0,1130	3,4813	+0,5417	0,1140	3,4430	+0,5369
0,1150	3,4053	+0,5322	0,1160	3,3683	+0,5274
0,1170	3,3320	+0,5227	0,1180	3,2963	+0,5180
0,1190	3,2612	+0,5134	0,1200	3,2267	+0,5088
0,1210	3,1927	+0,5042	0,1220	3,1594	+0,4996
0,1230	3,1265	+0,4951	0,1240	3,0943	+0,4906
0,1250	3,0625	+0,4861	0,1260	3,0313	+0,4816
0,1270	3,0005	+0,4772	0,1280	2,9703	+0,4728
0,1290	2,9405	+0,4684	0,1300	2,9112	+0,4641
0,1310	2,8823	+0,4597	0,1320	2,8539	+0,4554
0,1330	2,8259	+0,4512	0,1340	2,7983	+0,4469
0,1350	2,7712	+0,4427	0,1360	2,7445	+0,4385
0,1370	2,7181	+0,4343	0,1380	2,6922	+0,4301
0,1390	2,6666	+0,4260	0,1400	2,6414	+0,4218

Tables of the Kubelka-Munk-Function

I. (continued)

R_∞	$F(R_\infty)$	$\log F(R_\infty)$	R_∞	$F(R_\infty)$	$\log F(R_\infty)$
0,1410	2,6166	+0,4177	0,1420	2,5921	+0,4137
0,1430	2,5680	+0,4096	0,1440	2,5442	+0,4056
0,1450	2,5208	+0,4015	0,1460	2,4977	+0,3975
0,1470	2,4749	+0,3936	0,1480	2,4524	+0,3896
0,1490	2,4302	+0,3856	0,1500	2,4083	+0,3817
0,1510	2,3868	+0,3778	0,1520	2,3655	+0,3739
0,1530	2,3445	+0,3700	0,1540	2,3238	+0,3662
0,1550	2,3033	+0,3624	0,1560	2,2831	+0,3585
0,1570	2,2632	+0,3547	0,1580	2,2436	+0,3509
0,1590	2,2242	+0,3472	0,1600	2,2050	+0,3434
0,1610	2,1861	+0,3397	0,1620	2,1674	+0,3359
0,1630	2,1490	+0,3322	0,1640	2,1308	+0,3285
0,1650	2,1128	+0,3249	0,1660	2,0950	+0,3212
0,1670	2,0775	+0,3175	0,1680	2,0602	+0,3139
0,1690	2,0431	+0,3103	0,1700	2,0262	+0,3067
0,1710	2,0095	+0,3031	0,1720	1,9930	+0,2995
0,1730	1,9767	+0,2959	0,1740	1,9606	+0,2924
0,1750	1,9446	+0,2888	0,1760	1,9289	+0,2853
0,1770	1,9134	+0,2818	0,1780	1,8980	+0,2783
0,1790	1,8828	+0,2748	0,1800	1,8678	+0,2713
0,1810	1,8529	+0,2679	0,1820	1,8383	+0,2644
0,1830	1,8237	+0,2610	0,1840	1,8094	+0,2575
0,1850	1,7952	+0,2541	0,1860	1,7812	+0,2507
0,1870	1,7673	+0,2473	0,1880	1,7536	+0,2439
0,1890	1,7400	+0,2405	0,1900	1,7266	+0,2372
0,1910	1,7133	+0,2338	0,1920	1,7002	+0,2305
0,1930	1,6872	+0,2272	0,1940	1,6743	+0,2238
0,1950	1,6616	+0,2205	0,1960	1,6490	+0,2172
0,1970	1,6366	+0,2139	0,1980	1,6243	+0,2107
0,1990	1,6121	+0,2074	0,2000	1,6000	+0,2041
0,2010	1,5881	+0,2009	0,2020	1,5762	+0,1976
0,2030	1,5646	+0,1944	0,2040	1,5530	+0,1912
0,2050	1,5415	+0,1880	0,2060	1,5302	+0,1847
0,2070	1,5190	+0,1815	0,2080	1,5078	+0,1784
0,2090	1,4968	+0,1752	0,2100	1,4860	+0,1720
0,2110	1,4752	+0,1688	0,2120	1,4645	+0,1657
0,2130	1,4539	+0,1625	0,2140	1,4434	+0,1594
0,2150	1,4331	+0,1563	0,2160	1,4228	+0,1531
0,2170	1,4126	+0,1500	0,2180	1,4026	+0,1469
0,2190	1,3926	+0,1438	0,2200	1,3827	+0,1407

I. (continued)

R_∞	$F(R_\infty)$	$\log F(R_\infty)$	R_∞	$F(R_\infty)$	$\log F(R_\infty)$
0,2210	1,3729	+0,1377	0,2220	1,3633	+0,1346
0,2230	1,3537	+0,1315	0,2240	1,3441	+0,1284
0,2250	1,3347	+0,1254	0,2260	1,3254	+0,1223
0,2270	1,3161	+0,1193	0,2280	1,3070	+0,1163
0,2290	1,2979	+0,1132	0,2300	1,2889	+0,1102
0,2310	1,2800	+0,1072	0,2320	1,2712	+0,1042
0,2330	1,2624	+0,1012	0,2340	1,2538	+0,0982
0,2350	1,2452	+0,0952	0,2360	1,2366	+0,0922
0,2370	1,2282	+0,0893	0,2380	1,2198	+0,0863
0,2390	1,2116	+0,0833	0,2400	1,2033	+0,0804
0,2410	1,1952	+0,0774	0,2420	1,1871	+0,0745
0,2430	1,1791	+0,0716	0,2440	1,1712	+0,0686
0,2450	1,1633	+0,0657	0,2460	1,1555	+0,0628
0,2470	1,1478	+0,0599	0,2480	1,1401	+0,0570
0,2490	1,1325	+0,0541	0,2500	1,1250	+0,0512
0,2510	1,1175	+0,0483	0,2520	1,1101	+0,0454
0,2530	1,1028	+0,0425	0,2540	1,0955	+0,0396
0,2550	1,0883	+0,0367	0,2560	1,0811	+0,0339
0,2570	1,0740	+0,0310	0,2580	1,0670	+0,0282
0,2590	1,0600	+0,0253	0,2600	1,0531	+0,0225
0,2610	1,0462	+0,0196	0,2620	1,0394	+0,0168
0,2630	1,0326	+0,0139	0,2640	1,0259	+0,0111
0,2650	1,0193	+0,0083	0,2660	1,0127	+0,0055
0,2670	1,0062	+0,0027	0,2680	0,9997	−143.47
0,2690	0,9932	−0,0029	0,2700	0,9869	−0,0057
0,2710	0,9805	−0,0085	0,2720	0,9742	−0,0113
0,2730	0,9680	−0,0141	0,2740	0,9618	−0,0169
0,2750	0,9557	−0,0197	0,2760	0,9496	−0,0225
0,2770	0,9436	−0,0252	0,2780	0,9376	−0,0280
0,2790	0,9316	−0,0308	0,2800	0,9257	−0,0335
0,2810	0,9199	−0,0363	0,2820	0,9140	−0,0390
0,2830	0,9083	−0,0418	0,2840	0,9026	−0,0445
0,2850	0,8969	−0,0473	0,2860	0,8913	−0,0500
0,2870	0,8857	−0,0527	0,2880	0,8801	−0,0555
0,2890	0,8746	−0,0582	0,2900	0,8691	−0,0609
0,2910	0,8637	−0,0636	0,2920	0,8583	−0,0663
0,2930	0,8530	−0,0691	0,2940	0,8477	−0,0718
0,2950	0,8424	−0,0745	0,2960	0,8372	−0,0772
0,2970	0,8320	−0,0799	0,2980	0,8269	−0,0826
0,2990	0,8217	−0,0853	0,3000	0,8167	−0,0880

Tables of the Kubelka-Munk-Function

I. (continued)

R_∞	$F(R_\infty)$	$\log F(R_\infty)$	R_∞	$F(R_\infty)$	$\log F(R_\infty)$
0,3010	0,8116	−0,0906	0,3020	0,8066	−0,0933
0,3030	0,8017	−0,0960	0,3040	0,7967	−0,0987
0,3050	0,7918	−0,1014	0,3060	0,7870	−0,1040
0,3070	0,7822	−0,1067	0,3080	0,7774	−0,1094
0,3090	0,7726	−0,1120	0,3100	0,7679	−0,1147
0,3110	0,7632	−0,1174	0,3120	0,7586	−0,1200
0,3130	0,7539	−0,1227	0,3140	0,7494	−0,1253
0,3150	0,7448	−0,1280	0,3160	0,7403	−0,1306
0,3170	0,7358	−0,1332	0,3180	0,7313	−0,1359
0,3190	0,7269	−0,1385	0,3200	0,7225	−0,1412
0,3210	0,7181	−0,1438	0,3220	0,7138	−0,1464
0,3230	0,7095	−0,1491	0,3240	0,7052	−0,1517
0,3250	0,7010	−0,1543	0,3260	0,6967	−0,1569
0,3270	0,6926	−0,1595	0,3280	0,6884	−0,1622
0,3290	0,6843	−0,1648	0,3300	0,6802	−0,1674
0,3310	0,6761	−0,1700	0,3320	0,6720	−0,1726
0,3330	0,6680	−0,1752	0,3340	0,6640	−0,1778
0,3350	0,6600	−0,1804	0,3360	0,6561	−0,1830
0,3370	0,6522	−0,1856	0,3380	0,6483	−0,1882
0,3390	0,6444	−0,1908	0,3400	0,6406	−0,1934
0,3410	0,6368	−0,1960	0,3420	0,6330	−0,1986
0,3430	0,6292	−0,2012	0,3440	0,6255	−0,2038
0,3450	0,6218	−0,2064	0,3460	0,6181	−0,2090
0,3470	0,6144	−0,2115	0,3480	0,6108	−0,2141
0,3490	0,6072	−0,2167	0,3500	0,6036	−0,2193
0,3510	0,6000	−0,2218	0,3520	0,5965	−0,2244
0,3530	0,5929	−0,2270	0,3540	0,5894	−0,2296
0,3550	0,5860	−0,2321	0,3560	0,5825	−0,2347
0,3570	0,5791	−0,2373	0,3580	0,5756	−0,2398
0,3590	0,5723	−0,2424	0,3600	0,5689	−0,2450
0,3610	0,5655	−0,2475	0,3620	0,5622	−0,2501
0,3630	0,5589	−0,2527	0,3640	0,5556	−0,2552
0,3650	0,5524	−0,2578	0,3660	0,5491	−0,2603
0,3670	0,5459	−0,2629	0,3680	0,5427	−0,2654
0,3690	0,5395	−0,2680	0,3700	0,5364	−0,2706
0,3710	0,5332	−0,2731	0,3720	0,5301	−0,2757
0,3730	0,5270	−0,2782	0,3740	0,5239	−0,2808
0,3750	0,5208	−0,2833	0,3760	0,5178	−0,2858
0,3770	0,5148	−0,2884	0,3780	0,5118	−0,2909
0,3790	0,5088	−0,2935	0,3800	0,5058	−0,2960

I. (continued)

R_∞	$F(R_\infty)$	$\log F(R_\infty)$	R_∞	$F(R_\infty)$	$\log F(R_\infty)$
0,3810	0,5028	−0,2986	0,3820	0,4999	−0,3011
0,3830	0,4970	−0,3037	0,3840	0,4941	−0,3062
0,3850	0,4912	−0,3087	0,3860	0,4883	−0,3113
0,3870	0,4855	−0,3138	0,3880	0,4827	−0,3164
0,3890	0,4798	−0,3189	0,3900	0,4771	−0,3214
0,3910	0,4743	−0,3240	0,3920	0,4715	−0,3265
0,3930	0,4688	−0,3290	0,3940	0,4660	−0,3316
0,3950	0,4633	−0,3341	0,3960	0,4606	−0,3367
0,3970	0,4579	−0,3392	0,3980	0,4553	−0,3417
0,3990	0,4526	−0,3443	0,4000	0,4500	−0,3468
0,4010	0,4474	−0,3493	0,4020	0,4448	−0,3519
0,4030	0,4422	−0,3544	0,4040	0,4396	−0,3569
0,4050	0,4371	−0,3595	0,4060	0,4345	−0,3620
0,4070	0,4320	−0,3645	0,4080	0,4295	−0,3670
0,4090	0,4270	−0,3696	0,4100	0,4245	−0,3721
0,4110	0,4220	−0,3746	0,4120	0,4196	−0,3772
0,4130	0,4172	−0,3797	0,4140	0,4147	−0,3822
0,4150	0,4123	−0,3848	0,4160	0,4099	−0,3873
0,4170	0,4075	−0,3898	0,4180	0,4052	−0,3924
0,4190	0,4028	−0,3949	0,4200	0,4005	−0,3974
0,4210	0,3981	−0,4000	0,4220	0,3958	−0,4025
0,4230	0,3935	−0,4050	0,4240	0,3912	−0,4076
0,4250	0,3890	−0,4101	0,4260	0,3867	−0,4126
0,4270	0,3845	−0,4151	0,4280	0,3822	−0,4177
0,4290	0,3800	−0,4202	0,4300	0,3778	−0,4227
0,4310	0,3756	−0,4253	0,4320	0,3734	−0,4278
0,4330	0,3712	−0,4304	0,4340	0,3691	−0,4329
0,4350	0,3669	−0,4354	0,4360	0,3648	−0,4380
0,4370	0,3627	−0,4405	0,4380	0,3606	−0,4430
0,4390	0,3585	−0,4456	0,4400	0,3564	−0,4481
0,4410	0,3543	−0,4506	0,4420	0,3522	−0,4532
0,4430	0,3502	−0,4557	0,4440	0,3481	−0,4583
0,4450	0,3461	−0,4608	0,4460	0,3441	−0,4633
0,4470	0,3421	−0,4659	0,4480	0,3401	−0,4684
0,4490	0,3381	−0,4710	0,4500	0,3361	−0,4735
0,4510	0,3341	−0,4761	0,4520	0,3322	−0,4786
0,4530	0,3303	−0,4812	0,4540	0,3283	−0,4837
0,4550	0,3264	−0,4862	0,4560	0,3245	−0,4888
0,4570	0,3226	−0,4913	0,4580	0,3207	−0,4939
0,4590	0,3188	−0,4964	0,4600	0,3170	−0,4990

I. (continued)

R_∞	$F(R_\infty)$	$\log F(R_\infty)$	R_∞	$F(R_\infty)$	$\log F(R_\infty)$
0,4610	0,3151	−0,5016	0,4620	0,3133	−0,5041
0,4630	0,3114	−0,5067	0,4640	0,3096	−0,5092
0,4650	0,3078	−0,5118	0,4660	0,3060	−0,5143
0,4670	0,3042	−0,5169	0,4680	0,3024	−0,5195
0,4690	0,3006	−0,5220	0,4700	0,2988	−0,5246
0,4710	0,2971	−0,5271	0,4720	0,2953	−0,5297
0,4730	0,2936	−0,5323	0,4740	0,2919	−0,5348
0,4750	0,2901	−0,5374	0,4760	0,2884	−0,5400
0,4770	0,2867	−0,5425	0,4780	0,2850	−0,5451
0,4790	0,2833	−0,5477	0,4800	0,2817	−0,5503
0,4810	0,2800	−0,5528	0,4820	0,2783	−0,5554
0,4830	0,2767	−0,5580	0,4840	0,2751	−0,5606
0,4850	0,2734	−0,5632	0,4860	0,2718	−0,5657
0,4870	0,2702	−0,5683	0,4880	0,2686	−0,5709
0,4890	0,2670	−0,5735	0,4900	0,2654	−0,5761
0,4910	0,2638	−0,5787	0,4920	0,2623	−0,5813
0,4930	0,2607	−0,5839	0,4940	0,2591	−0,5865
0,4950	0,2576	−0,5891	0,4960	0,2561	−0,5917
0,4970	0,2545	−0,5943	0,4980	0,2530	−0,5969
0,4990	0,2515	−0,5995	0,5000	0,2500	−0,6021
0,5010	0,2485	−0,6047	0,5020	0,2470	−0,6073
0,5030	0,2455	−0,6099	0,5040	0,2441	−0,6125
0,5050	0,2426	−0,6151	0,5060	0,2411	−0,6177
0,5070	0,2397	−0,6203	0,5080	0,2383	−0,6230
0,5090	0,2368	−0,6256	0,5100	0,2354	−0,6282
0,5110	0,2340	−0,6308	0,5120	0,2326	−0,6335
0,5130	0,2312	−0,6361	0,5140	0,2298	−0,6387
0,5150	0,2284	−0,6414	0,5160	0,2270	−0,6440
0,5170	0,2256	−0,6466	0,5180	0,2243	−0,6493
0,5190	0,2229	−0,6519	0,5200	0,2215	−0,6546
0,5210	0,2202	−0,6572	0,5220	0,2189	−0,6598
0,5230	0,2175	−0,6625	0,5240	0,2162	−0,6651
0,5250	0,2149	−0,6678	0,5260	0,2136	−0,6705
0,5270	0,2123	−0,6731	0,5280	0,2110	−0,6758
0,5290	0,2097	−0,6784	0,5300	0,2084	−0,6811
0,5310	0,2071	−0,6838	0,5320	0,2058	−0,6864
0,5330	0,2046	−0,6891	0,5340	0,2033	−0,6918
0,5350	0,2021	−0,6945	0,5360	0,2008	−0,6972
0,5370	0,1996	−0,6998	0,5380	0,1984	−0,7025
0,5390	0,1971	−0,7052	0,5400	0,1959	−0,7079

I. (continued)

R_∞	$F(R_\infty)$	$\log F(R_\infty)$	R_∞	$F(R_\infty)$	$\log F(R_\infty)$
0,5410	0,1947	−0,7106	0,5420	0,1935	−0,7133
0,5430	0,1923	−0,7160	0,5440	0,1911	−0,7187
0,5450	0,1899	−0,7214	0,5460	0,1888	−0,7241
0,5470	0,1876	−0,7268	0,5480	0,1864	−0,7295
0,5490	0,1852	−0,7322	0,5500	0,1841	−0,7350
0,5510	0,1829	−0,7377	0,5520	0,1818	−0,7404
0,5530	0,1807	−0,7431	0,5540	0,1795	−0,7459
0,5550	0,1784	−0,7486	0,5560	0,1773	−0,7513
0,5570	0,1762	−0,7541	0,5580	0,1751	−0,7568
0,5590	0,1740	−0,7596	0,5600	0,1729	−0,7623
0,5610	0,1718	−0,7651	0,5620	0,1707	−0,7678
0,5630	0,1696	−0,7706	0,5640	0,1685	−0,7733
0,5650	0,1675	−0,7761	0,5660	0,1664	−0,7789
0,5670	0,1653	−0,7816	0,5680	0,1643	−0,7844
0,5690	0,1632	−0,7872	0,5700	0,1622	−0,7900
0,5710	0,1612	−0,7928	0,5720	0,1601	−0,7955
0,5730	0,1591	−0,7983	0,5740	0,1581	−0,8011
0,5750	0,1571	−0,8039	0,5760	0,1561	−0,8067
0,5770	0,1551	−0,8095	0,5780	0,1541	−0,8123
0,5790	0,1531	−0,8151	0,5800	0,1521	−0,8180
0,5810	0,1511	−0,8208	0,5820	0,1501	−0,8236
0,5830	0,1491	−0,8264	0,5840	0,1482	−0,8293
0,5850	0,1472	−0,8321	0,5860	0,1462	−0,8349
0,5870	0,1453	−0,8378	0,5880	0,1443	−0,8406
0,5890	0,1434	−0,8435	0,5900	0,1425	−0,8463
0,5910	0,1415	−0,8492	0,5920	0,1406	−0,8520
0,5930	0,1397	−0,8549	0,5940	0,1388	−0,8578
0,5950	0,1378	−0,8606	0,5960	0,1369	−0,8635
0,5970	0,1360	−0,8664	0,5980	0,1351	−0,8693
0,5990	0,1342	−0,8722	0,6000	0,1333	−0,8751
0,6010	0,1324	−0,8780	0,6020	0,1316	−0,8809
0,6030	0,1307	−0,8838	0,6040	0,1298	−0,8867
0,6050	0,1289	−0,8896	0,6060	0,1281	−0,8925
0,6070	0,1272	−0,8954	0,6080	0,1264	−0,8984
0,6090	0,1255	−0,9013	0,6100	0,1247	−0,9042
0,6110	0,1238	−0,9072	0,6120	0,1230	−0,9101
0,6130	0,1222	−0,9131	0,6140	0,1213	−0,9160
0,6150	0,1205	−0,9190	0,6160	0,1197	−0,9219
0,6170	0,1189	−0,9249	0,6180	0,1181	−0,9279
0,6190	0,1173	−0,9309	0,6200	0,1165	−0,9339

I. (continued)

R_∞	$F(R_\infty)$	$\log F(R_\infty)$	R_∞	$F(R_\infty)$	$\log F(R_\infty)$
0,6210	0,1157	−0,9368	0,6220	0,1149	−0,9398
0,6230	0,1141	−0,9428	0,6240	0,1133	−0,9458
0,6250	0,1125	−0,9488	0,6260	0,1117	−0,9519
0,6270	0,1109	−0,9549	0,6280	0,1102	−0,9579
0,6290	0,1094	−0,9609	0,6300	0,1087	−0,9640
0,6310	0,1079	−0,9670	0,6320	0,1071	−0,9701
0,6330	0,1064	−0,9731	0,6340	0,1056	−0,9762
0,6350	0,1049	−0,9792	0,6360	0,1042	−0,9823
0,6370	0,1034	−0,9854	0,6380	0,1027	−0,9884
0,6390	0,1020	−0,9915	0,6400	0,1013	−0,9946
0,6410	0,1005	−0,9977	0,6420	0,0998	−1,0008
0,6430	0,0991	−1,0039	0,6440	0,0984	−1,0070
0,6450	0,0977	−1,0101	0,6460	0,0970	−1,0133
0,6470	0,0963	−1,0164	0,6480	0,0956	−1,0195
0,6490	0,0949	−1,0227	0,6500	0,0942	−1,0258
0,6510	0,0935	−1,0290	0,6520	0,0929	−1,0321
0,6530	0,0922	−1,0353	0,6540	0,0915	−1,0385
0,6550	0,0909	−1,0416	0,6560	0,0902	−1,0448
0,6570	0,0895	−1,0480	0,6580	0,0889	−1,0512
0,6590	0,0882	−1,0544	0,6600	0,0876	−1,0576
0,6610	0,0869	−1,0608	0,6620	0,0863	−1,0641
0,6630	0,0856	−1,0673	0,6640	0,0850	−1,0705
0,6650	0,0844	−1,0738	0,6660	0,0838	−1,0770
0,6670	0,0831	−1,0803	0,6680	0,0825	−1,0835
0,6690	0,0819	−1,0868	0,6700	0,0813	−1,0901
0,6710	0,0807	−1,0934	0,6720	0,0800	−1,0967
0,6730	0,0794	−1,0999	0,6740	0,0788	−1,1033
0,6750	0,0782	−1,1066	0,6760	0,0776	−1,1099
0,6770	0,0771	−1,1132	0,6780	0,0765	−1,1165
0,6790	0,0759	−1,1199	0,6800	0,0753	−1,1232
0,6810	0,0747	−1,1266	0,6820	0,0741	−1,1300
0,6830	0,0736	−1,1333	0,6840	0,0730	−1,1367
0,6850	0,0724	−1,1401	0,6860	0,0719	−1,1435
0,6870	0,0713	−1,1469	0,6880	0,0707	−1,1503
0,6890	0,0702	−1,1537	0,6900	0,0696	−1,1572
0,6910	0,0691	−1,1606	0,6920	0,0685	−1,1640
0,6930	0,0680	−1,1675	0,6940	0,0675	−1,1709
0,6950	0,0669	−1,1744	0,6960	0,0664	−1,1779
0,6970	0,0659	−1,1814	0,6980	0,0653	−1,1849
0,6990	0,0648	−1,1884	0,7000	0,0643	−1,1919

I. (continued)

R_∞	$F(R_\infty)$	$\log F(R_\infty)$	R_∞	$F(R_\infty)$	$\log F(R_\infty)$
0,7010	0,0638	−1,1954	0,7020	0,0633	−1,1989
0,7030	0,0627	−1,2025	0,7040	0,0622	−1,2060
0,7050	0,0617	−1,2096	0,7060	0,0612	−1,2131
0,7070	0,0607	−1,2167	0,7080	0,0602	−1,2203
0,7090	0,0597	−1,2239	0,7100	0,0592	−1,2275
0,7110	0,0587	−1,2311	0,7120	0,0582	−1,2347
0,7130	0,0578	−1,2384	0,7140	0,0573	−1,2420
0,7150	0,0568	−1,2456	0,7160	0,0563	−1,2493
0,7170	0,0559	−1,2530	0,7180	0,0554	−1,2567
0,7190	0,0549	−1,2603	0,7200	0,0544	−1,2640
0,7210	0,0540	−1,2678	0,7220	0,0535	−1,2715
0,7230	0,0531	−1,2752	0,7240	0,0526	−1,2790
0,7250	0,0522	−1,2827	0,7260	0,0517	−1,2865
0,7270	0,0513	−1,2902	0,7280	0,0508	−1,2940
0,7290	0,0504	−1,2978	0,7300	0,0499	−1,3016
0,7310	0,0495	−1,3054	0,7320	0,0491	−1,3093
0,7330	0,0486	−1,3131	0,7340	0,0482	−1,3170
0,7350	0,0478	−1,3208	0,7360	0,0473	−1,3247
0,7370	0,0469	−1,3286	0,7380	0,0465	−1,3325
0,7390	0,0461	−1,3364	0,7400	0,0457	−1,3403
0,7410	0,0453	−1,3442	0,7420	0,0449	−1,3482
0,7430	0,0444	−1,3522	0,7440	0,0440	−1,3561
0,7450	0,0436	−1,3601	0,7460	0,0432	−1,3641
0,7470	0,0428	−1,3681	0,7480	0,0424	−1,3721
0,7490	0,0421	−1,3762	0,7500	0,0417	−1,3802
0,7510	0,0413	−1,3843	0,7520	0,0409	−1,3883
0,7530	0,0405	−1,3924	0,7540	0,0401	−1,3965
0,7550	0,0398	−1,4006	0,7560	0,0394	−1,4048
0,7570	0,0390	−1,4089	0,7580	0,0386	−1,4131
0,7590	0,0383	−1,4172	0,7600	0,0379	−1,4214
0,7610	0,0375	−1,4256	0,7620	0,0372	−1,4298
0,7630	0,0368	−1,4341	0,7640	0,0365	−1,4383
0,7650	0,0361	−1,4426	0,7660	0,0357	−1,4468
0,7670	0,0354	−1,4511	0,7680	0,0350	−1,4554
0,7690	0,0347	−1,4597	0,7700	0,0344	−1,4641
0,7710	0,0340	−1,4684	0,7720	0,0337	−1,4728
0,7730	0,0333	−1,4772	0,7740	0,0330	−1,4816
0,7750	0,0327	−1,4860	0,7760	0,0323	−1,4904
0,7770	0,0320	−1,4948	0,7780	0,0317	−1,4993
0,7790	0,0313	−1,5038	0,7800	0,0310	−1,5083

I. (continued)

R_∞	$F(R_\infty)$	$\log F(R_\infty)$	R_∞	$F(R_\infty)$	$\log F(R_\infty)$
0,7810	0,0307	−1,5128	0,7820	0,0304	−1,5173
0,7830	0,0301	−1,5219	0,7840	0,0298	−1,5264
0,7850	0,0294	−1,5310	0,7860	0,0291	−1,5356
0,7870	0,0288	−1,5402	0,7880	0,0285	−1,5449
0,7890	0,0282	−1,5495	0,7900	0,0279	−1,5542
0,7910	0,0276	−1,5589	0,7920	0,0273	−1,5636
0,7930	0,0270	−1,5684	0,7940	0,0267	−1,5731
0,7950	0,0264	−1,5779	0,7960	0,0261	−1,5827
0,7970	0,0259	−1,5875	0,7980	0,0256	−1,5923
0,7990	0,0253	−1,5972	0,8000	0,0250	−1,6021
0,8010	0,0247	−1,6070	0,8020	0,0244	−1,6119
0,8030	0,0242	−1,6168	0,8040	0,0239	−1,6218
0,8050	0,0236	−1,6268	0,8060	0,0233	−1,6318
0,8070	0,0231	−1,6368	0,8080	0,0228	−1,6418
0,8090	0,0225	−1,6469	0,8100	0,0223	−1,6520
0,8110	0,0220	−1,6571	0,8120	0,0218	−1,6623
0,8130	0,0215	−1,6674	0,8140	0,0213	−1,6726
0,8150	0,0210	−1,6778	0,8160	0,0207	−1,6831
0,8170	0,0205	−1,6883	0,8180	0,0202	−1,6936
0,8190	0,0200	−1,6990	0,8200	0,0198	−1,7043
0,8210	0,0195	−1,7097	0,8220	0,0193	−1,7151
0,8230	0,0190	−1,7205	0,8240	0,0188	−1,7259
0,8250	0,0186	−1,7314	0,8260	0,0183	−1,7369
0,8270	0,0181	−1,7424	0,8280	0,0179	−1,7480
0,8290	0,0176	−1,7536	0,8300	0,0174	−1,7592
0,8310	0,0172	−1,7649	0,8320	0,0170	−1,7705
0,8330	0,0167	−1,7762	0,8340	0,0165	−1,7820
0,8350	0,0163	−1,7877	0,8360	0,0161	−1,7935
0,8370	0,0159	−1,7994	0,8380	0,0157	−1,8052
0,8390	0,0154	−1,8111	0,8400	0,0152	−1,8171
0,8410	0,0150	−1,8230	0,8420	0,0148	−1,8290
0,8430	0,0146	−1,8351	0,8440	0,0144	−1,8411
0,8450	0,0142	−1,8472	0,8460	0,0140	−1,8534
0,8470	0,0138	−1,8595	0,8480	0,0136	−1,8657
0,8490	0,0134	−1,8720	0,8500	0,0132	−1,8783
0,8510	0,0130	−1,8846	0,8520	0,0129	−1,8909
0,8530	0,0127	−1,8973	0,8540	0,0125	−1,9038
0,8550	0,0123	−1,9103	0,8560	0,0121	−1,9168
0,8570	0,0119	−1,9233	0,8580	0,0118	−1,9299
0,8590	0,0116	−1,9366	0,8600	0,0114	−1,9433

I. (continued)

R_∞	$F(R_\infty)$	$\log F(R_\infty)$	R_∞	$F(R_\infty)$	$\log F(R_\infty)$
0,8610	0,0112	−1,9500	0,8620	0,0110	−1,9568
0,8630	0,0109	−1,9636	0,8640	0,0107	−1,9705
0,8650	0,0105	−1,9774	0,8660	0,0104	−1,9843
0,8670	0,0102	−1,9913	0,8680	0,0100	−1,9984
0,8690	0,0099	−2,0055	0,8700	0,0097	−2,0127
0,8710	0,0096	−2,0199	0,8720	0,0094	−2,0271
0,8730	0,0092	−2,0344	0,8740	0,0091	−2,0418
0,8750	0,0089	−2,0492	0,8760	0,0088	−2,0567
0,8770	0,0086	−2,0642	0,8780	0,0085	−2,0718
0,8790	0,0083	−2,0794	0,8800	0,0082	−2,0872
0,8810	0,0080	−2,0949	0,8820	0,0079	−2,1027
0,8830	0,0078	−2,1106	0,8840	0,0076	−2,1186
0,8850	0,0075	−2,1266	0,8860	0,0073	−2,1347
0,8870	0,0072	−2,1428	0,8880	0,0071	−2,1510
0,8890	0,0069	−2,1593	0,8900	0,0068	−2,1676
0,8910	0,0067	−2,1761	0,8920	0,0065	−2,1845
0,8930	0,0064	−2,1931	0,8940	0,0063	−2,2018
0,8950	0,0062	−2,2105	0,8960	0,0060	−2,2193
0,8970	0,0059	−2,2281	0,8980	0,0058	−2,2371
0,8990	0,0057	−2,2461	0,9000	0,0056	−2,2553
0,9010	0,0054	−2,2645	0,9020	0,0053	−2,2738
0,9030	0,0052	−2,2832	0,9040	0,0051	−2,2927
0,9050	0,0050	−2,3022	0,9060	0,0049	−2,3119
0,9070	0,0048	−2,3217	0,9080	0,0047	−2,3315
0,9090	0,0046	−2,3415	0,9100	0,0045	−2,3516
0,9110	0,0043	−2,3618	0,9120	0,0042	−2,3721
0,9130	0,0041	−2,3825	0,9140	0,0040	−2,3930
0,9150	0,0039	−2,4036	0,9160	0,0039	−2,4144
0,9170	0,0038	−2,4252	0,9180	0,0037	−2,4362
0,9190	0,0036	−2,4474	0,9200	0,0035	−2,4586
0,9210	0,0034	−2,4700	0,9220	0,0033	−2,4816
0,9230	0,0032	−2,4933	0,9240	0,0031	−2,5051
0,9250	0,0030	−2,5170	0,9260	0,0030	−2,5292
0,9270	0,0029	−2,5415	0,9280	0,0028	−2,5539
0,9290	0,0027	−2,5665	0,9300	0,0026	−2,5793
0,9310	0,0026	−2,5923	0,9320	0,0025	−2,6054
0,9330	0,0024	−2,6188	0,9340	0,0023	−2,6323
0,9350	0,0023	−2,6460	0,9360	0,0022	−2,6599
0,9370	0,0021	−2,6741	0,9380	0,0020	−2,6884
0,9390	0,0020	−2,7030	0,9400	0,0019	−2,7179

I. (continued)

R_∞	$F(R_\infty)$	$\log F(R_\infty)$	R_∞	$F(R_\infty)$	$\log F(R_\infty)$
0,9410	0,0018	−2,7329	0,9420	0,0018	−2,7482
0,9430	0,0017	−2,7638	0,9440	0,0017	−2,7796
0,9450	0,0016	−2,7957	0,9460	0,0015	−2,8121
0,9470	0,0015	−2,8288	0,9480	0,0014	−2,8458
0,9490	0,0014	−2,8632	0,9500	0,0013	−2,8808
0,9510	0,0013	−2,8988	0,9520	0,0012	−2,9172
0,9530	0,0012	−2,9359	0,9540	0,0011	−2,9551
0,9550	0,0011	−2,9746	0,9560	0,0010	−2,9946
0,9570	0,00097	−3,0150	0,9580	0,00092	−3,0359
0,9590	0,00088	−3,0573	0,9600	0,00083	−3,0792
0,9610	0,00079	−3,1016	0,9620	0,00075	−3,1246
0,9630	0,00071	−3,1483	0,9640	0,00067	−3,1725
0,9650	0,00064	−3,1974	0,9660	0,00060	−3,2230
0,9670	0,00056	−3,2494	0,9680	0,00053	−3,2766
0,9690	0,00050	−3,3046	0,9700	0,00046	−3,3336
0,9710	0,00043	−3,3635	0,9720	0,00040	−3,3944
0,9730	0,00038	−3,4264	0,9740	0,00035	−3,4596
0,9750	0,00032	−3,4942	0,9760	0,00030	−3,5301
0,9770	0,00027	−3,5675	0,9780	0,00025	−3,6065
0,9790	0,00022	−3,6474	0,9800	0,00020	−3,6902
0,9810	0,00018	−3,7352	0,9820	0,00016	−3,7826
0,9830	0,00015	−3,8327	0,9840	0,00013	−3,8858
0,9850	0,00011	−3,9423	0,9860	0,00010	−4,0027
0,9870	0,00008	−4,0675	0,9880	0,00007	−4,1374
0,9890	0,00006	−4,2134	0,9900	0,00005	−4,2967
0,9910	0,00004	−4,3886	0,9920	0,00003	−4,4914
0,9930	0,00002	−4,6078	0,9940	0,00002	−4,7421
0,9950	0,00001	−4,9009	0,9960	0,00001	−5,0952
0,9970	0,00000	−5,3455	0,9980	0,00000	−5,6981
0,9990	0,00000	−6,3006			

II. Tables of $\sinh^{-1} x$; $\cosh^{-1} x$; $\coth^{-1} x$

x	$\sinh^{-1} x$	$\cosh^{-1} x$	$\coth^{-1} x$	x	$\sinh^{-1} x$	$\cosh^{-1} x$	$\coth^{-1} x$
0,00	0,0000			**0,50**	0,4812		
0,01	0,0100			0,51	0,4901		
0,02	0,0200			0,52	0,4990		
0,03	0,0300			0,53	0,5079		
0,04	0,0400			0,54	0,5167		
0,05	0,0500			0,55	0,5255		
0,06	0,0600			0,56	0,5342		
0,07	0,0699			0,57	0,5429		
0,08	0,0799			0,58	0,5516		
0,09	0,0899			0,59	0,5602		
0,10	0,0998			**0,60**	0,5688		
0,11	0,1098			0,61	0,5774		
0,12	0,1197			0,62	0,5859		
0,13	0,1296			0,63	0,5944		
0,14	0,1395			0,64	0,6028		
0,15	0,1494			0,65	0,6112		
0,16	0,1593			0,66	0,6196		
0,17	0,1692			0,67	0,6279		
0,18	0,1790			0,68	0,6362		
0,19	0,1889			0,69	0,6445		
0,20	0,1987			**0,70**	0,6527		
0,21	0,2085			0,71	0,6608		
0,22	0,2183			0,72	0,6690		
0,23	0,2280			0,73	0,6771		
0,24	0,2378			0,74	0,6851		
0,25	0,2475			0,75	0,6931		
0,26	0,2572			0,76	0,7011		
0,27	0,2668			0,77	0,7091		
0,28	0,2765			0,78	0,7170		
0,29	0,2861			0,79	0,7248		
0,30	0,2957			**0,80**	0,7327		
0,31	0,3052			0,81	0,7405		
0,32	0,3148			0,82	0,7482		
0,33	0,3243			0,83	0,7559		
0,34	0,3338			0,84	0,7636		
0,35	0,3432			0,85	0,7712		
0,36	0,3526			0,86	0,7788		
0,37	0,3620			0,87	0,7864		
0,38	0,3714			0,88	0,7939		
0,39	0,3807			0,89	0,8014		
0,40	0,3900			**0,90**	0,8089		
0,41	0,3993			0,91	0,8163		
0,42	0,4085			0,92	0,8237		
0,43	0,4177			0,93	0,8310		
0,44	0,4269			0,94	0,8383		
0,45	0,4360			0,95	0,8456		
0,46	0,4452			0,96	0,8528		
0,47	0,4542			0,97	0,8600		
0,48	0,4633			0,98	0,8672		
0,49	0,4722			0,99	0,8743		
0,50	0,4812			**1,00**	0,8814		

II. (continued)

x	$\sinh^{-1}x$	$\cosh^{-1}x$	$\coth^{-1}x$	x	$\sinh^{-1}x$	$\cosh^{-1}x$	$\coth^{-1}x$
1,00	0,8814.	0,0000	∞	**1,50**	1,1948.	0,9624˙	0,8047˙
1,01	0,8884˙	0,1413	2,652.	1,51	1,2003	0,9713	0,7968
1,02	0,8954˙	0,1997.	2,308.	1,52	1,2058	0,9801	0,7891
1,03	0,9024˙	0,2443˙	2,107˙	1,53	1,2113	0,9888	0,7815˙
1,04	0,9094.	0,2819	1,966	1,54	1,2167˙	0,9974.	0,7742.
1,05	0,9163	0,3149˙	1,857.	1,55	1,2222.	1,0059.	0,7670.
1,06	0,9232.	0,3447	1,768	1,56	1,2276	1,0143.	0,7599
1,07	0,9300	0,3720˙	1,693˙	1,57	1,2330.	1,0226.	0,7530
1,08	0,9368˙	0,3974.	1,629	1,58	1,2383˙	1,0308	0,7463.
1,09	0,9436	0,4211˙	1,573.	1,59	1,2437.	1,0389˙	0,7396˙
1,10	0,9503˙	0,4436.	1,522˙	**1,60**	1,2490.	1,0470.	0,7332.
1,11	0,9571.	0,4648˙	1,477	1,61	1,2543.	1,0549˙	0,7268˙
1,12	0,9637˙	0,4851˙	1,436	1,62	1,2595˙	1,0628˙	0,7206
1,13	0,9704.	0,5045˙	1,398˙	1,63	1,2648.	1,0706˙	0,7145
1,14	0,9770	0,5232.	1,363˙	1,64	1,2700	1,0784.	0,7085˙
1,15	0,9836.	0,5411	1,331˙	1,65	1,2752	1,0860	0,7027.
1,16	0,9901	0,5584	1,301˙	1,66	1,2804.	1,0936.	0,6969˙
1,17	0,9966˙	0,5751˙	1,273˙	1,67	1,2855	1,1011	0,6913.
1,18	1,0031	0,5913˙	1,247	1,68	1,2906˙	1,1086.	0,6857˙
1,19	1,0096.	0,6071.	1,222˙	1,69	1,2957˙	1,1159˙	0,6803
1,20	1,0160.	0,6224.	1,199	**1,70**	1,3008˙	1,1232˙	0,6750.
1,21	1,0224.	0,6372.	1,177.	1,71	1,3059.	1,1305.	0,6697˙
1,22	1,0287	0,6517˙	1,156.	1,72	1,3109˙	1,1376˙	0,6646.
1,23	1,0350˙	0,6659.	1,136	1,73	1,3159˙	1,1448.	0,6595
1,24	1,0413˙	0,6797.	1,117.	1,74	1,3209˙	1,1518˙	0,6545˙
1,25	1,0476	0,6931˙	1,099.	1,75	1,3259	1,1588	0,6496˙
1,26	1,0538˙	0,7063˙	1,081˙	1,76	1,3308˙	1,1657˙	0,6448˙
1,27	1,0600˙	0,7192˙	1,065.	1,77	1,3358.	1,1726˙	0,6401
1,28	1,0662	0,7319	1,049.	1,78	1,3407	1,1794˙	0,6355.
1,29	1,0723˙	0,7443.	1,033˙	1,79	1,3456.	1,1862	0,6309.
1,30	1,0785.	0,7564˙	1,018˙	**1,80**	1,3504˙	1,1929	0,6264.
1,31	1,0845˙	0,7684.	1,004˙	1,81	1,3553	1,1996.	0,6220.
1,32	1,0906	0,7801.	0,9905	1,82	1,3601	1,2062.	0,6176
1,33	1,0966	0,7916.	0,9773.	1,83	1,3649˙	1,2127˙	0,6133
1,34	1,1026	0,8029.	0,9645.	1,84	1,3697	1,2192˙	0,6091.
1,35	1,1086.	0,8140	0,9521˙	1,85	1,3745.	1,2257.	0,6049˙
1,36	1,1145.	0,8249˙	0,9402.	1,86	1,3792	1,2321.	0,6008˙
1,37	1,1204˙	0,8357	0,9286.	1,87	1,3839˙	1,2384˙	0,5968
1,38	1,1263	0,8463	0,9173˙	1,88	1,3886˙	1,2447˙	0,5928
1,39	1,1322.	0,8567˙	0,9065.	1,89	1,3933˙	1,2510.	0,5889
1,40	1,1380.	0,8670	0,8959.	**1,90**	1,3980	1,2572	0,5850˙
1,41	1,1438.	0,8771˙	0,8856.	1,91	1,4026˙	1,2634.	0,5812˙
1,42	1,1496.	0,8871˙	0,8756˙	1,92	1,4073.	1,2695	0,5775.
1,43	1,1553	0,8970	0,8659˙	1,93	1,4119	1,2756.	0,5738
1,44	1,1610	0,9067	0,8565	1,94	1,4165.	1,2816	0,5701˙
1,45	1,1667	0,9163	0,8473	1,95	1,4210˙	1,2876	0,5665˙
1,46	1,1724.	0,9258.	0,8383˙	1,96	1,4256	1,2935.	0,5630
1,47	1,1780	0,9351	0,8296˙	1,97	1,4301˙	1,2995.	0,5595
1,48	1,1836˙	0,9443	0,8211	1,98	1,4347.	1,3053.	0,5561.
1,49	1,1892	0,9534˙	0,8128	1,99	1,4392.	1,3112.	0,5527.
1,50	1,1948.	0,9624˙	0,8047˙	**2,00**	1,4436˙	1,3170˙	0,5493

II. (continued)

x	$\sinh^{-1} x$	$\cosh^{-1} x$	$\coth^{-1} x$	x	$\sinh^{-1} x$	$\cosh^{-1} x$	$\coth^{-1} x$
2,00	1,4436	1,3170.	0,5493	**2,50**	1,6472	1,5668	0,4236
2,01	1,4481	1,3227	0,5460	2,51	1,6509	1,5712.	0,4218.
2,02	1,4525	1,3284	0,5427	2,52	1,6546	1,5755	0,4199.
2,03	1,4570.	1,3341	0,5395	2,53	1,6583	1,5798	0,4180
2,04	1,4614.	1,3397	0,5363	2,54	1,6620	1,5841	0,4162.
2,05	1,4658.	1,3454.	0,5332.	2,55	1,6656	1,5884.	0,4143
2,06	1,4702.	1,3509	0,5301.	2,56	1,6693	1,5926	0,4125
2,07	1,4745	1,3565.	0,5270	2,57	1,6729	1,5969.	0,4107
2,08	1,4789.	1,3620.	0,5240	2,58	1,6765	1,6011.	0,4090.
2,09	1,4832.	1,3674	0,5210	2,59	1,6801	1,6053.	0,4072
2,10	1,4875.	1,3729.	0,5180	**2,60**	1,6837	1,6094	0,4055.
2,11	1,4918.	1,3783.	0,5151	2,61	1,6873	1,6136	0,4037
2,12	1,4960	1,3836	0,5123.	2,62	1,6909	1,6177.	0,4020
2,13	1,5003	1,3890.	0,5094	2,63	1,6945.	1,6219.	0,4003
2,14	1,5045	1,3943.	0,5066	2,64	1,6980	1,6260.	0,3986
2,15	1,5088.	1,3995	0,5038	2,65	1,7015	1,6300	0,3970.
2,16	1,5130.	1,4048.	0,5011.	2,66	1,7051.	1,6341	0,3953
2,17	1,5172.	1,4100.	0,4984.	2,67	1,7086.	1,6382.	0,3937
2,18	1,5214.	1,4152.	0,4957	2,68	1,7121.	1,6422	0,3921.
2,19	1,5255	1,4203	0,4930	2,69	1,7156.	1,6462	0,3904
2,20	1,5297.	1,4254	0,4904	**2,70**	1,7191.	1,6502	0,3889.
2,21	1,5338	1,4305	0,4878	2,71	1,7225	1,6542.	0,3873.
2,22	1,5379	1,4356.	0,4853.	2,72	1,7260.	1,6581	0,3857
2,23	1,5420	1,4406	0,4827	2,73	1,7294	1,6621.	0,3841
2,24	1,5461	1,4456	0,4802	2,74	1,7329.	1,6660	0,3826
2,25	1,5502.	1,4506.	0,4778.	2,75	1,7363.	1,6699	0,3811.
2,26	1,5542	1,4555	0,4753.	2,76	1,7397	1,6738	0,3796.
2,27	1,5583.	1,4604	0,4729	2,77	1,7431	1,6777	0,3780
2,28	1,5623.	1,4653	0,4705	2,78	1,7465.	1,6816.	0,3766.
2,29	1,5663	1,4702	0,4681	2,79	1,7499.	1,6854	0,3751.
2,30	1,5703.	1,4750	0,4658.	**2,80**	1,7532	1,6892	0,3736
2,31	1,5743.	1,4799.	0,4635.	2,81	1,7566	1,6931.	0,3722.
2,32	1,5782	1,4846	0,4612.	2,82	1,7599	1,6969.	0,3707
2,33	1,5822.	1,4894	0,4589	2,83	1,7633.	1,7006	0,3693.
2,34	1,5861	1,4942.	0,4567.	2,84	1,7666	1,7044	0,3679.
2,35	1,5900	1,4989.	0,4544	2,85	1,7699	1,7082.	0,3664
2,36	1,5939	1,5036.	0,4522	2,86	1,7732	1,7119	0,3650
2,37	1,5978	1,5082	0,4501.	2,87	1,7765	1,7156	0,3637.
2,38	1,6017	1,5129.	0,4479	2,88	1,7798	1,7193	0,3623
2,39	1,6056	1,5175	0,4458.	2,89	1,7831	1,7230	0,3609
2,40	1,6094	1,5221.	0,4437.	**2,90**	1,7863	1,7267	0,3596.
2,41	1,6133.	1,5267.	0,4416.	2,91	1,7896	1,7304.	0,3582
2,42	1,6171	1,5312	0,4395	2,92	1,7928	1,7340	0,3569.
2,43	1,6209	1,5357	0,4374	2,93	1,7961.	1,7377.	0,3556.
2,44	1,6247	1,5402	0,4354	2,94	1,7993	1,7413	0,3542
2,45	1,6285	1,5447	0,4334	2,95	1,8025	1,7449	0,3529
2,46	1,6323.	1,5492.	0,4314	2,96	1,8057	1,7485	0,3516
2,47	1,6360	1,5536	0,4294	2,97	1,8089	1,7521.	0,3504.
2,48	1,6398.	1,5580.	0,4275	2,98	1,8121	1,7556.	0,3491
2,49	1,6435	1,5624	0,4256.	2,99	1,8153.	1,7592	0,3478
2,50	1,6472	1,5668	0,4236	**3,00**	1,8184	1,7627	0,3466.

II. (continued)

x	$\sinh^{-1}x$	$\cosh^{-1}x$	$\coth^{-1}x$	x	$\sinh^{-1}x$	$\cosh^{-1}x$	$\coth^{-1}x$
3,00	1,8184˙	1,7627˙	0,3466.	**3,50**	1,9657˙	1,9248˙	0,2939
3,01	1,8216	1,7663.	0,3453˙	3,51	1,9685.	1,9278˙	0,2930
3,02	1,8248.	1,7698	0,3441	3,52	1,9712	1,9308	0,2921˙
3,03	1,8279	1,7733.	0,3429.	3,53	1,9739˙	1,9338.	0,2913.
3,04	1,8310˙	1,7768	0,3416˙	3,54	1,9767.	1,9367	0,2904.
3,05	1,8341˙	1,7803.	0,3404˙	3,55	1,9794.	1,9396˙	0,2895˙
3,06	1,8373.	1,7837˙	0,3392˙	3,56	1,9821.	1,9426.	0,2887.
3,07	1,8404.	1,7872.	0,3380˙	3,57	1,9848.	1,9455	0,2878
3,08	1,8434˙	1,7906	0,3369.	3,58	1,9875.	1,9484	0,2870.
3,09	1,8465˙	1,7940˙	0,3357	3,59	1,9902.	1,9513	0,2861
3,10	1,8496	1,7975.	0,3345˙	**3,60**	1,9928˙	1,9542	0,2853.
3,11	1,8527.	1,8009.	0,3334.	3,61	1,9955.	1,9571	0,2844˙
3,12	1,8557˙	1,8042˙	0,3322˙	3,62	1,9982.	1,9600.	0,2836
3,13	1,8588.	1,8076˙	0,3311.	3,63	2,0008˙	1,9628˙	0,2828
3,14	1,8618	1,8110	0,3299˙	3,64	2,0035	1,9657	0,2820.
3,15	1,8648˙	1,8143˙	0,3288˙	3,65	2,0061˙	1,9686.	0,2812.
3,16	1,8679.	1,8177	0,3277	3,66	2,0088.	1,9714	0,2803˙
3,17	1,8709.	1,8210˙	0,3266	3,67	2,0114	1,9742˙	0,2795˙
3,18	1,8739.	1,8243˙	0,3255	3,68	2,0140˙	1,9771.	0,2787˙
3,19	1,8769.	1,8276˙	0,3244	3,69	2,0166˙	1,9799	0,2779˙
3,20	1,8799.	1,8309˙	0,3233	**3,70**	2,0193.	1,9827.	0,2772.
3,21	1,8828˙	1,8342˙	0,3222˙	3,71	2,0219.	1,9855.	0,2764.
3,22	1,8858	1,8375	0,3212.	3,72	2,0245.	1,9883.	0,2756
3,23	1,8888.	1,8408.	0,3201	3,73	2,0271.	1,9911.	0,2748
3,24	1,8917˙	1,8440	0,3190˙	3,74	2,0296˙	1,9939.	0,2740˙
3,25	1,8947.	1,8472˙	0,3180	3,75	2,0322˙	1,9966˙	0,2733.
3,26	1,8976	1,8505.	0,3170.	3,76	2,0348	1,9994	0,2725
3,27	1,9005˙	1,8537.	0,3159˙	3,77	2,0374.	2,0021˙	0,2717˙
3,28	1,9035.	1,8569.	0,3149	3,78	2,0399˙	2,0049	0,2710
3,29	1,9064.	1,8601	0,3139.	3,79	2,0425.	2,0076˙	0,2702˙
3,30	1,9093.	1,8633.	0,3129.	**3,80**	2,0450˙	2,0104.	0,2695
3,31	1,9122.	1,8665.	0,3118˙	3,81	2,0476.	2,0131	0,2688.
3,32	1,9151.	1,8696˙	0,3108˙	3,82	2,0501	2,0158	0,2680˙
3,33	1,9179˙	1,8728.	0,3098˙	3,83	2,0526˙	2,0185	0,2673
3,34	1,9208	1,8759	0,3089.	3,84	2,0552.	2,0212˙	0,2666.
3,35	1,9237.	1,8790˙	0,3079.	3,85	2,0577.	2,0239	0,2658˙
3,36	1,9265˙	1,8822.	0,3069	3,86	2,0602.	2,0266	0,2651
3,37	1,9294.	1,8853.	0,3059˙	3,87	2,0627.	2,0293.	0,2644
3,38	1,9322˙	1,8884.	0,3050.	3,88	2,0652.	2,0319˙	0,2637.
3,39	1,9351.	1,8915.	0,3040˙	3,89	2,0677.	2,0346	0,2630.
3,40	1,9379.	1,8946.	0,3031.	**3,90**	2,0702.	2,0373.	0,2623.
3,41	1,9407.	1,8976˙	0,3021˙	3,91	2,0727.	2,0399.	0,2616.
3,42	1,9435.	1,9007	0,3012	3,92	2,0751˙	2,0426.	0,2609.
3,43	1,9463.	1,9037˙	0,3003.	3,93	2,0776	2,0452	0,2602.
3,44	1,9491	1,9068	0,2993˙	3,94	2,0801.	2,0478˙	0,2595.
3,45	1,9519	1,9098˙	0,2984	3,95	2,0825	2,0504˙	0,2588
3,46	1,9547.	1,9128˙	0,2975	3,96	2,0850.	2,0531.	0,2581
3,47	1,9574˙	1,9159.	0,2966	3,97	2,0874.	2,0557.	0,2574˙
3,48	1,9602	1,9189.	0,2957.	3,98	2,0899.	2,0583.	0,2568.
3,49	1,9630.	1,9219.	0,2948	3,99	2,0923.	2,0609.	0,2561.
3,50	1,9657˙	1,9248˙	0,2939	**4,00**	2,0947	2,0634˙	0,2554

II. (continued)

x	$\sinh^{-1}x$	$\cosh^{-1}x$	$\coth^{-1}x$	x	$\sinh^{-1}x$	$\cosh^{-1}x$	$\coth^{-1}x$
4,00	2,0947	2,0634·	0,2554	**4,50**	2,2093·	2,1846·	0,2260
4,01	2,0971·	2,0660	0,2547·	4,51	2,2115	2,1869·	0,2255.
4,02	2,0996.	2,0686	0,2541	4,52	2,2137.	2,1892	0,2250.
4,03	2,1020.	2,0712.	0,2534·	4,53	2,2158·	2,1915.	0,2244·
4,04	2,1044.	2,0737	0,2528.	4,54	2,2180	2,1937·	0,2239·
4,05	2,1068.	2,0763.	0,2521·	4,55	2,2201·	2,1960.	0,2234·
4,06	2,1092.	2,0788	0,2515.	4,56	2,2223.	2,1982·	0,2229·
4,07	2,1116.	2,0813·	0,2508·	4,57	2,2244·	2,2005.	0,2224
4,08	2,1139·	2,0839.	0,2502	4,58	2,2266.	2,2027	0,2219
4,09	2,1163	2,0864	0,2496.	4,59	2,2287	2,2049·	0,2214
4,10	2,1187	2,0889·	0,2489·	**4,60**	2,2308	2,2072.	0,2209
4,11	2,1211.	2,0914·	0,2483	4,61	2,2329·	2,2094	0,2204·
4,12	2,1234	2,0939·	0,2477.	4,62	2,2351.	2,2116·	0,2199·
4,13	2,1258.	2,0964·	0,2470·	4,63	2,2372.	2,2138·	0,2194·
4,14	2,1281·	2,0989·	0,2464	4,64	2,2393.	2,2160·	0,2190.
4,15	2,1305.	2,1014	0,2458	4,65	2,2414.	2,2182·	0,2185.
4,16	2,1328	2,1039	0,2452.	4,66	2,2435.	2,2204·	0,2180.
4,17	2,1351·	2,1064.	0,2446.	4,67	2,2456.	2,2226·	0,2175
4,18	2,1375.	2,1088·	0,2440.	4,68	2,2477.	2,2248·	0,2170·
4,19	2,1398	2,1113	0,2434.	4,69	2,2498.	2,2270	0,2165·
4,20	2,1421	2,1137·	0,2428.	**4,70**	2,2518·	2,2292	0,2161.
4,21	2,1444·	2,1162	0,2422.	4,71	2,2539·	2,2314.	0,2156
4,22	2,1467·	2,1186·	0,2416.	4,72	2,2560	2,2335·	0,2151·
4,23	2,1490·	2,1211.	0,2410.	4,73	2,2581.	2,2357	0,2147.
4,24	2,1513·	2,1235	0,2404.	4,74	2,2601·	2,2379.	0,2142
4,25	2,1536·	2,1259·	0,2398	4,75	2,2622	2,2400·	0,2137·
4,26	2,1559	2,1283·	0,2392	4,76	2,2643.	2,2422.	0,2133.
4,27	2,1582	2,1308.	0,2386·	4,77	2,2663	2,2443·	0,2128
4,28	2,1605.	2,1332.	0,2380·	4,78	2,2684.	2,2465.	0,2123·
4,29	2,1627·	2,1356.	0,2375.	4,79	2,2704	2,2486	0,2119.
4,30	2,1650·	2,1380.	0,2369	**4,80**	2,2724·	2,2507·	0,2114·
4,31	2,1673.	2,1403·	0,2363·	4,81	2,2745.	2,2529.	0,2110.
4,32	2,1695·	2,1427·	0,2358.	4,82	2,2765	2,2550.	0,2105·
4,33	2,1718	2,1451	0,2352	4,83	2,2785·	2,2571	0,2101.
4,34	2,1740·	2,1475.	0,2346·	4,84	2,2806.	2,2592	0,2096·
4,35	2,1763.	2,1498·	0,2341.	4,85	2,2826	2,2613·	0,2092
4,36	2,1785·	2,1522	0,2335	4,86	2,2846.	2,2634·	0,2087·
4,37	2,1808.	2,1546.	0,2330.	4,87	2,2866·	2,2655·	0,2083
4,38	2,1830.	2,1569	0,2324	4,88	2,2886·	2,2676·	0,2079.
4,39	2,1852	2,1592·	0,2319.	4,89	2,2906·	2,2697	0,2074·
4,40	2,1874·	2,1616.	0,2313	**4,90**	2,2926·	2,2718	0,2070
4,41	2,1896·	2,1639	0,2308.	4,91	2,2946·	2,2739	0,2066.
4,42	2,1918·	2,1662·	0,2302·	4,92	2,2966·	2,2760.	0,2061·
4,43	2,1940·	2,1686.	0,2297	4,93	2,2986·	2,2780·	0,2057
4,44	2,1962·	2,1709.	0,2292.	4,94	2,3006	2,2801	0,2053.
4,45	2,1984·	2,1732.	0,2286·	4,95	2,3026	2,2822.	0,2048·
4,46	2,2006·	2,1755	0,2281	4,96	2,3046.	2,2842·	0,2044
4,47	2,2028·	2,1778.	0,2276.	4,97	2,3065·	2,2863	0,2040
4,48	2,2050	2,1801.	0,2270·	4,98	2,3085·	2,2883·	0,2036.
4,49	2,2072.	2,1824.	0,2265	4,99	2,3105.	2,2904	0,2032.
4,50	2,2093·	2,1846·	0,2260	**5,00**	2,3124·	2,2924·	0,2027·

II. (continued)

x	$\sinh^{-1} x$	$\cosh^{-1} x$	$\coth^{-1} x$	x	$\sinh^{-1} x$	$\cosh^{-1} x$	$\coth^{-1} x$
5,0	2,3124˙	2,2924˙	0,2027˙	7,5	2,7125.	2,7036.	0,1341˙
5,1	2,3319.	2,3126˙	0,1987.	7,6	2,7256	2,7169˙	0,1323˙
5,2	2,3509˙	2,3324˙	0,1947˙	7,7	2,7386.	2,7301˙	0,1306
5,3	2,3696˙	2,3518˙	0,1910.	7,8	2,7514.	2,7431˙	0,1289
5,4	2,3880	2,3709.	0,1873˙	7,9	2,7640	2,7560.	0,1273.
5,5	2,4061.	2,3895˙	0,1839.	**8,0**	2,7765.	2,7687.	0,1257.
5,6	2,4238	2,4078˙	0,1805	8,1	2,7888	2,7812.	0,1241
5,7	2,4412˙	2,4258˙	0,1773.	8,2	2,8010.	2,7935˙	0,1226.
5,8	2,4584.	2,4435	0,1742.	8,3	2,8130	2,8058.	0,1211.
5,9	2,4752	2,4608˙	0,1711˙	8,4	2,8249	2,8178˙	0,1196
6,0	2,4918.	2,4779	0,1682˙	8,5	2,8367.	2,8297˙	0,1182
6,1	2,5081	2,4946˙	0,1654˙	8,6	2,8483.	2,8415	0,1168
6,2	2,5241˙	2,5111˙	0,1627	8,7	2,8598.	2,8532.	0,1155.
6,3	2,5399˙	2,5273˙	0,1601	8,8	2,8711	2,8647.	0,1141˙
6,4	2,5555	2,5433	0,1575˙	8,9	2,8823˙	2,8760˙	0,1128˙
6,5	2,5708	2,5590.	0,1551.	**9,0**	2,8934˙	2,8873.	0,1116.
6,6	2,5859	2,5744˙	0,1527	9,1	2,9044˙	2,8984	0,1103˙
6,7	2,6008.	2,5896˙	0,1504.	9,2	2,9153	2,9094	0,1091˙
6,8	2,6154˙	2,6046.	0,1481˙	9,3	2,9260˙	2,9203˙	0,1079˙
6,9	2,6299.	2,6194.	0,1460.	9,4	2,9367.	2,9310	0,1068
7,0	2,6441˙	2,6339	0,1438˙	9,5	2,9472	2,9417.	0,1057.
7,1	2,6582.	2,6482˙	0,1418	9,6	2,9576	2,9522	0,1045˙
7,2	2,6720	2,6624.	0,1398	9,7	2,9679˙	2,9626	0,1035.
7,3	2,6857.	2,6763.	0,1379.	9,8	2,9781˙	2,9729	0,1024
7,4	2,6992.	2,6900˙	0,1360.	9,9	2,9882˙	2,9831˙	0,1014.
7,5	2,7125.	2,7036.	0,1341˙	**10,0**	2,9982˙	2,9932˙	0,1003˙

Subject Index

absolute and relative measurements 137 ff.
— — — reflectance values 274
— diffuse reflectance of magnesium oxide 138
— reflectance according to Melamed 167
— —, according to *Shibata* 143
— — — for quantitative measurements 183
— — — from transmittance values 142
— — — of a series of white standards 148
— — — — freshly prepared magnesium oxide 137
— — — — prepared MgO according to various authors 146
— — — — prepared MgO, table 147
— — — magnesium oxide in the infrared 145
absorption coefficient according to *Kubelka-Munk* 112
— — as a function of particle size 59
— — definition 21, 108
— coefficients of a colored glass in transmittance and reflectance 195
— — of aerosils 207
— — of glass powder 208
— — of graphite from reflection measurements on aerosil-graphite mixtures 210
— cross section according to *Mie* 256
— index, definition 21
— maximum of chlorophyll in living cells 307
— — relative shift as function of the half width and of the wave number dependence of S 212
— parameter, definition 315
— spectra of $K(Mn, Cl)O_4$ mixed crystals 187

absorption sprectrum of monocellular algae in aqueous suspension 307
acid-base reaction between the adsorbed substance and the adsorbent 257
activation energies, calculation from the rate constants 283
adsorption from solution 243
— from the gas phase 242
— isotherm of p-dimethylamino-azobenzene on $BaSO_4$ 259
— isotherms of the chemisorption on NaCl of different grain sizes 272
— of I_2 of Br_2 on highly dried oxides 263
aerosil, diffuse reflectance 149
— goniophotometric curves 52
—, grain size dependence of scattering coefficients 203
ageing standards 149
albedo, definition 28
analytical photometric measurements in reflexion 290ff.
— procedure according *Ringbom* and *Giovanelli* 292
angle of polarization, definition 8
angular distribution, isotropic at multiple scattering 99
— — of Rayleigh scattering 78
— — of the scattered intensity after single and twofold scattering on dielectric spheres 99
— — of the scattered radiation at a homogeneous sphere 84
anthraquinone oxidation on surfaces 278
anisotropic materials, investigation by ATR 330
apparent absorbance of powders 59
applications of ATR 329 ff.
approximate equations of the Kubelka-Munk-theory 120

aromatic hydrocarbons adsorbed on cracking catalysts 265
ATR spectra of aqueous Eosin-Y solutions 334
attachment for attenuated total reflection according to *Fahrenfort* 326
— for commercial spectrophotometers for measuring internal reflection spectra 327
attenuated total reflection 309 ff.
— — — in the visible and UV spectral regions 332
— — — on a phase boundary as function of the angle of incidence with different absorption index of the rarer medium 311
— — —, sensitivity to changes in $\varkappa$ 311
average path length of radiation in a scattering medium 107

Babinet compensator 18
ball mills 237
biological materials, reflectance measurements 306
Bouguer-Beer Law in ATR-spectra 333
Bouguer elementary mirror hypothesis 52
— hypothesis of elementary mirrors 29 ff.
Brewster Law 8, 58
bulk dyes in the textile, laquers, pigments and plastics industries 301

calibration curves for determination of concentrations 290
characteristic absorption of adsorbents 297
— function 158
—, linear of amplifier and detector 226
characteristics of the rotation ellipsoid 247
charge-transfer-band of the complex pyrene-s-trinitrobenzene 275
charge transfer complexes on surfaces 262 ff.
chemisorption and physical adsorption 272
—, spectral changes 257

chromatogram-spectrophotometer 225
circularly diffuse reflection 29
cleavage reactions, reversible on surfaces 266
color curve, typical 186
— matching 303
— measurement 301
— measuring instruments 231
complexes, formation on surfaces 261
— $OH^- I_2$ and $OH^- Br_2$ 203
complex refractive index 21, 310
continuum theory of absorption and scattering of tightly packed particles 103
contrast ratio, ideal 112
covering glasses, effect on scattering and transmission of a sample 135
— —, influence on reflectance spectra 189
cracking catalysts, adsorption 265
critical angle of total reflection 13
crystal spectra and solution spectra of complexes 285

deformation and shift of bands at internal reflection 319
dehydration of phenylbenzyl carbinol to stilbene on Al_2O_3 as catalyst 283
depolarization of linear polarized light at diffuse reflection 37
depth of penetration in the rarer medium at toal reflection 316
— — of radiation in powders 59
— —, relative and effective layer thickness at ATR-measurements 318
deviations between Kubelka-Munk theory and rigorous theory 162
— from the Lambert Cosine Law 38 ff., 173
— from the Lambert Law in the region of selective absorption 55
dielectric polarization 80
differential equations, simultaneous of isotropic scattering 104, 109

Subject Index

differential equations, substitution for the radiation transfer equation 157
diffraction phenomena at total reflection 20
diffuse reflectance at absorbing materials 55
— — on aerosil 44
— reflection, definition 26
dilution method 175 ff.
— — in measuring reflectance spectra 215
— series with the same standard 180
dimethylaminoazobenzene, reflection spectrum in adsorbed state 257
dipole moment, definition 75
discontinuum theories 163 ff.
— theory according to *Melamed* 166
— — — to *Melamed*, experimental test 216
discussion of errors 250
dispersion of the optically rarer medium at ATR measurements 319
— of the refractive index, influence on scattering coefficients 206
dissociation equilibrium of molecular complexes on surfaces 276
distribution function for the grain sizes of ground glas 192
— — for the Bouguer elementary mirrors 32, 45
double-beam spectrophotometer, attachments for reflection measurements 226
— spectrophotometer for determination of reflection spectra 328
double-pass elements for internal reflection 322
dyeing recipes 304
dynamic reflectance spectroscopy 288

effective layer thickness at films 318
— — —, defination 316
efficiency factor according to *Mie* for spherical particles 90
— —, definition 89
— factors of gold sols according to *Mie* 93

electric vector of the radiation 6
elementary mirror hypothesis according to *Bouguer* 32
elements for internal reflection 324
elliptical polarization at total reflection 18
Elrepho of Zeiss 222
equilibria at surfaces 273
errors, relative of Kubelka-Munk function 250
external and internal field 80
extinction $E_\perp$ and $E_\parallel$ of an aqueous solution of Eosin B as a function of concentration 335

fading reaction of the tautomeric 2-(2′.4′-dinitrobenzyl)-pyridine, adsorbed on SiO_2 282
Fahrenfort method for determination of n and κ 309 ff.
forward and backward scattering 88, 95
Fresnel equations 8
— — for complex refractive index 310
— — for the reflection of absorbing materials 21
— — for total reflection 18
— formulas, development in series 332
— reflectance of TiO_2 (rutile) MgO, $BaSO_4$ and SiO_2 46
Freundlich isotherm 263

gas chromatograph fraction, spectra by ATR 331
geophysical problems, investgation with reflectance measurements 307
germanium detector 248
glass powder, scattering coefficients as function of wavenumber 206
— powders, absorption coefficients 207
— —, reflectance as function of wavelength 62
gloss, definition 173
glossy peaks as a Fresnel effect 38, 44

glove boxes 236
goniophotometer for testing Lamberts cosine law 217
— of Carl Zeiss 218
—, recording 219
goniophotometric curves 35
— — of $BaSO_4$, Rutil, MgO, Aerosil 43
— measurements on K_2CrO_4 powder at different wavelengths 56
— — testing the Lambert Cosine Law 38
grain size dependence of $F(R_\infty)$ 214
— — — of the scattering coefficients of aerosil 203
— — distribution of ground Didym filter glass 192
grinding time influence on the reflection spectra 238

half cylinder for 5-fold reflection according to *Fahrenfort* 321
— — for measurements of the attenuated total reflection 320
Hardy spectrophotometer for reflectance measurements 228
heavy metal complexes, analytical determination 293
Helmholtz Reciprocity Law 10, 39, 172
hiding power 112
H-integrals 158
hohlraum reflectance spectrophotometer for the infrared 246
hyperbolic functions 116
— —, expressed in series 121

ideal diffuse reflector 28
indicatrix, definition 29
— for the reflection of infrared radiation of germanium powder 36
— — on drawing paper 37
— of a Lambert radiator 30
— of a Seeliger radiator 30
infinite layer thickness 213

influence of regular reflection on $F(R_\infty)$ 174
influence of various parameters on the validity of the Lambert Law 38
infrared reflectance spectrometers 247
— — spectrometer with rotation ellipsoid 248, 249
— spectra of liquid dibutyl phthalate in transmission, normal reflection and in attenuated total reflection 314
inhomogeneous layers, reflectance and transmittance according to *Kubelka* 123
— media, treated by *Giovanelli* 163
inhomogenieties in chromatogram spots 300
integrating sphere according to *Taylor* 139, 219ff.
— — for the vacuum UV 220
— —, radiation flux 219
— —, sphere error (table) 221
interference method for the measurement of phase difference 24
— of scattered waves 83
— spectroscopy 323
"internal" regular reflection according to *Ryde* 131
— — — of diffuse radiation 132
— reflectance at phase boundaries 15
— —, dependence on the extinction modulus of the optically rarer medium 315
— — of a phase boundary as function of the angle of incidence 310
— reflection coefficient 168
— —, definition 14
— — spectroscopy 313ff.
intrinsic absorption of reference standards 183
iodides of heavy metals adsorbed on alkali or alkaline earth halides 260
ion-exchange resins 293
iron oxide pigments 253
isomeric matched coloration 305
isotropic diffuse reflection 32
— distribution of multiple scattering 100

Subject Index

KBr-method in the infrared 2
kinetic measurements on surfaces 281
K_R/K_T ratio 195
Kubelka-Munk coefficients R_∞, R_0, K, S, a, b (table) 119
— function 111
— — and absorbance as function of R_∞ and T respectively 179
— —, concentration dependence 178, 181
— —, additivity in mixtures 303
— — of a rutile-anatase mixture, measured against pure anatase as standard 291
— — of stilbene, adsorbed on active Al_2O_3, concentration dependence 284
— —, particle-size dependence 213
— —, relativ errors 250
— —, testing by means of transmittance and reflectance measurements on the same mixed crystals 188
—, limiting straight line of pyrene, adsorbed on NaCl 182
— theory, application on papers 126
— —, applicability on papers 199
— — at directed irradiation 127
— —, comparison with rigorous theory 160
— —, exponential solution 107
— —, hyperbolic solution 116

lactone of malachite-green-o-carboxylic acid, cleavage reaction 266
Lambert Cosine Law 28
— — —, experimental test 34, 217
— Law and Bouguer hypothesis 50
— —, deviations according to *Pokrowski* 31
— —, influence of the azimuth 35
Langmuir adsorption isotherm 260, 271
layer thickness for the measurement of R_∞ 239
— — for measurments of R_0 240

Legendre polynomials 157
low temperature cells 230
luminescence of scattering samples 150
— reabsorption in scattering media 154
—, true in scattering media 156
luminescent radiation flux in scattering media 153
lyophilisation 238

magnesium oxide as reflectance standard 137
— — primary standard according to Zeiss 147
major angle of incidence, definition 22
— azimuth, definition 22
mean path length of diffuse irradiation 108
measurement of the regular reflectance of plane surfaces 228
measurements in the infrared 245
— with linearly polarized radiation 231
Melamed, discontinuum theory 166
— statistical theory of diffuse reflectance, experimental test 216
metameric colorations 304
methods, ATR-measurements ff. 319
— for determination of n and $\varkappa$ 22
Mie functions 89
— scattered radiation, variation of the wavelength exponents 92
Mie-theory 84, 85
— for absorbing spheres 93
mirror reflection 5
mixture of colored pigments, quantitative analysis 302
moisture, influence on reflectance spectra 234
molecular complex hexamethylbenzene-s-trinitrobenzene, adsorbed on silicagel 275
multiple reflection of a plate bounded by parallel planes 10
— scattering 94

natural extinction modulus, definition 21
normal spectral values for the equal-energy spectrum 304

numerical quadrature according to *Gauss* 157

opacity 112
optical clearing agents 233
— constants n and $\varkappa$ 309ff.
— contact between sample and reflection element 325
— geometry of reflectometers 227
— — of the measuring arrangements 170ff.
— projection of a surface ΔF onto a smaller surface $\Delta F'$ 245
— — of the sample onto the detector by means of a rotation ellipsoid 248
— thickness 163
— —, definition 101
orientation of adsorbed molecules on surfaces 277
— of elementary mirrors to the macroscopic surface 46

packing density 201, 239
paper chromatograms, consideration of the spot size 298
— —, dependence of $F(R_{ads})/n$ on the amount of material present 300
— —, quantitative evaluation 298
papers, applicability of Kubelka-Munk function 197
particle size dependence of absorption at reflectance measurements 214
— — — of light absorption in heterogeneous systems 61
— — — of the absorption at adsorbed substances 216
Perkin-Elmer 12 C spectrometer, converted to a reflection spectrometer with rotation ellipsoid 249
phase difference in case of scattering at two centers with a given distance 82
— function, definition 101
— functions, different 102
— shift at reflection on metals 24
— — at total reflection 17
phenomenological theories of absorption and scattering 103

photochemical reaction of anthraquinone, adsorbed on Al_2O_3 278
— reactions on surfaces 278
photography in the IR 256
photometric determination of heavy metal ions 293
piezochromism and thermochromism of bianthrone and bixanthylene 284
pigments 253
plates for multiple internal reflection 322
polar diagram of the dipole scattering of non-polarized light 79
polarizability of dissolved molecules 80
— of molecules, definition 74
polarization of natural radiation on diffuse reflection at absorbing materials 58
polarized radiation for determination of optical constants 325
powders, surface area determination 270
preparation of samples for measurement 237
primary standard according to Zeiss 147

quantitative analysis of mixed powders 290
— determination of water by reflectance in the infrared 293
quantum efficiency of luminescence from reflectance measurements 152

radiation characteristic of a dipole 75, 77
— densities, normalized according to *v. Seeliger* 53
— —, — of powders of different refractive index 51
— density according to *Lambert* and to *v. Seeliger* 28
— —, definition 27
— —, methods of measurement 29
— — of white standards, λ-dependence 52

radiation flux of dipole scattering 76
— — per unit area, definition 5
— intensities of different scattering order in case of multiple scattering 98
— loss due to scattering 2
radiation-transfer equation 100
— —, attempts for a rigorous solution 156
— —, general form 101
radical ions of aromatic hydrocarbons 265
rate constants, calculation from reflectance measurements 281
Rayleigh scattering 73
— — a limiting case of the Mie theory 88
— — formula for the dilute gas 78
reactions in mixed solid phases 282
reactivity of powdered pharmaceutical preparations 307
reciprocity law 224
redox reactions on surfaces 265
red-shift of bandmaxima of typical color curves 211
— of bandmaxima by wavenumber dependence of S 211
reflectance according to Giovanelli for semi-infinite media 160
— and dyestuff concentration 301
— and transmittance of filter papers 198
— — of opal glass plates 131
— — of samples with covering glasses 136
— — of several layers 123
— — — layers one behind the other 197
—, concentration dependence 178
— dependent on the refractive index of the reflecting medium 45
—, diffuse, definition 110
—, — of a layer with black background 111
— for diffuse incident radiation 11
— function for isotropic scattering 105

reflectance of a semi-infinite medium in a matrix 301
— — — with non-isotropic scattering 158
— — scattering medium embedded in an matrix according to Fresnel equations 134
— of magnesium oxide, dependence on different parameters 142
— of non-homogeneous scattering layers 123
— of transparent papers from transmittance measurements 144
— oximetry 306
— R_∞, calculated according to *Giovanelli* and *Kubelka-Munk* 161
—, relative of a mixture of Cr_2O_3 with impure CaF_2 in excess 184
— spectra at extremely high or low temperatures 229
— — obtained by attenuated total reflection 309ff.
— — of allotropes 289
— — of heavy metal complexes on ion-exchange resins 293
— standards, different 146
— —, intrinsic absorption 149
reflection coefficients, definition 8
— — in the case of reflection at the rarer medium 14
— cuvette for measurements at low temperatures 230
—, diffuse of a layer of finite thickness with black background 113
— losses at phase boundaries 1
— measurements as a function of the optical geometry 170
— of infrared radiation at powdered germanium 35
— of linearly-polarized radiation at powders of various particle sizes 66
— spectra of adsorbed substances 256
— — of anthracene on silica gel, influence of drieing 241
— — of anthraquinone, undiluted or adsorbed on NaCl 177

reflection spectra of Cr_2O_3 + excess of CaF_2 in dependence on water vapor pressure 235
— — ferric oxide pigments 255
— — of green MnS and of red MnS 288
— — of HgI_2 260
— — of iodine adsorbed on dry NaI 262
— — — vapor on undried MgO 264
— — of $MnCl_2 \cdot 4H_2O$ 287
— — of p-dimethylamino-azobenzene. adsorbed at different concentrations 259
— — of rutile and anatase 291
— — of substances which have been adsorbed from solution 244
— — — which have been adsorbed from the gaseous phase 242
— — of the molecular compound pyrene-s-trinitrobenzene 274
— spectrometer, model 13/205 from Perkin-Elmer 247
— spectrum and transmission spectrum of a Didym filter glass 189
— — of a naphtospirane adsorbed on NaCl 271
— — of BiI_3 adsorbed on dry KI 262
— — of Cr_2O_3/MgO without and with quartz cover plate 190
— — of perylene, adsorbed on a SiO_2-Al_2O_3 cracking catalysts 266
reflectometers 221 ff.
regular reflectance at strongly absorbing media 21
— — for diffuse and perpendicular incidence, table 12
— — of absorbing materials 23
— — of a plane parallel plate 10
— — — parallel plate for diffuse incidence 13
— — of metals (table) 25
— — of non-absorbing media 9
— reflection at phase-boundaries 130
— — of linearly polarized radiation, method 67

relations between the coefficients of the Kubelka-Munk theory 120
relative error $\Delta F(R'_\infty)/F(R'_\infty)$ with $R_{\infty\,standard}$ as parameter 185
— — in K, ratio to the relative error in R_∞ 251
— — in S, ratio to the relative error in R_∞ 251
— — of the Kubelka-Munk function in dependence on R_∞ 251
remission attachment for measurements between crossed polarization prisms 233
— — for the evaluation of thin layer chromatograms 224
— — RA 2 and RA 3 for the Zeiss PMQ II 223, 224
— curves dependence on particle size 58
— — of glass powder of different grain sizes 62
— — of mixed crystals consisting of $KMnO_4$ and $KClO_4$ 63
— measurements of fluorescent samples 232
— spectra of pure $CuSO_4$-$5H_2O$ with polarized radiation or with natural radiation 68
— spectrum of pure $KMnO_4$ 64, 70
— — — $K_3[Fe(CN)_6]$ as a function of particle size 66
residual radiation method in the infrared 23
Richter "Gloss Number" 39
ring cleavage, influence of drying of the adsorbent 269
— — of lactones 267
— —, reversibility 268
Ryde-theory of non-isotropic scattering 127

sample preparation 237
— surface, distance from the image-forming lens 239
— —, effect on the measured reflectance spectrum 239
scattering angle, definition 77

Subject Index

scattering coefficient according to *Kubelka-Munk* 112
— —, definition 73
— —, effective 162
— — of the Kubelka-Munk-theory 109
— —, specific of papers 198
— coefficients and absorption coefficients of the Kubelka-Munk-theory 191
— —, dependence on wavenumber and grain size 196
— —, derived from R_∞ and T 197
— — from measurements of R_0 and R_∞ or of T and R_∞ 239
— — from transmittance spectra 192
— — in the forward and reverse directions for parallel and perpendicular irradiation 128
— —, linear dependence on wavenumber 204
— — of aerosil, dependent on the packing density 202
— — of aerosils of different specific surface 202
— — of glass filter powder, dependence on grain size and wavenumber 193
— — of MgO 204
— — of NaCl, influence of dispersion of refraction index 203
— — of pure aerosil and of aerosil + graphite 209
— — of quartz powder 206
— — of the Kubelka-Munk theory (Table) 200
— — of white standards as function of wavenumber, grain size, and refractive index 199
— — of strongly absorbing substances 240
— diagram according to Mie for dielectric spheres 89
— intensity for spherical particles 86
— of a light wave on a dipole 74

scattering of large isotropic spherical particles 81
— power according to *Kubelka-Munk* 111
— — as function of R_0 and R_∞ 195
— — — of the reflection in front of a black background 115
— — at variable packing density 122
— –, derived from R_∞ and T 197
Schuster equation for isotropic scattering 103
scintillation method 228
selecting of standards 241
self-absorption of standards, elimination 149
semi-infinite medium, definition 102
separation of scattering from absorption 3
shadowing effects in rough surfaces 52
— — of loose powder surfaces 54
single scattering, definition 72
Snell Law of Refraction 6
source function, definition 101
sources of error at reflectance measurements 224
spectra of adsorbed materials 176
— of crystalline powders 285
— of powered materials by ATR 329
— of slightly soluble substances 253
spectrophotometers for color measurements 231
spectrum from attenuated total reflection of the 1035 cm^{-1} band of benzene and the optical constants n and $\varkappa$ 312
specular reflection 7
— — of small crystals 228
sphere error of the integrating sphere 220
— factor of the integrating sphere 220
spiropyranes, adsorption as cation or zwitterion 270
—, ring cleavage on surfaces 269
spreading on surfaces 176, 261
standardizing of industrial products 3
state of polarization of "diffusely" reflected radiation 37

surface area determination of powders 270
— brightness, definition 27
— —, normalized of powders of different refraction index 49
— spectrum by means of internal reflection 331

v. Seeliger law for diffuse reflection 28, 53
V-form element for internal reflection 323
vibrator 238
vulcanic dust, influence on the reflectance of snow in Antarctica 308

Tappi-opacity 112
tautomeric complexes 277
— transformation of 2-(2′,4′-dinitrobenzyl)pyridine 279
Taylor sphere 138
temperature induced emission of a scattering and absorbing medium 150
thermal transformation and decomposition reactions, investigation by reflectance measurements 289
thin films, total reflection 317
thin-layer chromatography 295
total reflection 13
totally reflecting phase boundary 16
transfer complexes of aromatic hydrocarbons 265
transmission coefficients, definition 8
— — in case of transition to an rarer medium 15
— of a plane parallel plate 11
— of a thin layer for diffuse radiation 227
— spectrum according to the KBr method and reflection derived from measurements of the attenuated total reflection 330

transmittance, definition 1
— of a non-homogeneous layer from reflectance measurements 125
— of a scattering layer according to Kubelka-Munk 118
—, true or internal 1
triphenylmethyl cation on surfaces 266
tristimulus values, normalized 304
turbidity as measure of the total scattering power of gases 79
— as "scattering surface" 91
—, definition 73
— of solutions 81
two-constant theories 3
— — of scattering 100
— theory of Kubelka and Munk 106
typical color curve, additivity for mixtures 256
— — —, definition 188
— — curves, distortion by the wavenumber dependence of the scattering coefficients 210
— — — of chromatogram spots 297
— — — of $Cu(H_2O)_4^{2+}$-ions, enriched on air-dried Dowex 294
— — — of iron oxide pigments 254
— — — of powders 199
— — —, relative intensities of the individual bands 210

water content of solids, quantitative determination in the infrared 295
Wright's measurements on validity of Lambert Law 34

Zeiss chromatogram spectrophotometer 225
— color measuring instrument 231
— Elrepho 232
— remission attachment RA 2 with built-in polarization foils 232